Molecular Biology of RNA Viruses

Molecular Biology of RNA Viruses

Editors

Yiping Li
Yuliang Liu

Basel • Beijing • Wuhan • Barcelona • Belgrade • Novi Sad • Cluj • Manchester

Editors
Yiping Li
Institute of Human Virology,
Zhongshan School of
Medicine, Key Laboratory of
Tropical Disease Control,
Ministry of Education, Sun
Yat-sen University
Guangzhou
China

Yuliang Liu
Department of Veterinary
Diagnosis, China Animal
Disease Control Center
Beijing
China

Editorial Office
MDPI
St. Alban-Anlage 66
4052 Basel, Switzerland

This is a reprint of articles from the Special Issue published online in the open access journal *Viruses* (ISSN 1999-4915) (available at: https://www.mdpi.com/journal/viruses/special_issues/molecular_biology_rna_viruses).

For citation purposes, cite each article independently as indicated on the article page online and as indicated below:

Lastname, A.A.; Lastname, B.B. Article Title. *Journal Name* **Year**, *Volume Number*, Page Range.

ISBN 978-3-7258-0521-1 (Hbk)
ISBN 978-3-7258-0522-8 (PDF)
doi.org/10.3390/books978-3-7258-0522-8

Contents

Article

Development and Characterization of a Genetically Stable Infectious Clone for a Genotype I Isolate of Dengue Virus Serotype 1

Mingyue Hu [1,2], Tiantian Wu [2], Yang Yang [2], Tongling Chen [2], Jiawei Hao [2], Youchuan Wei [1], Tingrong Luo [1], De Wu [3,*] and Yi-Ping Li [2,4,*]

1 College of Animal Sciences and Veterinary Medicine, Guangxi University, Nanning 530004, China
2 Institute of Human Virology, Department of Pathogen Biology and Biosecurity, Key Laboratory of Tropical Disease Control of Ministry of Education, Zhongshan School of Medicine, Sun Yat-sen University, Guangzhou 510080, China
3 Institute of Pathogenic Microbiology, Center for Disease Control and Prevention of Guangdong, Guangzhou 511430, China
4 Department of Infectious Diseases, The Third Affiliated Hospital of Sun Yat-sen University, Guangzhou 510630, China
* Correspondence: wu_de68@cdcp.org.cn (D.W.); lyiping@mail.sysu.edu.cn (Y.-P.L.)

Abstract: Dengue virus (DENV) is primarily transmitted by the bite of an infected mosquito of *Aedes aegypti* and *Aedes albopictus*, and symptoms caused may range from mild dengue fever to severe dengue hemorrhagic fever and dengue shock syndrome. Reverse genetic system represents a valuable tool for the study of DENV virology, infection, pathogenesis, etc. Here, we generated and characterized an eukaryotic-activated full-length infectious cDNA clone for a DENV serotype 1 (DENV-1) isolate, D19044, collected in 2019. Initially, nearly the full genome was determined by sequencing overlapping RT-PCR products, and was classified to be genotype I DENV-1. D19044 wild-type cDNA clone (D19044_WT) was assembled by four subgenomic fragments, in a specific order, into a low-copy vector downstream the CMV promoter. D19044_WT released the infectious virus at a low level (1.26×10^3 focus forming units per milliliter [FFU/mL]) following plasmid transfection of BHK-21 cells. Further adaptation by consecutive virus passages up to passage 37, and seven amino acid substitutions (7M) were identified from passage-recovered viruses. The addition of 7M (D19044_7M) greatly improved viral titer (7.5×10^4 FFU/mL) in transfected BHK-21 culture, and virus infections in 293T, Huh7.5.1, and C6/36 cells were also efficient. D19044_7M plasmid was genetically stable in transformant bacteria after five transformation-purification cycles, which did not change the capacity of producing infectious virus. Moreover, the D19044_7M virus was inhibited by mycophenolic acid in a dose-dependent manner. In conclusion, we have developed a DNA-launched full-length infectious clone for a genotype I isolate of DENV-1, with genetic stability in transformant bacteria, thus providing a useful tool for the study of DENV-1.

Keywords: dengue virus; reverse genetics; infectious clone; amino acid substitution; genetically stable

Citation: Hu, M.; Wu, T.; Yang, Y.; Chen, T.; Hao, J.; Wei, Y.; Luo, T.; Wu, D.; Li, Y.-P. Development and Characterization of a Genetically Stable Infectious Clone for a Genotype I Isolate of Dengue Virus Serotype 1. *Viruses* **2022**, *14*, 2073. https://doi.org/10.3390/v14092073

Academic Editor: Qiang Ding

Received: 2 September 2022
Accepted: 16 September 2022
Published: 18 September 2022

Publisher's Note: MDPI stays neutral with regard to jurisdictional claims in published maps and institutional affiliations.

1. Introduction

Dengue virus (DENV) belongs to the genus *Flavivirus* of the family *Flaviviridae*. It is the most widespread arbovirus mainly transmitted by *Aedes aegypti* and *Aedes albopictus*. DENV infection can cause a spectrum of diseases varying from mild dengue fever to severe diseases, such as dengue haemorrhagic fever (DHF) and dengue shock syndrome (DSS) [1,2]. With global warming and active commercial and traveling flows, the vector mosquitoes have expanded unexpectedly, increasing the risk of global DENV prevalence. In the past 50 years, the incidence of dengue fever has soared 30 times [3]. The World Health Organization (WHO) reports that DENV infection has been identified in 128 countries, mostly in tropical and subtropical regions [1,4,5]. In 2019, the WHO announced dengue as one of the top ten threats to global health [6].

Four serotypes of DENV (DENV-1-4) have transmission cycles in humans, and serotypes differ from each other by 30–35% amino acid (aa) sequence [7,8]. Antibody-dependent enhancement (ADE) and cross-reactive T cell-mediated cytotoxicity may occur when a secondary infection caused by a heterogenous serotype virus [9] or other flaviviruses [10–13]. ADE is believed as an important risk factor of DHF and DSS [10]. To date, only one vaccine was licensed for DENV, however the issues of effectiveness and safety remain, and it has not been widely used [14]. Moreover, there is no an antiviral drug available for DENV infection.

DENV-1 has five genotypes, I-V, varying by approximately 3% at amino acid level and 6% at nucleotide level [7,8,15]. The emergence of a highly divergent isolate suggested the existence of genotype VI DENV-1, and it may have circulated in multiple regions in China and other countries [16]. Genotype I is the most prevalent genotype of DENV-1 responsible for several major outbreaks in south China since 2000 [1], however only few DENV-1 and one genotype I infectious clones (DENV1-3_C_0268-96, Thailand, 1996) have been developed previously [17]. Reverse genetic systems representative of different serotypes and prevalent isolates will provide valuable tools for the studies of molecular virology, virus-host interactions, viral pathogenesis, cross-species transmission, vaccine and antiviral drugs. Given the antigenic and genetic diversity of DENV, an infectious cDNA clone of clinical isolate from recent outbreaks will be highly relevant, especially for genotype I of DENV-1. However, some technical challenges exist in the construction of a genome length cDNA clone for DENV and other flaviviruses, such as Zika virus (ZIKV), into a plasmid vector, as the viral cDNA sequence may contain cryptic prokaryotic promoters leading to unexpected transcription and expression of proteins toxic or lethal to the transformant recipient bacteria [18–20]. As a consequence, the replication of plasmid becomes unstable in the transformant *E. coli* bacteria, and it remains a challenge to overcome the instability of genomic cDNA in bacteria.

In this study, we constructed and adapted a full-length infectious clone of DENV-1 genotype I isolate, D19044, for efficient virus production in cultured cells. The genetic stability in transformant bacteria and responsiveness to antiviral compound were also proven. D19044 was isolated in 2019 from an endemic in Jieyang city, Guangdong province, China; thus, this clone will be highly relevant for the study of DENV-1 in prevalence and those close-related endemics.

2. Materials and Methods

2.1. Cells and DENV-Infected Serum

Huh7.5.1 cells were provided by Dr. Francis Chisari (The Scripps Research Institute, La Jolla, CA, USA) and Dr. Jin Zhong (Institut Pasteur of Shanghai, Shanghai, China). C6/36 cells were provided by Dr. Ping Zhang (Sun Yat-sen University, Guangzhou, China). BHK-21, A549, 293T, and Vero (Vero, originally refers to ATCC No. CCL-81) cells were maintained in our laboratory. BHK-21, A549, 293T, Huh7.5.1, and Vero cells were cultured in a high-glucose Dulbecco's modified Eagle's medium (DMEM) supplemented with 10% fetal bovine serum (FBS) and 1% penicillin/streptomycin, and incubated at 37 °C with 5% CO_2. C6/36 cells were maintained in a high-glucose DMEM supplemented with 10% FBS, 1% penicillin/streptomycin, and 1 × Non-Essential Amino Acids Solution (Gibco, CA, USA), and incubated at 28 °C with 5% CO_2. The cells used for experiments were tested routinely by mycoplasma-specific PCR to ensure free of mycoplasma contamination. The DENV-1-infected serum was collected from a patient (no. D19044) by Guangdong Provincial Center for Disease Prevention and Control, China, for the purpose of disease surveillance and prevention, during an endemic in Jieyang city, Guangdong, China, in 2019.

2.2. Determination of Viral Genome Sequence and Construction of Full-Length cDNA Clone

The patient serum (no. D19044) was inoculated to C6/36 cells and the culture supernatant from day 6 was divided into two parts. One part of supernatant was assigned for sequencing analysis and subjected to RNA extraction using MagZol Reagent (Magen, Guangzhou, China). The nearly full-length genomic cDNA was synthesized by reverse transcription using SuperScript III Reverse Transcriptase (Invitrogen, CA, USA) (primers R6260*/R10654*) (Table 1). Four overlapping PCRs covering the ORF and partial 5′UTR and 3′UTR sequences were performed using primers F17/R3347, F2724/R4897, F4755/R8347, F7485/R10544, F-5′UTR/R-5′UTR, and F-3′UTR/R-3′UTR (Table 1). The PCR fragments were sequenced by Sanger sequencing method, and nearly full-length genome (nucleotides [nts] 50 to 10508; GenBank no. OP315653) was determined and confirmed by aligning to DENV-1 sequences from GenBank (MT076934.1, KY057373.1, MG679801.1, KY672944.1, KY926849.1, MF405201.1, KX225491.1, KP406801.1, AF226685.2, AY732483.1, and NC_001477.1).

Table 1. Primers were used for RT-PCR of DENV-1 isolate, D19044.

Primers	Position	Sequence (5′-3′)	Product Length (nt)
F17	17–37	GGACCGACAAGAACAGTTTCG	3331
R3347	3324–3347	CATGAATTATCTTCCCTGTGACTG	
F2724	2724–2743	AAATGATTAGGCCACAACCC	2174
R4897	4877–4897	AATGGCTCCAACTTCACCTTC	
F4755	4755–4778	ATGGAGGAGGTTGGAGGCTTCAAG	3593
R8347	8324–8347	GCCTAGGTCCACGTCTCTTTCATA	
F7485	7485–7510	TTTCCATGGCAAACATTTTCAGAGGA	3060
R10544	10,524–10,544	GTTGTGTCTTGGGAGGGGTCT	
F-5′UTR	Vector-15	gagctcgtttagtgaaccgtcAGTTGTTAGTCTACG	70
R-5′UTR	15–49	AAGCTTCCGATTTGAAACTGTTCTTGTCGGTCCAC	
F-3′UTR	10,509–10,529	ACTAGTGGTTAGAGGAGACCC	277
R-3′UTR	Vector	ttcggatgcccaggtcggaccgcgaggaggtggagat	
R6260 *	6239–6260	CCACGTCCATGTTCTCCTCCAA	
R10654 *	10,633–10,654	GGTCTCTCCCAGCGTCAATATG	

*, the primers were used for reverse transcription. Lowercases indicate the sequences priming to the vector sequence.

To assemble full-length D19044 wild-type cDNA clone (D19044_WT), nts 1–49 and 10,509–10,735 were synthesized identical to a DENV-1 isolate, GZ2002 (GenBank accession number, JN205310.1) [21]. Amino acid substitutions were engineered into the D19044_WT by fusion PCR or gene synthesis, such as D19044_4M and D19044_7M (GenBank no. OP315654). Using partial 5′UTR and 3′UTR sequences from isolate GZ2002 avoided performing rapid amplification of cDNA 5′ ends (5′RACE) and 3′RACE to determine the terminus sequences of D19044 genome. Although the secondary and tertiary RNA structures in the 5′UTR and 3′UTR of DENV and other flaviviruses play fundamental roles in viral replication and translation [22], the sequences from GZ2002 (nts 1–49 and 10,509–10,735) were located in the conserved regions of DENV genome [23]; D19044 had no difference with GZ2002 in the 5′UTR nts 50–94 and only differed by 11 nts in the 3′UTR nts 10,274–10,508, which were identified in this study. D19044 shared 91.36% and 96.93% sequence identity with GZ2002 at nucleotide and amino acid levels, respectively. In addition, GZ2002 was isolated also from Guangdong province, China, in 2002 [21]. To construct the full-length cDNA clone, we divided the D19044 genome into four subgenomic fragments: fragment A (nts 1–2805, *Sac*I-*Bsr*GI), fragment B (nts 2724–4897, *Bsr*GI-*Bsr*GI), fragment C (nts 4755–8347, *Bsr*GI-*Cla*I), and fragment D (nts 7485–10,735, *Cla*I-*Rsr*II). These four fragments were assembled into pTight vector by in-fusion cloning using a One Step Cloning Kit (Vazyme, Nanjing, China), and surprisingly the cloning succeeded only when fragments were assembled in the order of D, C, A, B. All restriction endonucleases were purchased (New England Biolabs, NEB, MA, USA). The pTight vector was originally modified from pBR322, which contains seven repeated tetracycline-response elements (7 × TRE) located upstream of the minimal cytomegalovirus (CMVmin) promoter sequence, and it was used previously for cloning a DENV-2, DENV-2-pTight (provided by Dr. Andrew Yueh, National Health Research Institutes, Taiwan, China) [19].

2.3. Transformation and Purification of Plasmids

Full-length D19044 cDNA clone plasmid was transformed into competent Turbo bacteria (NEB, MA, USA) and plated on solid 2 × YT medium plates containing ampicillin (15 μg/mL) at 28 °C for 24 h. The colonies were selected and cultured in liquid 2 × YT medium with ampicillin (15 μg/mL) at 28 °C for 20 h. The plasmid was extracted by HiPure Plasmid Mini kit (Magen, Guangzhou, China) and confirmed by Sanger sequencing analysis.

2.4. Phylogenetic Analysis of D19044 and Other DENV Sequences

The sequence of D19044 virus was generated by Sanger sequencing four overlapping RT-PCR products (above). Twelve DENV-1 genotypes I–V isolates with complete genome sequences available in GenBank and other eight sequences from DENV-2, -3, and -4 (two sequences for each) were selected for phylogenetic analysis. Sequences representative of DENV serotypes 2–4 were selected, especially for those with infectious clones developed. The phylogenetic tree was made based on the ORF nucleotide sequences using the neighbor-joining method in the Molecular Evolutionary Genetics Analysis (MEGA, version 6, Kimura 2-parameter). The Basic Local Alignment Search Tool (BLAST, version +2.13.0) was used to analyze the sequences homology to D19044.

2.5. Virus Stock Preparation, Infection Experiment, and Virus Passage

Virus stock was made by transfection and infection of BHK-21 cells. Briefly, BHK-21 cells were seeded to 12-well plates 16–20 h prior to transfection and allowed to grow to 80% confluence for transfection. One microgram of D19044_7M plasmid (or D19044_4M; D19044_WT) and 0.5 μg of pTet-off plasmid were mixed and incubated in 200 μL of Opti-MEM medium containing 2.25 μL of ExFect transfection reagent (Vazyme, Nanjing, China) for 20 min. The plasmid mix was transfected to BHK-21 cells and cultured at 37 °C for 6 h, and then transfection medium was replaced with DMEM supplemented with 2% FBS and left for 2 days. The transfected culture was split every 2 days and supernatant was harvested at days 2, 4, 6, and 8, filtered (0.2 μm), and stored at −80 °C for later use. To expand the virus stock or perform infection experiment, transfection-derived supernatant virus was inoculated to naïve BHK-21 cells or other cell lines for 2 h, and then infection supernatant was replaced with DMEM with 2% FBS. The cultured supernatant was harvested at time points as required and stored at −80 °C.

To adapt the virus for efficient infection in culture, we continuously passaged the virus to naïve BHK-21 cells. Briefly, naïve BHK-21 cells were inoculated with transfection-derived virus supernatant in DMEM with 2% FBS for 72 h. Culture supernatant was collected, filtered (0.2 μm), and designated as first passage virus. The first passage virus was inoculated to naïve BHK-21 cells for 72 h to generate second passage virus. We consecutively passaged the virus until a fast virus spread and more intense fluorescence DENV-positive cells were detected through IFA in the culture. The passage-recovered viruses were stored at −80 °C and viruses from passages 21, 36, and 37 were subjected to RNA extraction using MagZol Reagent (Magen, Guangzhou, China). The cDNA was synthesized by reverse transcription using SuperScript III Reverse Transcriptase (Invitrogen, CA, USA) and amplified by four overlapping fragments (Section 2.2 and Table 1). The PCR products were analyzed by Sanger sequencing to identify if any substitution had occurred.

2.6. Virus Titration and Indirect Immunofluorescence Foci Assay (IFA)

The DENV titer was determined by indirect immunofluorescence foci assay (IFA) using BHK-21 cells [24,25] and expressed as focus forming unit per ml (FFU/mL). Briefly, BHK-21 cells were seeded in 96-well plates (4×10^3 cells/well) and grown for 16–20 h. The virus supernatant was prepared as 10-fold serial dilutions in DMEM (from 10^0 to 10^6 dilutions), inoculated to BHK-21 cells, and left at 37 °C for 2 h. The cells were then washed once with DMEM, and incubated in DMEM containing 2% FBS for 48 h. Then, the culture supernatant was removed, and the cells were fixed using methanol (−20 °C) for 10–15 min. The fixed cells were incubated with DENV NS3 antibody (1:2000 dilution; GTX124252,

GeneTex, CA, USA) at 4 °C overnight. To visualize DENV infected cells, the secondary antibody Alex Fluor 488 Goat Anti-Mouse IgG (H + L) (1:1000; Invitrogen, CA, USA) and cell nucleus staining dye Hoechst 33258 (1:1000; Invitrogen, CA, USA) were added at room temperature and incubated for 1 h. The numbers of FFU were manually counted using a fluorescence microscope (Leica Microsystems, DMI 4000B, Weizla, Germany).

2.7. Cytotoxicity Assay

BHK-21 cells were seeded in a 96-well plate (4×10^3 cells/well) and incubated for 16–20 h. The cells were incubated with different concentrations of drugs for 48 h, and 10 µL of Cell Counting Kit-8 (CCK8, Dojindo, Kumamoto, Japan) was added to each well, gently shaken and incubated at 37 °C for 90 min. The optical density (OD_{450}) was measured by a microplate reader (BioTek ELx800, Vermont, Germany).

2.8. Antiviral Drug Treatment

Mycophenolic acid (MPA, T1335, TargetMol, Shanghai, China) was dissolved to 10 mM using DMSO, aliquoted, and stored at −80 °C. Moxifloxacin hydrochloride (Tianjin Chasesun Pharmaceutical, Tianjin, China) was preserved at room temperature. Before treatment experiment, drugs were diluted to different concentrations (below cytotoxicity concentration) required for experiments using DMEM with 2% FBS. For antiviral treatment, BHK-21 cells were seeded in a 96-well plate (4×10^3 cells/well) and cultured for 16–20 h. The cells were infected with virus dilution at multiplicity of infection (MOI) of 0.1 containing drug of different concentration for 2 h. The treatment medium was removed, and the cells were washed with DMEM and incubated in DMEM with 2% FBS containing drugs for 48 h. The cells were fixed and subjected to IFA, and the FFU were enumerated for each well.

3. Results

3.1. Genome Sequence of DENV-1 Genotype I Isolate, D19044

DENV-1-infected serum D19044 was collected from a patient in Jieyang city, Guangdong province, China, in 2019. The patient serum was inoculated to C6/36 cells, and culture supernatant was collected at day 6 post infection. A part of C6/36 culture supernatant was inoculated to Vero cells and consecutively passaged up to 16 passages, followed by 13 passages in BHK-21 cells; unfortunately, the virus did not spread up in the cultures, without detectable viral titers. The other part of supernatant was subjected to RNA extraction and amplified by four overlapping DENV-1-specific RT-PCRs (Table 1). The PCR products were analyzed by Sanger sequencing, and the sequence covering genome nucleotides (nts) 50-10508 was obtained and designated D19044 wild-type (WT) sequence (OP315653). The D19044 WT sequence was phylogenetically analyzed against twelve sequences of DENV-1 genotypes I-V and eight sequences of DENV-2, -3, and -4 (two sequences for each) from GenBank (Figure 1). The results showed that D19044 was genetically close to the strain UI13412 (KC172835.1), which belongs to genotype I of DENV-1 (Figure 1). BLAST analysis revealed that D19044 was highly homologous to GZ8H/2019180/2019/I (MW261838.1), a clinical serum from Guangzhou in 2019, differing by 13 nucleotides and 2 amino acids (nt T1650C in E sequence with aa change I519T; G9452A in NS5, V3120I). Additionally, D19044 shared 97.8% and 99.49% identity at nucleotide and amino acids levels with infectious clone of genotype I isolate DENV1-3 C_0268-96 (OK605755.1), respectively [17]. Thus, D19044 was a genotype I isolate of DENV-1 recently prevalent in south China.

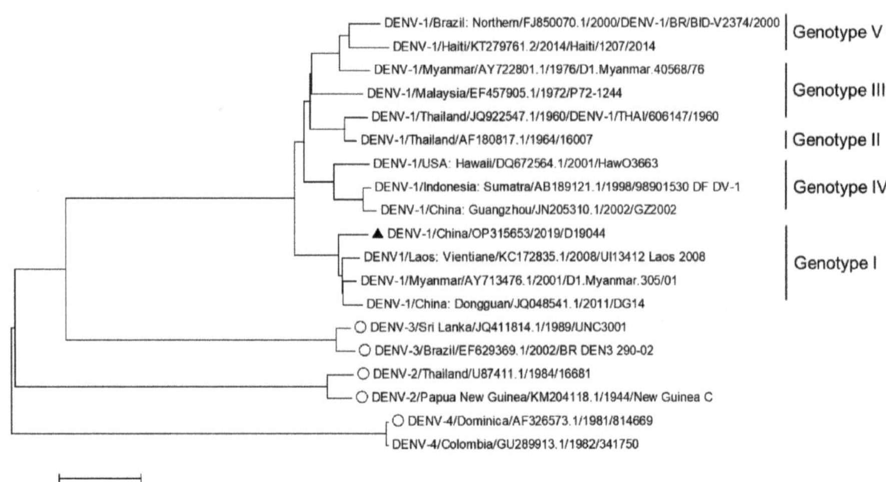

Figure 1. Phylogenetic analysis of D19044 and other DENV isolates. The isolates of genotypes I-V DENV-1 with complete genome sequences available from GenBank were selected to make phylogenetic analysis. Of DENV-2, -3, and -4, two sequences representative of each serotype were selected, especially for those with infectious clones developed. The phylogenetic tree is made based on the ORF nucleotide sequences using neighbor-joining method. A solid triangle (▲) highlights the D19044 isolate of this study. A circle (○) indicates that the infectious clone of the isolate has been developed.

3.2. Construction of Full-Length D19044 cDNA Clone

To overcome the instability of DENV cDNA clone in transformant *E. coli* bacteria, we selected a low-copy plasmid pTight, which was modified from pBR322 [19,20]. We and others have previously used this vector for development of infectious clones of JEV, DENV, and ZIKV [18,19]. To construct a D19044 full-length clone, we divided its genomic cDNA into four fragments A, B, C, and D, by restriction sites (Figure 2A); fragment A was fused from sub-fragments A1 and A2, while D was a fusion of D1 and D2 (Figure 2A and Section 2). Partial sequences of 5′UTR (nts 1–49, mainly A1 fragment) and 3′UTR (nts 10,509–10,735, D2 fragment) were obtained from isolate GZ2002 (JN205310) [21]. Inexplicably, we found that the cloning was restricted and selective by the order of fragment cloning into the pTight vector. When fragments were cloned in the order of C → D → A, namely fragment A was first cloned into the vector to get pTight-A, followed by adding D to pTight_A to make pTight_A-D, then C was added in to get pTight_A-C-D, all 70 colonies from three independent ligation-transformation experiments carried multiple nucleotides and/or amino acid substitutions or deletions. In subsequent attempts, we tested different orders of fragment cloning and found that D19044 full-length clone could be constructed only in the order B → A → C → D, namely D was first cloned to get pTight_D, followed by adding C to make pTight_C-D, and then adding A to obtain pTight_A-C-D. Finally, B was added to generate full-length D19044 wild-type cDNA clone (D19044_WT). All subclones and final D19044_WT clone were confirmed by Sanger sequencing (Figure 2A).

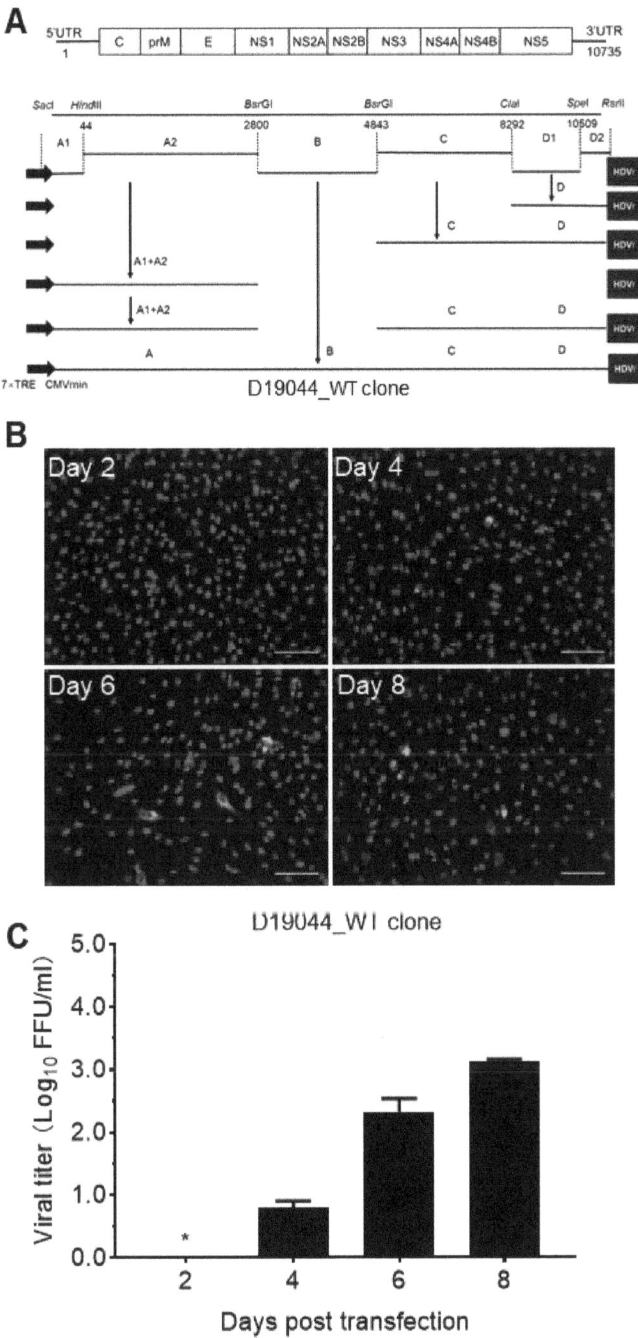

Figure 2. Construction of D19044 clone and its capacity to generate infectious virus particles. (**A**) Schematic diagram of construction of full-length D19044 cDNA clone. The full-length genome was illustrated and divided into four overlapping subgenomic fragments (A–D) for the convenience

of cloning, of which fragment (A) was subdivided into A1 and A2, and (D) was subdivided into D1 and D2. The key restriction sites and nucleotide positions are indicated for fragment A (nts 1-2805, *SacI-BsrGI*), B (nts 2724-4897, *BsrGI-BsrGI*), C (nts 4755-8347, *BsrGI-ClaI*) and D (nts 7485-10735, *ClaI-RsrII*). The pTight vector contains 7 × TRE and CMVmin, as well as hepatitis D virus ribosome (HDVr). The order of cloning fragments was as the followings: B → A → C → D (see Results), as indicated by solid lines in the constructs. Finally, a full-length D19044 wild-type (D19044_WT) clone was made. (**B**) D19044_WT clone released infectious virus particles from BHK-21 cells. The D19044_WT clone plasmid was transfected into BHK-21 cells, and DENV-1 positive cells were monitored by immunofluorescence assay (IFA) for DENV NS3 protein (green) and cell nuclei (blue). The images were taken from one transfection experiment. Bar, scale 100 μm. (**C**) The viral titer of D19044_WT after transfection of BHK-21 cells. The culture supernatant collected at different time points of post transfection (p.t.) and subjected to FFU assay. The mean ± standard error of the mean (SEM) from three independent experiments are shown. An asterisk (*) indicates the viral titer was below the limit of detection (LOD; < 0.69 $\log_{10}$FFU/mL).

3.3. Transfection of D19044 Clone Plasmid Produced Infectious DENV Particles

To test the capacity of producing infectious particles for D19044_WT clone, we transfected D19044_WT plasmid together with pTet-off plasmid into five different types of cells, A549, 293T, Huh7.5.1, Vero, and BHK-21 cells. The plasmid of pTet-off expresses the tet-responsive transcriptional activator (tTA), which will bind to a tet-responsive element (TRE) in the pTight plasmid to activate the CMVmin promoter-controlled transcription to produce viral genomic RNA, thus initiating translation and replication of D19044_WT RNA [18,19]. The results showed that only the BHK-21 culture showed a greater number of DENV positive cells with intense fluorescence by the IFA method (Figure 2B), and other cell cultures showed weak or undistinguishable fluorescence signals. Thus, we selected BHK-21 cells for subsequent experiments, without further optimizing the transfection for other cell lines. Noting that, we did not observe apparent cytopathic effect or cell death in infected culture and did not succeed for plaque forming assay, as described previously [18], thus we monitored D19044 virus infection by NS3-specific immunofluorescence method in this study. The culture supernatant was collected at days 2, 4, 6, and 8 post transfection (p.t.), and viral titers were determined by inoculating supernatant to naïve BHK-21 cells for FFU assay. The results showed that D19044_WT transfection of BHK-21 cells released infectious virus, however, the viral titer was low (1.26 × 10^3 FFU/mL; day 8 p.t.) (Figure 2C).

To further adapt the D19044_WT virus, we continuously passaged the virus to naïve BHK-21 cells, determined the viral titer regularly, and found that the viral titer rose slightly at passage 13 (47 days from transfection; ~8 × 10^3 FFU/mL), claiming a further adaptation. When reaching passage 34, the viral titer rose significantly (3.9 × 10^5 FFU/mL), thus we expanded this virus stock in the following passages and terminated at passage 37. To identify adaptive acid substitution that may have acquired for the enhanced viral titer, we sequenced D19044 ORF region for passages 21, 36, and 37 by Sanger sequencing. Seven amino acid substitutions (7M) were identified, including G2084A (aa change G664K in E), C4410T (T1439M in NS2B), and G6143A (D2017N in NS3) in the 21th passage virus, T7161C (V2356A in NS4B) and C7689T (S2532F in NS5) in the 36th passage virus, and A1125C (K344T in E) and C1416T (T441I in E) in the 37th passage virus (Table 2).

3.4. Efficient Virus Production of D19044 Clone Carrying Amino Acid Substitutions

To test the adaptation effect of passage-derived amino acid substitutions, we engineered them back into a D19044_WT clone. Since V2356A was partially changed in passage 21 (1 × 10^3 FFU/mL) and became completely changed in passage 36, we initiated to test the effect of four amino acid substitutions (4M, G664K/T1439M/D2017N/V2356A) and 7M in D19044_WT and designated D19044_4M and D19044_7M, respectively (Figure 3A). Both plasmids were transformed into competent Turbo *E. coli* bacteria and transformant colonies appeared after 24 h incubation. D19044_7M acquired G3384T (aa R1097L in NS1), which may stabilize its sequence in bacteria, and we kept this change in D19044_7M clone.

Table 2. Amino acid substitutions identified in the recovered viruses after serial passages.

Number of Passage	Nucleotide Position	Nucleotide Change *	Gene Region	Amino Acid Position	Amino Acid Change
21	2084	G–A	E	664	Glu–Lys
21	4410	C–T	NS2B	1439	Thr–Met
21	6143	G–A	NS3	2017	Asp–Asn
36	7161	T–C/t	NS4B	2356	Val–Ala
36	7689	C–C/t	NS5	2532	Ser–Phe
37	1125	A–C	E	344	Lys–Thr
37	1416	C–T/c	E	441	Thr–Ile

*, Capital letters indicate complete change (or >50% change) in sequencing reads; lowercases indicate a minor change (<50%) in sequencing reads.

To determine whether amino acid substitutions enhance virus production, we transfected D19044_4M, D19044_7M, and D19044_WT clones into BHK-21 cells and examined the viral titers of culture supernatant at different time points post transfection (p.t.) (Figure 3B,C). D19044_4M and D19044_7M produced DENV, with detectable titers at day 2 p.t., while D19044_WT was undetectable. At day 4 p.t., virus spread and titers of D19044_7M were higher than that of D19044_4M and D19044_WT by 5.1~8-fold; at days 6 and 8, D19044_4M and D19044_7M were higher than D19044_WT by 16.5~20.7 and 29.8~60.1 fold, respectively (Figure 3C). D19044_7M reached peak titer (7.5×10^4 FFU/mL) at day 8 p.t. and higher than D19044_4M (2.24×10^4 FFU/mL) (Figure 3C). The recovered viruses were confirmed by sequencing analysis, and all amino acid substitutions engineered were maintained. Thus, passage-recovered amino acid substitutions promoted virus production of D19044 clone, and D19044_7M represented most efficient clone for subsequent experiments.

Since DENV-2 clones were used in the field more frequently, we compared virus production of D19044_7M to a DENV-2 infectious clone, 16681, by plasmid transfection of BHK-21 cells [19,26,27]. The results showed that D19044_7M produced lower viral titers than 16681 clone at days 2, 4, and 6, but surpassed 16681 at day 8 (Figure 3D). Noting that, 16681 tended to cause cell death from day 6 p.t., whereas D19044_7M did not apparently cause cell death, which may contribute to a higher titer for D19044_7M at day 8 (Figure 3D).

3.5. D19044_7M Clone Was Genetically Stable in Bacteria after Multiple Transformation-Purification Cycles

It is an unsolicited issue that flavivirus cDNA encounters instability in transformant recipient bacteria [20,28,29], thus we proceed to test the bacterial stability of D19044_7M clone by repeating transformation-purification cycles (TPC) as previous described [18]. Briefly, D19044_7M plasmid was transformed into Turbo competent bacteria, single colonies were picked and cultured in 2 × YT broth, and plasmid DNA was extracted and transformed into competent Turbo bacteria again. We repeated five rounds of the transformation-purification procedure and confirmed the sequence of plasmid. One amino acid substitution was identified (NS4B, nt C7158G, aa A2355G) in the recovered plasmid and designated D19044_7M/5TPC.

Next, we tested the capacity of producing viral particles for D19044_7M/5TPC in comparison to D19044_7M. Both clones released infectious viruses with comparable viral titers at days 2, 4, 6, and 8 following plasmid transfection of BHK-21 cells. Peak viral titers peaked at day 8 (1×10^5 FFU/mL), indicating that A2355G did not affect production of infectious virus (Figure 4). Moreover, no other deletion and amino acid substitution was found in the genome, especially in substitution- and deletion-prone region, prM-E-NS1 [18,30]. Thus, we considered D19044_7M recombinant was genetically stable in bacteria, feasible for molecular manipulation.

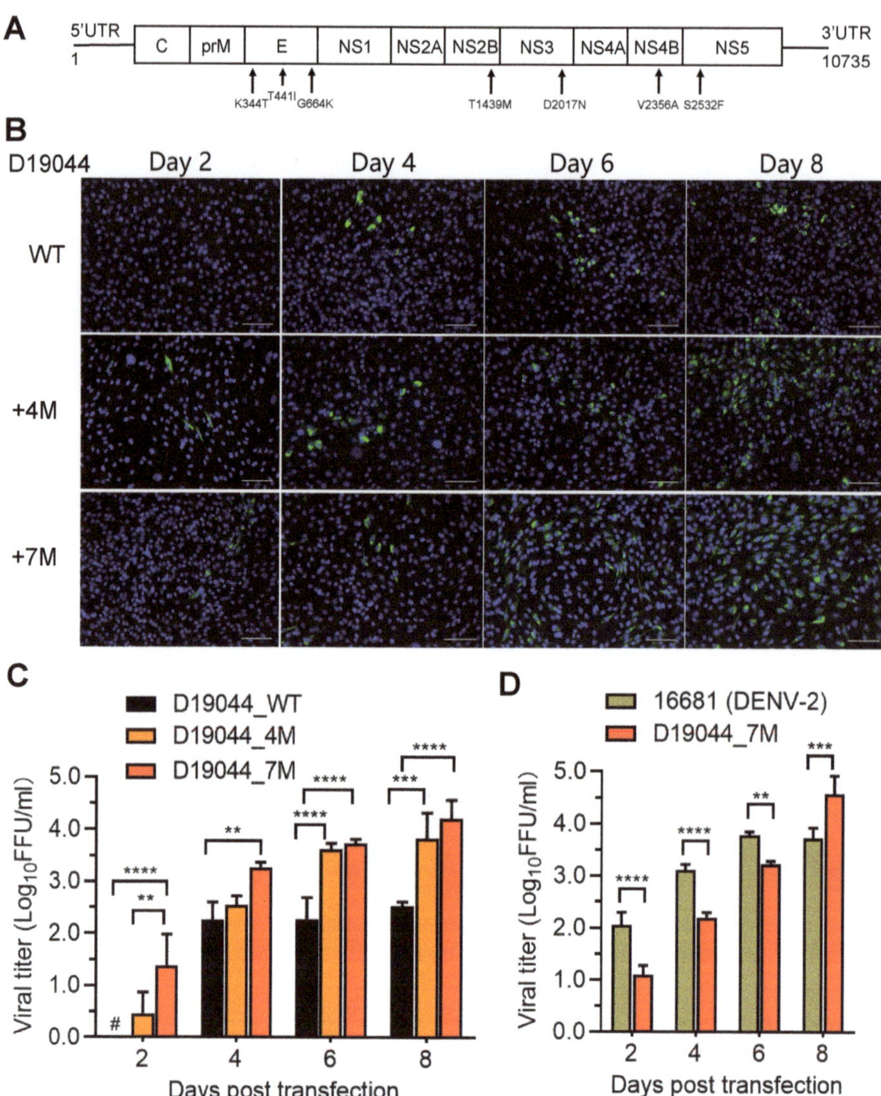

Figure 3. Amino acid substitutions improved the virus production of D19044 clone. (**A**) Diagram of DENV-1 genome with seven amino acid substitutions (7M) identified from passage-recovered viruses. (**B**) IFA of D19044_WT, D19044_4M, and D19044_7M after plasmid transfection of BHK-21 cells. DENV-1 NS3 protein (green) and cell nuclei (blue) were visualized. Bar, scale 100 μm. (**C**) The DENV titer of culture supernatant. Three D19044 clones (D19044_WT, _4M, and _7M) were transfected into BHK-21 cells, and the supernatant were collected for FFU assay at time points as indicated. The means ± SEM of three independent experiments are shown. A "#" indicates the titer was below LOD (<0.69 $\log_{10}$FFU/mL). (**D**) Comparison of viral titers released by D19044_7M and DENV-2 clone 16681. BHK-21 cells were transfected with D19044_7M and DENV-2 16681 plasmids, and the supernatant were collected for FFU assay at time points as indicated. The means ± SEM of three independent experiments are shown. In panels **C** and **D**, statistics analyses were performed using multiple comparison test in a two-way ANOVA. ** $p < 0.01$; *** $p < 0.001$; **** $p < 0.0001$.

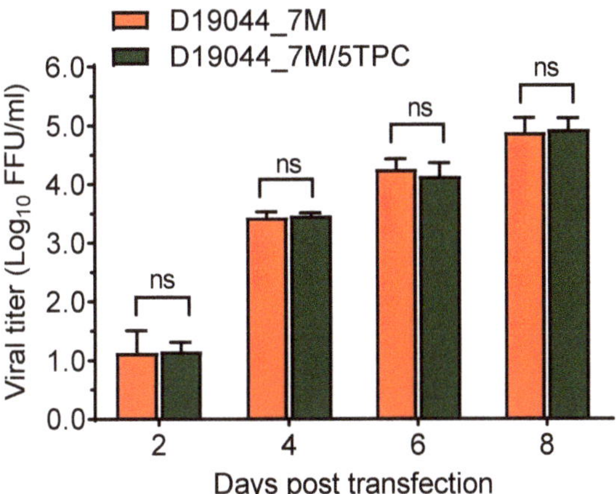

Figure 4. D19044_7M clone were genetically stable in the transformant bacteria. D19044_7M clone plasmid that have gone through five rounds of transformation-purification cycle (5TPC) produced infectious virus comparable to D19044_7M clone. The viral titers were determined by FFU assay. The means ± SEM of three independent experiments are shown. Statistics analysis were performed using multiple comparison test in a two-way ANOVA. ns, no significance.

3.6. Differential Efficiency of D19044_7M and D19044_WT in Various Vell Lines

Since D19044_7M was generated by adapting D19044_WT in BHK-21 cells through serial passages and it carried seven amino acid substitutions, we wanted to test whether it was capable of infecting other cell lines. We infected BHK-21, Huh7.5.1, 293T, and C6/36 cells with same MOI of viruses (MOI = 0.01; ~70% culture confluence) and determined the viral titers released at days 2, 4, 6, and 8 p.i. (Figure 5). The results showed that all cell lines were infected by both viruses, however D19044_7M was more efficient in BHK-21, Huh7.5.1, and C6/36 cells (peak titers, $1 \times 10^{5.3}$–$10^{5.9}$ FFU/mL) than in 293T cells ($1 \times 10^{4.0}$ FFU/mL) (Figure 5A–D). D19044_WT was apparently less efficient than D19044_7M by 75~268-fold in BHK-21, Huh7.5.1, and 293T cells at days 4 and 6, however similar to D19044_7M in C6/36 cells (~2-folds) (Figure 5A–D). Together, D19044_7M was more adapted for infection of vertebrate cell lines, without loss infectivity to the mosquito cell line.

3.7. D19044_7M Virus Was Inhibited by Antiviral Compound

An infectious clone provides a valuable tool for the study of various aspects of DENV virology, of which one most expected may be its feasibility for testing antivirals. We proceeded to test its responsiveness against mycophenolic acid (MPA). MPA inhibits DENV infection by preventing accumulation of positive- and negative-strand RNAs [2] and has been shown to suppress ZIKV infection in different cell types [31,32]. Moxifloxacin hydrochloride was selected as the negative control, which is used for treatment of pulmonary infections and has no inhibition to flaviviruses [33]. The antiviral treatment was performed below the cytotoxic concentration of the drugs (BHK-21 cells, MPA CC_{50} = 1.087 μM) (Figure 6A,B). BHK-21 cells in 96-well plates were infected with D19044_WT and D19044_7M (MOI = 0.1) and then treated with drugs of serial dilution. The results showed that D19044_7M (IC_{50} = 0.04827 μM) and D19044_WT (IC_{50} = 0.04657 μM) were inhibited by MPA in a dose-dependent manner; the IC_{50} comparable for both viruses indicate that amino acid substitutions (7M) did not affect viral response to MPA treatment (Figure 6C). Moxifloxacin hydrochloride did not inhibit virus infection (Figure 6D). Together, these data demonstrate that D19044_7M could serve as a tool for the study of anti-DENV drugs.

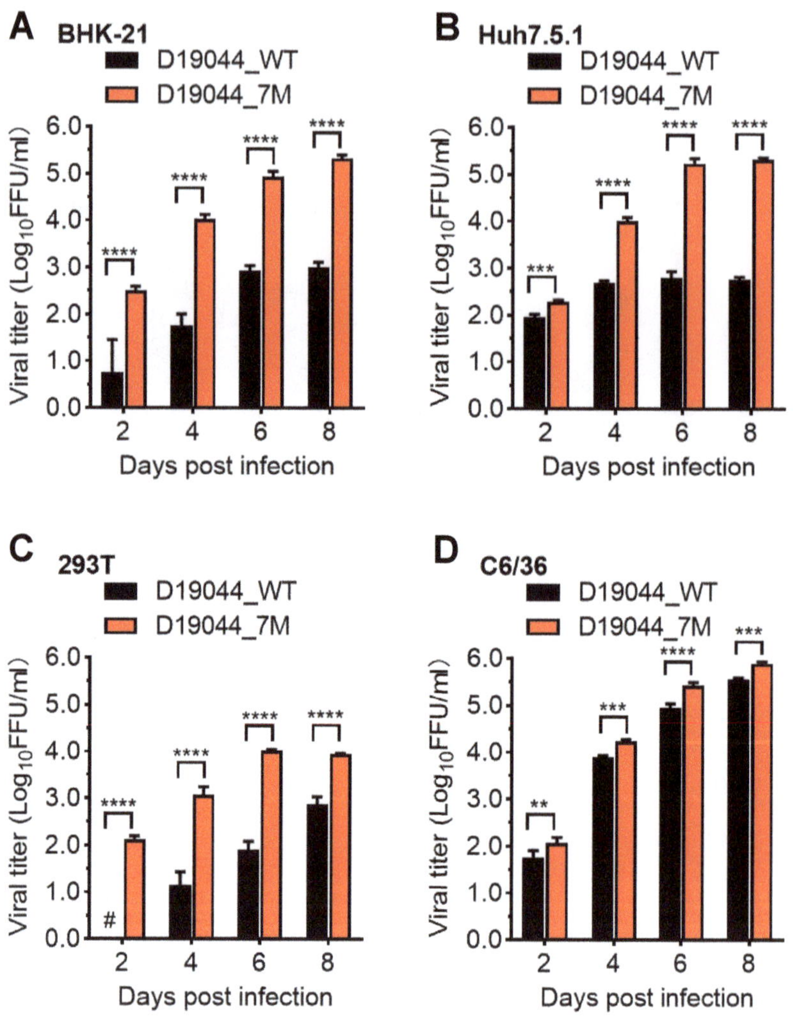

Figure 5. Virus production of D19044_WT and D19044_7M in various cell lines. BHK-21 cells (**A**), Huh7.5.1 cells (**B**), 293T cells (**C**), and C6/36 cells (**D**) were infected with D19044_WT and D19044_7M (MOI = 0.001). The viral titers released in supernatant were determined by FFU assay. A "#" indicates the titer was below LOD (<0.69 $\log_{10}$FFU/mL). The means $\pm$ SEM of three independent experiments are shown. Statistics analysis were performed using multiple comparison test in a two-way ANOVA. ** $p < 0.01$; *** $p < 0.001$; **** $p < 0.0001$.

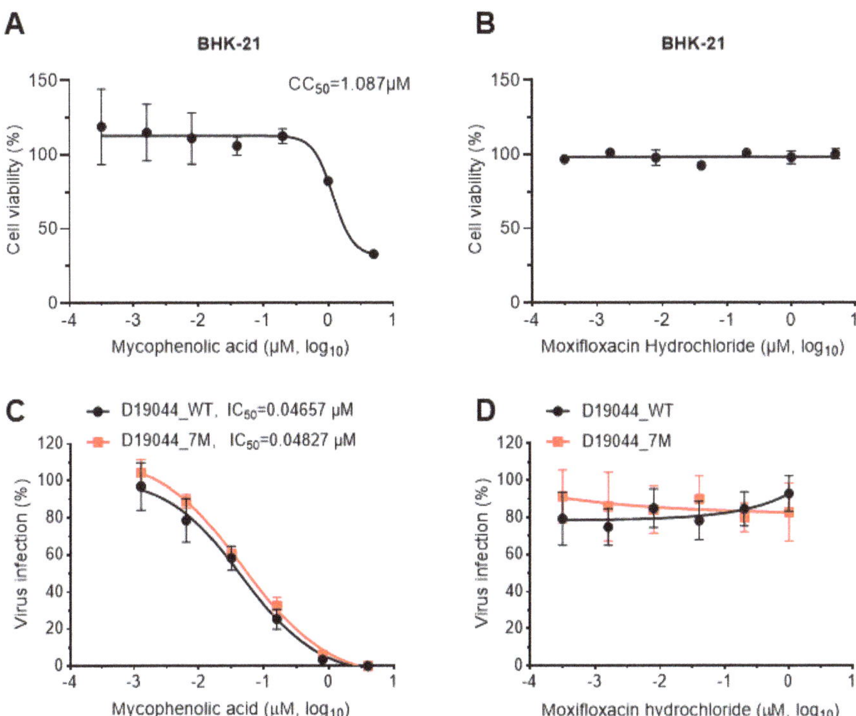

Figure 6. D19044 was inhibited antiviral compound in a dose-dependent manner. (**A**,**B**) The cell viability of BHK-21 cells under the treatment of mycophenolic acid (MPA) and moxifloxacin hydrochloride. (**C**) MPA inhibited D19044_7M and D19044_WT in a dose-dependent manner with comparable IC_{50} (0.04827 µM and 0.04657 µM). (**D**) Moxifloxacin hydrochloride did not inhibit D19044_7M and D19044_WT viruses. In **C** and **D**, BHK-21 cells were infected with D19044_7M and D19044_WT with MOI of 0.1 for 2 h, and then incubated with drugs of different concentrations for 48 h. The infection rates were examined by IFA and normalized to the nontreated mock controls. The means ± SEM of four determinations are shown.

4. Discussion

In this study, we constructed a DNA-launched full-length infectious clone for a DENV-1 genotype I isolate, D19044. Through serial passages in cell culture, seven amino acid substitutions (7M) were identified and subsequently demonstrated to promote efficient virus production in vitro. The final D19044_7M clone was able to release infectious viruses with peak titers of 7.5×10^4 FFU/mL and was genetically stable in transformant bacteria, thus enabling its usefulness to the field. Both D19044_WT and D19044_7M virus showed infections to BHK-21, 293T, Huh7.5.1, and C6/36 cells, and were inhibited by MPA drug in a dose-dependent manner. The development of a full-length infectious cDNA clone for a clinical isolate provides a useful tool for the study of DENV, especially for the serotype I DENV-1.

There are different ways to develop infectious clones for DENV isolates, and vector selection plays a key role for assembling a full-length cDNA clone with genetic stability in transformant bacteria. Different plasmid vectors have been used, including high copy number plasmid vector for a DENV-2 isolate [26], low copy number plasmid vectors for DENV-1 [34], DENV-2 [27,35], and DENV-4 [36], bacterial artificial chromosomes for DENV-1 [37] and DENV-2 [38]. Yeast artificial chromosomes [39] and shuttle vector [40–42] have also been used. Recently, 20 infectious clones covering DENV-1-4 were constructed using the circular polymerase extension reaction method [17,43], thus wisely bypassing propagation of flavivirus cDNA in bacteria and yeast, which was previously used for developing infectious clones of flaviviruses. In addition to vector selection, other strategies to stabilize the cDNA clones include inserting intron sequences to interrupt the putative *E. coli* promoters (ECP) [28,44] or introducing silent substitutions into the putative ECP sequence [18,20]. Moreover, methods avoiding transformation step have also been applied, such as long PCR [45], infectious-subgenomic-amplicons [46], and subgenomic plasmid recombination method (SuPReMe) [47].

We selected pTight vector to construct the D19044 full-length clone, as this vector is a low copy number plasmid originally modified from pBR322 [19], containing 7 × TRE and CMVmin promoter upstream of the cloned viral cDNA sequence. The 7 × TRE could inhibit the potential ECP activity that a viral cDNA sequence may have, thus permitting plasmid replication in bacteria and making laboratory manipulation feasible [18,19]. The CMVmin promoter-controlled vector allows a DNA-launched production of viral genomic RNA transcripts by plasmid transfection of eukaryotic cells, thus avoiding in vitro transcription and RNA manipulation. In addition, the CMVmin promoter vector also showed advantages in maintaining the stability of flavivirus genome in transformant bacteria, such as C41 and Turbo *E. coli* bacteria [18,19]. Nevertheless, although these elements noticeably reduced the ECP activity, we identified deletions, neucleotide and/or amino acid substitutions, or recombination in transformation-recovered plasmids, mainly in prM-E-NS1 region, when we ligated four fragments (A, B, C, and D) in the order of C → D → A. Thus, we tried different orders of assembling and only succeeded when these fragments were cloned in the order of B → A → C → D (Figure 2A). Noted that, colonies with mutagenesis events generally showed abnormal manifestation, such as a faster growth and a larger colony size; mutagenesis events may have eliminated or reduced toxic proteins producing in bacteria, thus bacteria gained a growth benefit. In contrast, bacteria colonies carrying correct viral cDNA usually grew slower, became identifiable after 24 h incubation, and had a smaller size. Through the use of the pTight vector and a specific order of fragment cloning, we generated a D19044_7M clone with genetic stability in the transformant bacteria. A full-length clone genetically stable in bacteria is vitally important and allows its utility for molecular manipulation.

Adaptive amino acid substitutions are often required to improve virus production of infectious clones [18,48,49]. D19044_WT clone showed a slow virus spread and low-level infectious virus after transfection (Figures 2 and 3), but became efficient after a number of virus passages (e.g., up to 34-37 passages, 3.9×10^5 FFU/mL). Seven amino acid substitutions (7M) were identified (Table 2) and proven to greatly improve virus production (1×10^5 FFU/mL). Therefore, the acquisition of adaptive amino acid substitutions by serial viral passage in the culture is a strategy recommendable for making an efficient infectious clone for DENV and other RNA viruses. Here, it is interesting that addition of 7M improved the virus infection greatly in the cell lines from human and Hamster Syrian, but to a less extent in mosquito C6/36 cells (Figure 5D). D19044_WT virus infected mosquito C6/36 cells more efficiently than other cell lines, suggesting that this virus may be more adaptive to C6/36 cells or mosquitos' phycological environment than in other species. These results may be due to the fact that the D19044_WT virus was generated from the cDNA clone constructed from infected-C6/36 cells after 6 days. It is unknown whether there was adaptive substitution(s) had been introduced into the viral genome for an efficient infection of C6/36 cells (without patient serum adequate for sequencing). However, in this case, 7M had proven a potential in culture adaptation of cross-species, thus the functional roles of 7M in the viral life cycle and cross-species adaptation in vitro and in vivo warrant future investigations.

The wild-type virus used in this study was generated from D19044_WT cDNA clone that was made from infection supernatant of C3/36 cells (day 6), due to a limited volume of patient serum available. The C6/36-deirived supernatant showed very low infection in Vero and BHK-21 cells after a number of passages without detectable viral titers, thus we were unable to compare the infection of D19044_7M and original serum virus in cultured cells. In addition, the functional roles of 7M and potential effect of partial 5′UTR and 3′UTR sequences from GZ2002 used in D19044_7M clone were not explored. These limitations of this study warrant future investigations towards an in-depth characterization and utility of D19044_7M clone.

In summary, we have developed an efficient infectious clone for a genotype I DENV-1 isolate that was recently prevalent in south China. Given that genotype I virus is predominantly circulating in different regions and is responsible for several major outbreaks, D19044_7M will be highly relevant for the study of DENV-1.

Author Contributions: Conceptualization, Y.-P.L. and T.L.; investigation, M.H., T.W., Y.Y., T.C., J.H., D.W.; data analysis, M.H., T.W., Y.Y., Y.W., and Y.-P.L.; resources, D.W. and Y.-P.L.; data curation, M.H. and Y.-P.L.; original draft preparation, M.H., T.W. and Y.-P.L.; review and editing, Y.-P.L.; supervision, Y.-P.L.; funding acquisition, Y.-P.L. All authors have read and agreed to the published version of the manuscript.

Funding: This work was supported by National Key R&D Program of China (No. 2020YFC1200100).

Institutional Review Board Statement: Not applicable.

Informed Consent Statement: Not applicable.

Data Availability Statement: All experimental data are present in the manuscript. Sequence data has been deposited in GenBank. Plasmids are available upon request.

Acknowledgments: We thank Andrew Yueh (National Health Research Institutes, Taiwan, China) for providing the modified pTight-DENV-2 clone, Frank Chisari (The Scripps Research Institute, CA, USA) and Jin Zhong (Institut Pasteur of Shanghai, Shanghai, China) for providing Huh7.5.1 cells, and Ping Zhang (Sun Yat-sen University, Guangzhou, China) for providing C6/36 cells. Hongying Chen (Northwest A&F University, Xianyang, China) for providing Turbo competent bacteria.

Conflicts of Interest: Authors declare no conflict of interest exists.

References

1. Wu, T.; Wu, Z.; Li, Y.P. Dengue fever and dengue virus in the People's Republic of China. *Rev. Med. Virol.* **2022**, *32*, e2245. [CrossRef] [PubMed]
2. Diamond, M.S.; Zachariah, M.; Harris, E. Mycophenolic Acid Inhibits Dengue Virus Infection by Preventing Replication of Viral RNA. *Virology* **2002**, *304*, 211–221. [CrossRef] [PubMed]
3. Guo, C.; Zhou, Z.; Wen, Z.; Liu, Y.; Zeng, C.; Xiao, D.; Ou, M.; Han, Y.; Huang, S.; Liu, D.; et al. Global Epidemiology of Dengue Outbreaks in 1990–2015: A Systematic Review and Meta-Analysis. *Front. Cell. Infect. Microbiol.* **2017**, *7*, 317. [CrossRef] [PubMed]
4. Zhang, Z.; Rong, L.; Li, Y.-P. *Flaviviridae* Viruses and Oxidative Stress: Implications for Viral Pathogenesis. *Oxidative Med. Cell. Longev.* **2019**, *2019*, 1409582. [CrossRef]
5. Wilder-Smith, A. Dengue vaccine development: Status and future. *Bundesgesundheitsblatt Gesundh. Gesundh.* **2020**, *63*, 40–44. [CrossRef]
6. WHO. Ten Threats to Global Health in 2019. WHO Spotlight 2019. Available online: https://www.who.int/news-room/spotlight/ten-threats-to-global-health-in-2019 (accessed on 29 May 2022).
7. Holmes, E.C.; Twiddy, S.S. The origin, emergence and evolutionary genetics of dengue virus. *Infect. Genet. Evol.* **2003**, *3*, 19–28. [CrossRef]
8. Rico-Hesse, R. Molecular evolution and distribution of dengue viruses type 1 and 2 in nature. *Virology* **1990**, *174*, 479–493. [CrossRef]
9. Guzman, M.G.; Harris, E. Dengue. *Lancet* **2015**, *385*, 453–465. [CrossRef]
10. Khetarpal, N.; Khanna, I. Dengue Fever: Causes, Complications, and Vaccine Strategies. *J. Immunol. Res.* **2016**, *2016*, 6803098. [CrossRef]
11. Langerak, T.; Mumtaz, N.; Tolk, V.I.; Van Gorp, E.C.M.; Martina, B.E.; Rockx, B.; Koopmans, M.P.G. The possible role of cross-reactive dengue virus antibodies in Zika virus pathogenesis. *PLOS Pathog.* **2019**, *15*, e1007640. [CrossRef]
12. Mishra, N.; Boudewijns, R.; Schmid, M.A.; Marques, R.E.; Sharma, S.; Neyts, J.; Dallmeier, K. A Chimeric Japanese Encephalitis Vaccine Protects against Lethal Yellow Fever Virus Infection without Inducing Neutralizing Antibodies. *mBio* **2020**, *11*, e02494-19. [CrossRef] [PubMed]
13. Shukla, R.; Ramasamy, V.; Shanmugam, R.K.; Ahuja, R.; Khanna, N. Antibody-Dependent Enhancement: A Challenge for Developing a Safe Dengue Vaccine. *Front. Cell. Infect. Microbiol.* **2020**, *10*, 572681. [CrossRef] [PubMed]
14. Halstead, S.B.; Dans, L.F. Dengue infection and advances in dengue vaccines for children. *Lancet Child Adolesc. Health* **2019**, *3*, 734–741. [CrossRef]

15. Thai, K.T.D.; Henn, M.R.; Zody, M.C.; Tricou, V.; Nguyet, N.M.; Charlebois, P.; Lennon, N.J.; Green, L.; de Vries, P.J.; Hien, T.T.; et al. High-Resolution Analysis of Intrahost Genetic Diversity in Dengue Virus Serotype 1 Infection Identifies Mixed Infections. *J. Virol.* **2012**, *86*, 835–843. [CrossRef] [PubMed]

16. Pyke, A.T.; Moore, P.R.; Taylor, C.T.; Hall-Mendelin, S.; Cameron, J.N.; Hewitson, G.R.; Pukallus, D.S.; Huang, B.; Warrilow, D.; van den Hurk, A.F. Highly divergent dengue virus type 1 genotype sets a new distance record. *Sci. Rep.* **2016**, *6*, 22356. [CrossRef] [PubMed]

17. Tamura, T.; Zhang, J.; Madan, V.; Biswas, A.; Schwoerer, M.P.; Cafiero, T.R.; Heller, B.L.; Wang, W.; Ploss, A. Generation and characterization of genetically and antigenically diverse infectious clones of dengue virus serotypes 1–4. *Emerg. Microbes Infect.* **2022**, *11*, 227–239. [CrossRef]

18. Chen, Y.; Liu, T.; Zhang, Z.; Chen, M.; Rong, L.; Ma, L.; Yu, B.; Wu, D.; Zhang, P.; Zhu, X.; et al. Novel genetically stable infectious clone for a Zika virus clinical isolate and identification of RNA elements essential for virus production. *Virus Res.* **2018**, *257*, 14–24. [CrossRef]

19. Pu, S.-Y.; Wu, R.-H.; Tsai, M.-H.; Yang, C.-C.; Chang, C.-M.; Yueh, A. A novel approach to propagate flavivirus infectious cDNA clones in bacteria by introducing tandem repeat sequences upstream of virus genome. *J. Gen. Virol.* **2014**, *95*, 1493–1503. [CrossRef] [PubMed]

20. Pu, S.-Y.; Wu, R.-H.; Yang, C.-C.; Jao, T.-M.; Tsai, M.-H.; Wang, J.-C.; Lin, H.-M.; Chao, Y.-S.; Yueh, A. Successful Propagation of Flavivirus Infectious cDNAs by a Novel Method to Reduce the Cryptic Bacterial Promoter Activity of Virus Genomes. *J. Virol.* **2011**, *85*, 2927–2941. [CrossRef] [PubMed]

21. Fang, X.; Zhang, J.; Zhu, J.; Hu, Z.; Liu, J.; Yang, J.; Yuan, W.; Shang, W.; Rao, X. Genomic phylogenetic analysis of one Dengue virus type 1 strain GZ2002 isolated in Guangzhou. *China J. Mod. Med.* **2012**, *22*, 25–33.

22. de Borba, L.; Villordo, S.M.; Iglesias, N.G.; Filomatori, C.V.; Gebhard, L.G.; Gamarnik, A.V. Overlapping Local and Long-Range RNA-RNA Interactions Modulate Dengue Virus Genome Cyclization and Replication. *J. Virol.* **2015**, *89*, 3430–3437. [CrossRef] [PubMed]

23. Ng, W.C.; Soto-Acosta, R.; Bradrick, S.S.; Garcia-Blanco, M.A.; Ooi, E.E. The 5′ and 3′ Untranslated Regions of the Flaviviral Genome. *Viruses* **2017**, *9*, 137. [CrossRef]

24. Payne, A.F.; Binduga-Gajewska, I.; Kauffman, E.B.; Kramer, L.D. Quantitation of flaviviruses by fluorescent focus assay. *J. Virol. Methods* **2006**, *134*, 183–189. [CrossRef] [PubMed]

25. Schoepp, R.J.; Beaty, B.J. Titration of dengue viruses by immunofluorescence in microtiter plates. *J. Clin. Microbiol.* **1984**, *20*, 1017–1019. [CrossRef]

26. Sriburi, R.; Keelapang, P.; Duangchinda, T.; Pruksakorn, S.; Maneekarn, N.; Malasit, P.; Sittisombut, N. Construction of infectious dengue 2 virus cDNA clones using high copy number plasmid. *J. Virol. Methods* **2001**, *92*, 71–82. [CrossRef]

27. Kinney, R.M.; Butrapet, S.; Chang, G.-J.J.; Tsuchiya, K.R.; Roehrig, J.; Bhamarapravati, N.; Gubler, D.J. Construction of Infectious cDNA Clones for Dengue 2 Virus: Strain 16681 and Its Attenuated Vaccine Derivative, Strain PDK-53. *Virology* **1997**, *230*, 300–308. [CrossRef] [PubMed]

28. Li, G.; Jin, H.; Nie, X.; Zhao, Y.; Feng, N.; Cao, Z.; Tan, S.; Zhang, B.; Gai, W.; Yan, F.; et al. Development of a reverse genetics system for Japanese encephalitis virus strain SA14-14-2. *Virus Genes* **2019**, *55*, 550–556. [CrossRef] [PubMed]

29. Mishin, V.P.; Cominelli, F.; Yamshchikov, V.F. A 'minima' approach in design of flavivirus infectious DNA. *Virus Res.* **2001**, *81*, 113–123. [CrossRef]

30. Schwarz, M.C.; Sourisseau, M.; Espino, M.M.; Gray, E.S.; Chambers, M.T.; Tortorella, D.; Evans, M.J. Rescue of the 1947 Zika Virus Prototype Strain with a Cytomegalovirus Promoter-Driven cDNA Clone. *mSphere* **2016**, *1*, e00246-16. [CrossRef] [PubMed]

31. Barrows, N.J.; Campos, R.K.; Powell, S.T.; Prasanth, K.R.; Schott-Lerner, G.; Soto-Acosta, R.; Galarza-Muñoz, G.; McGrath, E.L.; Urrabaz-Garza, R.; Gao, J.; et al. A Screen of FDA-Approved Drugs for Inhibitors of Zika Virus Infection. *Cell Host Microbe* **2016**, *20*, 259–270. [CrossRef] [PubMed]

32. Dong, S.; Kang, S.; Dimopoulos, G. Identification of anti-flaviviral drugs with mosquitocidal and anti-Zika virus activity in Aedes aegypti. *PLoS Negl. Trop. Dis.* **2019**, *13*, e0007681. [CrossRef]

33. Shandil, R.K.; Jayaram, R.; Kaur, P.; Gaonkar, S.; Suresh, B.L.; Mahesh, B.N.; Jayashree, R.; Nandi, V.; Bharath, S.; Balasubramanian, V. Moxifloxacin, ofloxacin, sparfloxacin, and ciprofloxacin against Mycobacterium tuberculosis: Evaluation of in vitro and pharmacodynamic indices that best predict in vivo efficacy. *Antimicrob. Agents Chemother.* **2007**, *51*, 576–582. [CrossRef]

34. Gallichotte, E.N.; Menachery, V.D.; Yount, B.L., Jr.; Widman, D.G.; Dinnon, K.H., III; Hartman, S.; de Silva, A.M.; Baric, R.S. Epitope Addition and Ablation via Manipulation of a Dengue Virus Serotype 1 Infectious Clone. *mSphere* **2017**, *2*, e00380-16. [CrossRef] [PubMed]

35. Davidson, A.D. Development and Application of Dengue Virus Reverse Genetic Systems. *Methods Mol. Biol.* **2014**, *1138*, 113–130. [CrossRef] [PubMed]

36. Lai, C.J.; Zhao, B.T.; Hori, H.; Bray, M. Infectious RNA transcribed from stably cloned full-length cDNA of dengue type 4 virus. *Proc. Natl. Acad. Sci. USA* **1991**, *88*, 5139–5143. [CrossRef] [PubMed]

37. Suzuki, R.; de Borba, L.; Duarte dos Santos, C.N.; Mason, P.W. Construction of an infectious cDNA clone for a Brazilian prototype strain of dengue virus type 1: Characterization of a temperature-sensitive mutation in NS1. *Virology* **2007**, *362*, 374–383. [CrossRef] [PubMed]

38. Usme-Ciro, J.A.; Lopera, J.A.; Enjuanes, L.; Almazán, F.; Gallego-Gomez, J.C. Development of a novel DNA-launched dengue virus type 2 infectious clone assembled in a bacterial artificial chromosome. *Virus Res.* **2014**, *180*, 12–22. [CrossRef] [PubMed]
39. Polo, S.; Ketner, G.; Levis, R.; Falgout, B. Infectious RNA transcripts from full-length dengue virus type 2 cDNA clones made in yeast. *J. Virol.* **1997**, *71*, 5366–5374. [CrossRef]
40. Kelly, E.P.; Puri, B.; Sun, W.; Falgout, B. Identification of mutations in a candidate dengue 4 vaccine strain 341750 PDK20 and construction of a full-length cDNA clone of the PDK20 vaccine candidate. *Vaccine* **2010**, *28*, 3030–3037. [CrossRef]
41. Puri, B.; Polo, S.; Hayes, C.G.; Falgout, B. Construction of a full length infectious clone for dengue-1 virus Western Pacific, 74 strain. *Virus Genes* **2000**, *20*, 57–63. [CrossRef] [PubMed]
42. Santos, J.; Cordeiro, M.T.; Bertani, G.R.; Marques, E.; Gil, L.H.V.G. Construction and characterisation of a complete reverse genetics system of dengue virus type 3. *Mem. Inst. Oswaldo Cruz* **2013**, *108*, 983–991. [CrossRef] [PubMed]
43. Quan, J.; Tian, J. Circular Polymerase Extension Cloning of Complex Gene Libraries and Pathways. *PLoS ONE* **2009**, *4*, e6441. [CrossRef] [PubMed]
44. Guo, J.; He, Y.; Wang, X.; Jiang, B.; Lin, X.; Wang, M.; Jia, R.; Zhu, D.; Liu, M.; Zhao, X.; et al. Stabilization of a full-length infectious cDNA clone for duck Tembusu virus by insertion of an intron. *J. Virol. Methods* **2020**, *283*, 113922. [CrossRef]
45. Govindarajan, D.; Guan, L.; Meschino, S.; Fridman, A.; Bagchi, A.; Pak, I.; ter Meulen, J.; Casimiro, D.R.; Bett, A.J. A Rapid and Improved Method to Generate Recombinant Dengue Virus Vaccine Candidates. *PLoS ONE* **2016**, *11*, e0152209. [CrossRef] [PubMed]
46. Aubry, F.; Nougairède, A.; De Fabritus, L.; Querat, G.; Gould, E.A.; De Lamballerie, X. Single-stranded positive-sense RNA viruses generated in days using infectious subgenomic amplicons. *J. Gen. Virol.* **2014**, *95*, 2462–2467. [CrossRef] [PubMed]
47. Driouich, J.-S.; Ali, S.M.; Amroun, A.; Aubry, F.; De Lamballerie, X.; Nougairède, A. SuPReMe: A rapid reverse genetics method to generate clonal populations of recombinant RNA viruses. *Emerg. Microbes Infect.* **2018**, *7*, 40. [CrossRef] [PubMed]
48. Chen, M.; Zheng, F.; Yuan, G.; Duan, X.; Rong, L.; Liu, J.; Feng, S.; Wang, Z.; Wang, M.; Feng, Y.; et al. Development of an Infectious Cell Culture System for Hepatitis C Virus Genotype 6a Clinical Isolate Using a Novel Strategy and Its Sensitivity to Direct-Acting Antivirals. *Front. Microbiol.* **2018**, *9*, 2950. [CrossRef]
49. Li, J.; Zhou, Q.; Rong, L.; Rong, D.; Yang, Y.; Hao, J.; Zhang, Z.; Ma, L.; Rao, G.; Zhou, Y.; et al. Development of cell culture infectious clones for hepatitis C virus genotype 1b and transcription analysis of 1b-infected hepatoma cells. *Antivir. Res.* **2021**, *193*, 105136. [CrossRef] [PubMed]

Article

Allosteric Inhibitors of Zika Virus NS2B-NS3 Protease Targeting Protease in "Super-Open" Conformation

Ittipat Meewan [1,2,†], Sergey A. Shiryaev [3,†], Julius Kattoula [2], Chun-Teng Huang [3], Vivian Lin [3], Chiao-Han Chuang [3], Alexey V. Terskikh [3,*] and Ruben Abagyan [2,*]

[1] Institute of Molecular Biosciences, Mahidol University, Nakhon Pathom 73170, Thailand
[2] Skaggs School of Pharmacy and Pharmaceutical Sciences, University of California San Diego, La Jolla, CA 92093, USA
[3] Sanford-Burnham-Prebys Medical Discovery Institute, La Jolla, CA 92037, USA
* Correspondence: terskikh@sbpdiscovery.org (A.V.T.); rabagyan@health.ucsd.edu (R.A.)
† These authors equally contributed to this work.

Abstract: The Zika virus (ZIKV), a member of the Flaviviridae family, is considered a major health threat causing multiple cases of microcephaly in newborns and Guillain-Barré syndrome in adults. In this study, we targeted a transient, deep, and hydrophobic pocket of the "super-open" conformation of ZIKV NS2B-NS3 protease to overcome the limitations of the active site pocket. After virtual docking screening of approximately seven million compounds against the novel allosteric site, we selected the top six candidates and assessed them in enzymatic assays. Six candidates inhibited ZIKV NS2B-NS3 protease proteolytic activity at low micromolar concentrations. These six compounds, targeting the selected protease pocket conserved in ZIKV, serve as unique drug candidates and open new opportunities for possible treatment against several flavivirus infections.

Keywords: Zika virus protease inhibitors; allosteric inhibitors; Zika virus NS2B-NS3 protease; super-open conformation

Citation: Meewan, I.; Shiryaev, S.A.; Kattoula, J.; Huang, C.-T.; Lin, V.; Chuang, C.-H.; Terskikh, A.V.; Abagyan, R. Allosteric Inhibitors of Zika Virus NS2B-NS3 Protease Targeting Protease in "Super-Open" Conformation. *Viruses* **2023**, *15*, 1106. https://doi.org/10.3390/v15051106

Academic Editor: Ronald N. Harty

Received: 21 February 2023
Revised: 26 April 2023
Accepted: 27 April 2023
Published: 30 April 2023

1. Introduction

The Zika virus (ZIKV) is an emerging, global pathogen declared as a Public Health Emergency of International Concern by the World Health Organization. Approximately 84,000 cumulative cases have been reported globally since 2016 [1–4]. The virus is transmitted by the mosquito species *Aedes aegypti* and *Aedes albopictus*, as well as through human contact [1,5–7]. The main symptoms of a ZIKV infection, including fever, rash, and conjunctivitis, are relatively mild and frequently asymptomatic in adults. However, ZIKV has been associated with microcephaly in children born to mothers infected with ZIKV during pregnancy, as evidenced by the four-times increase in reported microcephaly cases from the end of January to mid-November of 2016 compared to the same period in 2015 [8–11]. Moreover, there is also evidence that ZIKV infection may be linked with Guillain-Barré syndrome and other neuroinflammatory disorders, which cause autoimmune degeneration of the myelin sheath in the peripheral neurons of adults since these conditions were observed to rise during the ZIKV epidemic in Colombia during 2015–2017 [12–15]. Currently, there are no effective drugs against ZIKV or any other flavivirus infection [16]. Furthermore, vaccines against several Flavivirus infections may be problematic due to antibody-dependent enhancement phenomena, with the notable exceptions of Japanese Encephalitis and Yellow Fever [17,18].

The two-part ZIKV protease is essential for viral replication, designating it as a potential target for treatment. ZIKV nonstructural NS3 protein is a multi-functional protein with protease and helicase activities. The C-terminal 440 residue region of NS3 encodes NS3 helicase (NS3hel), which participates in RNA replication and RNA capping [19–23]. A central cytosolic ~50 residue fragment of the transmembrane NS2B protein forms a functional

NS2B-NS3pro complex with the N-terminal ~170 residues of NS3, responsible for the cleavage of ZIKV polyprotein (Figure 1). The cleavage results in activating capsid (C), pre-membrane (pr), membrane (M), and envelope (E) structural, and nonstructural (NS1, NS2A, NS2B, NS3, NS4A, NS4B, and NS5) proteins [24,25]. The NS2B cofactor is required for the protease activity of NS3pro. ZIKV NS2B-NS3pro possesses high proteolytic activity but becomes enzymatically inactive if the NS2B cofactor is deleted [25,26]. In addition to the viral polyprotein processing, the presence of multiple copies of active viral protease inside the host cell cytoplasm may lead to irreversible damage to numerous host cell proteins. Thus, NS2B-NS3pro is a promising drug target for the treatment of ZIKV infection [16,26,27]. Targeting the viral protease has been shown as a successful therapeutic strategy for ZIKV as well as the infections caused by other members of the Flaviviridae family, including Dengue, West Nile, and Hepatitis C viruses [28–33].

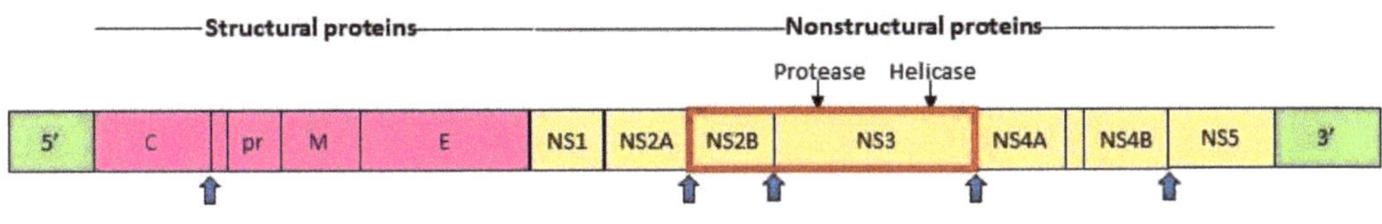

Figure 1. Schematic representation of the ZIKV genome. ZIKV polyprotein precursor includes structural proteins, shown in magenta, and nonstructural proteins, shown in yellow. ZIKV NS2B-NS3pro, highlighted by a red box, plays a significant role in viral protein activation. Its cleavage sites are represented by blue arrows.

Designing or screening for effective competitive inhibitors for flavivirus proteases has been challenging since strong and specific binding to S1 and S2 active sub-pockets requires hydrophilic and electrostatic interactions, usually associated with low membrane permeabilities [29]. Therefore, in our research, we sought to identify a novel type of non-competitive inhibitor against NS2B-NS3pro that prevent the formation of the active ZIKV NS2B-NS3pro complex. We chose to target a recently identified transient pocket present in the crystal structure of WNV NS2B-NS3pro, termed the "super-open" conformation [34]. The structural analysis of WNV NS2B-NS3pro reveals that there are two distinct conformations based on the placement of NS2B cofactor: the "closed conformation", in which the NS2B wraps around NS3 and is the active conformation, and the "super-open" conformation, in which the NS2B chain turns and binds to a small area behind the active site, resulting in its inactivation (see Figure 2). Here, we employed a large-scale Molsoft-ICM docking screen of approximately seven million compounds to find effective non-covalent inhibitors for the allosteric pocket of ZIKV NS2B-NS3pro, with six selected top candidates evaluated experimentally, and they were confirmed to inhibit the enzymatic activity of the viral protease.

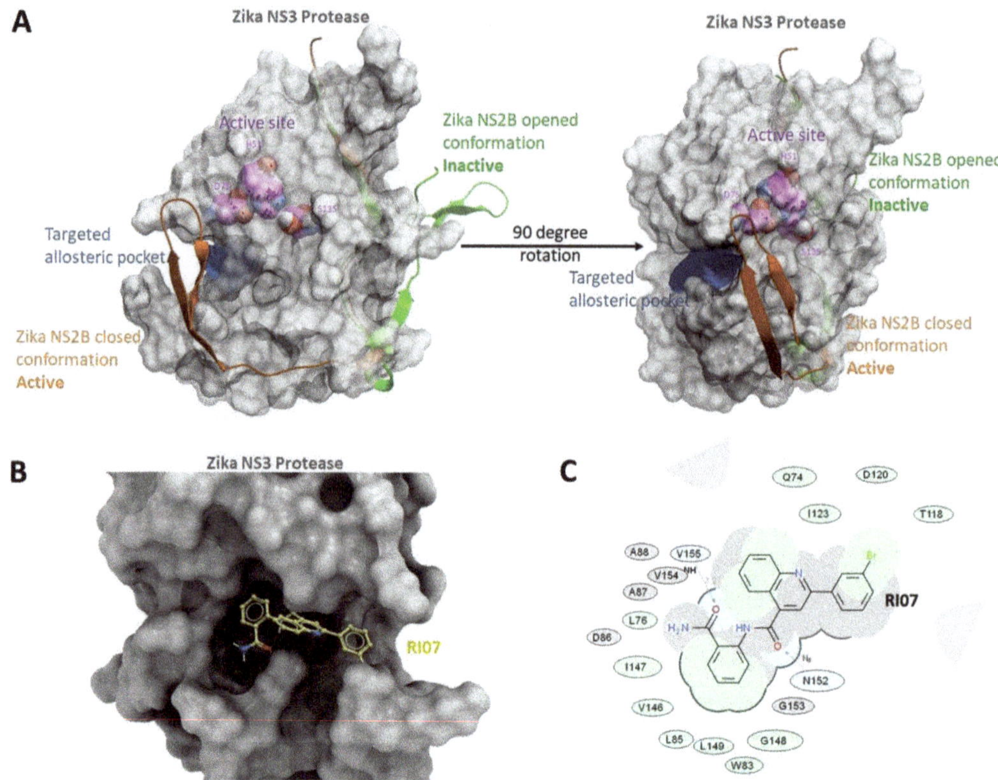

Figure 2. Targeting a hidden pocket exposed in the "super-open" conformation of ZIKV NS2B-NS3 protease. (**A**) ZIKV NS3 subunit in grey and active site in purple. Beta hairpin of NS2B shown in closed conformation opens a transient (allosteric) pocket by moving away into an open (green ribbon) conformation. Targeted allosteric pocket is in blue, and catalytic triad of NS3, S135, H51, and D75 are labeled and shown in CPK; (**B**) predicted 3D binding pose of RI07 targeting the super-open conformation of NS2B-NS3 protease; and (**C**) 2D interaction diagram for predicted pose of RI07 compound (Q74, D75, L76, W83, L85, D86, A87, A88, T118, D120, I123, V146, I147, G148, L149, N152, G153, V154, and V155). Hydrogen bonding interactions are shown by grey dotted lines.

2. Materials and Methods

2.1. Reagents

Routine laboratory reagents were purchased from Thermo Fisher Scientific, US. The tested compounds were purchased from Chembridge Corp., San Diego, CA, USA.

2.2. Virtual Ligand Screening of the Compound Library

The structure of NS2-NS3pro was obtained from the X-ray crystal structure (PDB ID 5TFN) [34]. The "super-open" conformation and the transient pocket exposed in that conformation were identified by comparing 5LC0 (closed) with five other structures (PDB ID: 5TFN, 5TFO, 6UM3, 5T1V, and 2GGV) [34–37]. The single chain construct in 5TFN contained both NS2 and NS3 domains, and the NS2 domain was marked to map possible interactions of the selected inhibitors with the NS3 chain. A docking screen was performed against approximately 7 million small molecules from the eMolecules catalog [38], commercially available compounds that have not been reported as ZIKV protease inhibitors and are predicted to have low toxicity using Molsoft ICM 3.9-1e software [39–41]. The scored binding poses of small molecules were predicted by the Biased-Probability Monte

Carlo optimizer in Molsoft ICM software, and binding affinity of small molecules to the receptor was ranked based on force field-based docking score extended with additional free-energy contributions. All scoring functions and pharmacokinetic properties prediction were performed using Molsoft ICM-Pro v3.9 [39–41].

2.3. Cloning and Purification of ZIKV NS2B-NS3pro Construct

The recombinant construct expressing ZIKV NS2B-NS3 protease with a 6xHisTag on its N-terminus was used to transform competent *E. coli* BL21 (DE3) Codon Plus cells obtained from Stratagene. Transformed cells were grown at 30 °C in LB broth containing carbenicillin (0.1 mg/mL). Cultures were induced with 0.6 mM IPTG for 16 h at 18 °C. Cells were collected by centrifugation, re-suspended in Tris-HCl buffer, pH 8.0, containing 1 M NaCl, and disrupted by sonication (30 s pulse and 30 s intervals; 8 pulses) on ice. The pellet was removed by centrifugation ($40,000\times g$; 30 min). The construct was then purified from the supernatant fraction on a Ni-NTA Sepharose, equilibrated with 20 mM Tris-HCl buffer, pH 8.0, supplemented with 1 M NaCl. After washing out the impurities using the same buffer supplemented with 35 mM imidazole, the bound material was eluted using a 35–500 mM gradient of imidazole. The fractions containing the recombinant protein were combined, dialyzed against 20 mM Tris-HCl, pH 8.0, containing 150 mM NaCl, and stored at −80 °C until use. The purity of the material was tested by SDS gel-electrophoresis (12% NuPAGE-MOPS, Invitrogen, Waltham, MA, USA), followed by Coomassie staining and by Western blotting with anti-HisTag antibodies.

2.4. Protease Activity Assay with Fluorescent Peptide

The peptide cleavage activity assay with the purified ZIKV NS2B-NS3pro samples was performed in 0.2 mL 20 mM Tris-HCl buffer, pH 8.0, containing 20% glycerol and 0.005% Brij-35. The cleavage peptide pyroglutamic acid Pyr-Arg-Thr-Lys-Arg-7-amino-4-methylcoumarin (Pyr-RTKR-AMC) and enzyme concentrations were 20 µM and 10 nM, respectively [26]. Reaction velocity was monitored continuously at $\lambda ex = 360$ nm and $\lambda ex = 465$ nm on a Tecan fluorescence spectrophotometer (Tecan Group Ltd., Männedorf, Switzerland). All assays were performed in triplicate wells of a 96-well plate. Dose–response curves and IC_{50} values of each compound were calculated accordingly using SciPy and Matplotlib Python packages.

2.5. Determination of the IC_{50} Values of the Inhibitory Compounds

The ZIKV NS2B-NS3pro construct (20 nM) was pre-incubated for 30 min at 20 °C with increasing concentrations of the individual compounds in 0.1 mL 20 mM Tris-HCl buffer, pH 8.0, containing 20% glycerol and 0.005% Brij 35. The Pyr-RTKR-AMC substrate (20 µM) was then added in 0.1 mL of the same buffer. All assays were performed in triplicate wells of a 96-well plate. IC_{50} values were calculated by determining the concentrations of the compounds needed to inhibit 50% of the NS2B-NS3pro activity against Pyr-RTKR-AMC. GraphPad Prism was used as fitting software.

3. Results

3.1. Identification of an Allosteric Site on ZIKV NS2B-NS3 Protease

The active site of ZIKV NS2B-NS3 protease has been a favorable protein target for designing small molecules against ZIKV infection. However, a recent analysis of the NS2B-NS3 protease structure suggested that the catalytic triad of ZIKV NS2B-NS3pro is conserved in various human serine proteinases, and the corresponding pocket is shallow. Thus, designing specific and potent competing small molecules against the active pocket of ZIKV NS2B-NS3pro is quite challenging. To identify a new promising druggable pocket, we analyzed four structures of ZIKV NS2B-NS3 protease that were recently deposited in the Protein Data Bank (PDB: 5TFN, 5TFO, 6UM3, and 5T1V) [25]. The protease in all four structures contained the novel "super-open" pocket. The active "closed" conformation observed in PDB ID 5LC0 [36] undergoes a transition to the inactive super-open conformation and reveals a targetable pocket. Therefore, we considered that interference between the NS3 domain and its cofactor, NS2B (Figure 2A), might be a superior drug target to stabilize the inactive "super-open" conformation in the design of allosteric novel highly potent and specific small molecules for ZIKV infection treatment. The list of contact residues in the newly identified transient pocket includes W83, L85, V146, A87, A88, I147, G148, D86, L149, V154, V155, I123, D120, T118, G153, N152, Q74, D75, and L76. The transient pocket was then used as the receptor to find strong binders using the virtual docking screen.

3.2. Virtual Docking Screen of Seven Million Compounds Targeting the Allosteric Hidden Site of ZIKV NS2B-NS3 Protease

The resulting model of the "super-open" state of ZIKV protease (PDB 5TFN) was docked and scored against approximately seven million small molecules from the eMolecules database. Chemicals were first filtered by removing compounds with high toxicity propensity based on over 1000 structural alerts [41]. The binding free energy of each compound to the allosteric pocket was estimated by a docking score computed via the MolSoft ICM-Pro package [39,40]. The ICM scoring function includes van der Waals potential for a hydrogen atom probe; van der Waals potential for a heavy-atom probe (generic carbon of 1.7 Å radius); optimized electrostatic term; hydrophobic terms; and lone-pair-based potential for approximation of the intermolecular interaction between the receptor and ligand. Ten different compounds with the top ICM-Docking scores were initially tested in the cell-based assay to assess their inhibition activity against ZIKV NS2B-NS3 protease (see Table S1).

3.3. Selection of Core Functional Group

Ten protease inhibitor candidates suggested from predicted docking poses and binding scores were purchased from an available vendor (ChemBridge Corp, 11199 Sorrento Valley Rd., Suite 206, San Diego, CA, USA) and tested against 10 nM of ZIKV NS2B-NS3 in proteolytic activity assays with 20 μM of the fluorogenic peptide substrate, Pyr-RTKR-AMC, in order to validate viral protease inhibition. The structures and docking scores of the initial ten protease inhibitor candidates found from the virtual screening can be found in Table S1. We found that compounds with phenylquinoline and aminobenzamide groups, shown in Figure 3, demonstrated desirable activity, indicating a potential scaffold for further optimization of allosteric inhibitors of ZIKV NS2B-NS3 protease. The binding conformation of RI07, the representative of phenylquinoline and aminobenzamide substituent compounds, in the deep allosteric pocket of the "super-open" conformation of NS2B-NS3 is shown in Figure 2B,C, suggesting strong binding affinity between aminobenzamide and hydrophobic residues including W83, L85, D86, A87, A88, V146, and L149, in the identified allosteric pocket of ZIKV NS2B-NS3.

General structure of R07, RI22, RI23, RI24, RI27, and RI28.

Figure 3. *Cont.*

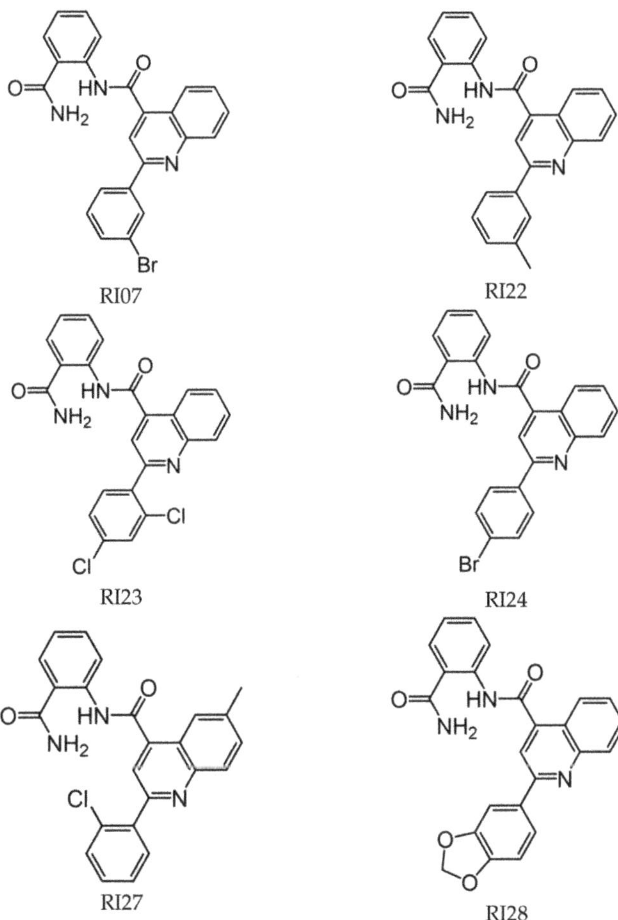

Figure 3. The figure presents the chemical structures of a general scaffold and six specific phenylquinoline and aminobenzamide compounds investigated in this study. Phenylquinoline substructures are highlighted in blue, and aminobenzamide substructures are highlighted in red.

3.4. Variation of Phenylquinoline and Aminobenzamide Substituents and Enzymatic Activities against ZIKV NS2B-NS3 Protease

Phenylquinoline and aminobenzamide-containing compounds showed high affinity to ZIKV NS2B-NS3 in proteolytic activity assays. This prompted us to search for more inhibitors from the same chemical class that are commercially available. We found five compounds from the available vendor with structures related to RI07, including the presence of aminobenzamide and phenylquinoline with varied substituents. These compounds, labeled as RI22, RI23, RI24, RI27, and RI28, were tested for their inhibition activity in the enzymatic assay. The structures of RI07 and its derivatives can be found in Figure 3.

The novel compounds in this series were evaluated for their ability to inhibit the proteolytic activity of the ZIKV protease. The inhibitor candidates, with IC_{50} values ranging in low micromolar concentrations from 3.8 to 14.4 µM, were identified (see Table 1 and Figure 4). Table 1 lists the docking scores, molecular weight, and IC_{50} values of all candidates as well as their relevant pharmacokinetic and toxicity properties, including water solubility, cell permeability, propensity for pan-assay interference, potassium channel blocking activity, polar surface area, and toxicity. These properties were evaluated using appropriate Molscreen models from the MolSoft ICM-Pro package [39–41]. The predicted parameters for these potential candidates fell within acceptable ranges. According to the docking simulation, RI07, the best compound in this series, was predicted to interact with several residues in the novel "super-open" pocket of the ZIKV NS2B-NS3 protease, including W83, L85, V146, A87, A88, I147, G148, D86, L149, V154, V155, I123, D120, T118, G153, N152, Q74, D75, and L76. The docking poses of all six compounds in this series, found in Figure S1, demonstrate consistent binding conformations in the identified allosteric pocket. The results suggest that targeting the allosteric pocket, which is necessary for ZIKV NS2B-NS3 complex formation, with compounds containing phenylquinoline and aminobenzamide groups has a high potential for further research on the development of anti-flaviviral agents (Table 2).

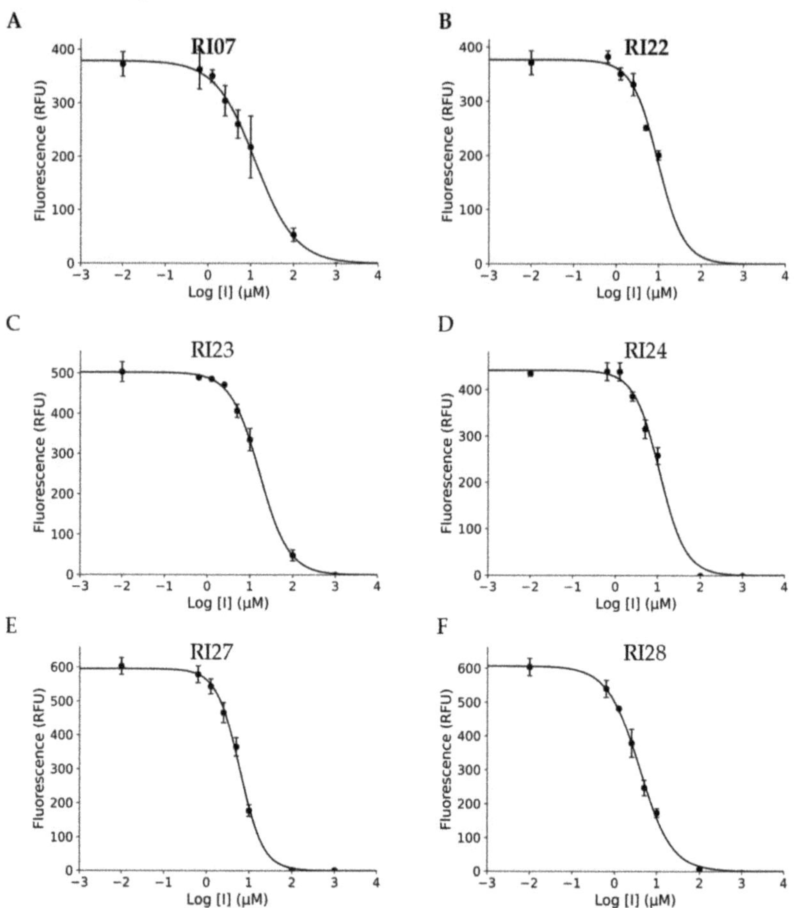

Figure 4. Six experimentally validated inhibitors of catalytic activity of Zika virus protease and their dose–response curves. The dose–response inhibition of NS2B-NS3 protease by RI07, RI22, RI23, RI24, RI27, and RI28 and the IC_{50} values of the compounds against ZIKV NS2B-NS3 protease were measured in protease cleavage assay with Pyr-Arg-Thr-Lys-Arg-MCA fluorescent peptidic substrate (Peptides International-MRP-3159-v) [26]. SciPy and Matplotlib Python packages were used for curve fitting (**A–F**). The X-axis shows decimal logarithms of compound concentrations in micromolar units versus the peptidic substrate cleavage in RFUs.

Table 1. IC_{50} values, binding score against ZIKV NS2B-NS3 protease, and pharmacokinetic parameters for compounds in this study.

Compounds ID	IC_{50} (μM)	Molecular Weight (g/mol)	ICM Binding Score [a]	Predicted Pharmacological Properties [b]								
				Solubility molLogS [c]	Lipophobicity molLogP [d]	Druglikeness [e]	molCACO2 [f]	molPAMPA [g]	molHERG [h]	molPAINS [i]	Polar Surface Area (molPSA) [j]	Tox_Score [k]
RI07	3.8	445	−37	−5.12	5.07	0.57	−5.09	−4.71	0.13	0.05	115	0
RI22	6.2	381	−31	−4.50	4.59	0.59	−4.99	−4.71	0.11	0.01	110	0
RI23	11.9	435	−31	−5.92	5.28	0.70	−5.08	−4.80	0.37	0.06	108	0.42
RI24	17.4	445	−30	−5.54	5.17	0.93	−5.07	−4.70	0.09	0.04	116	0
RI27	10.5	415	−33	−4.65	5.16	0.78	−5.10	−4.63	0.09	0.04	104	0.42
RI28	14.4	411	−29	−4.45	4.14	0.40	−5.26	−4.61	0.07	0.02	156	0

[a] Binding score was calculated using the Dockscan function in ICM-Pro [39,40]. The binding scores are calculated based on the binding free energy between the receptor and ligand. Lower scores suggest strong complex binding. [b] The prediction of relevant pharmacological properties was calculated using the Chemical Properties prediction function in ICM-Pro [41]. [c] Water solubility (molLogS) in 10-based logarithms of the solubility in M. [d] Lipophilicity (LogP) is the logarithm of the ratio of the compound concentration in octanol and water. [e] Druglikeness, value ranging between −1 and 1. A higher number indicates more drug-like properties. [f] CACO-2 permeability; value over −5 indicates high permeability. [g] PAMPA permeability, value over −5 indicates high permeability. [h] hERG inhibition; value over 0.5 indicates high probability of being an hERG inhibitor that may block potassium ion channels in the heart. [i] Pan Assay Interference Compound (PAINS), value over 0.5 indicates high probability of being a PAIN compound. [j] Polar surface area in square angstroms. [k] Toxicity scores calculated based on the presence of known toxic substructures within the compounds, value over 1 indicates that the molecule contains unfavorable substructures or substituents, suggesting a potentially toxic compound.

Table 2. IC_{50} values of the compounds reported to target the active site of ZIKV NS2B-NS3 protease and the compounds tested against the "super-open" pocket of ZIKV NS2B-NS3 protease in this study.

Compounds ID	IC_{50}	Comments	Targets	Ref
Compound 1	0.2	Dipeptide inhibitor	Active site	[25,42]
Compound 3	4.1	Nonpeptidic small molecule inhibitor contains sulfonamide and benzothiazole groups	Active site	[25,42]
Bromocriptine	21.6	Repurposing dopamine receptor agonist as a ZIKV protease inhibitor	Active site	[43]
RI07	3.8	445	"Super-open" pocket	This study
RI22	6.2	381	"Super-open" pocket	This study
RI23	11.9	435	"Super-open" pocket	This study
RI24	17.4	445	"Super-open" pocket	This study
RI27	10.5	415	"Super-open" pocket	This study
RI28	14.4	411	"Super-open" pocket	This study

4. Discussion

ZIKV NS2B-NS3 protease is a two-component chymotrypsin-like serine protease consisting of the NS2B cofactor and the NS3 protease domain, similar to other Flaviviridae members. The active site of ZIKV NS2B-NS3pro includes three conserved amino acid residues forming a classic catalytic triad (His51, Asp75, and Ser135 in ZIKV) [18]. Due to its importance in the virus life cycle and propagation, NS2B-NS3pro is a promising target for antiviral drug design. Unfortunately, because of its shallow and solvent-exposed pocket and high structural homology among the active centers of multiple cellular serine proteases and the viral NS2B-NS3pro, the development of inhibitors with high cellular permeability, stability, and selectivity targeting the active site is challenging [44]. Thus, through the design of inhibitors targeting new allosteric sites on ZIKV NS2B-NS3pro, we can bypass this obstacle and develop effective therapeutic approaches.

The analysis of recently deposited protein structures found two conformations of ZIKV NS2B-NS3pro, which are the inactive "super-open" conformation and the active "closed" conformation. All structural rearrangements of ZIKV protease in "super-open" conformation are incompatible with the protease's catalytic activity. However, this catalytic activity of NS2B-NS3pro can be completely restored by transitioning back to the "closed" conformation. Transitioning from ZIKV NS2B-NS3pro closed conformation to super-open conformation reveals a transient pocket at the interface between NS2B and NS3 in the closed conformation. Our purpose was to show that the major reorganization of the NS3pro C-terminal loop creates a transient novel druggable pocket that can be used for the development of specific binding, disrupting the protease activity allosterically (Figure 2A). This transient pocket may be present in proteases of other flaviviruses (e.g., PDB ID 2GGV for the West Nile virus protease) [34]. These novel structures demonstrate the detachment of the C-terminal part of NS2B from a deep pocket. The contact residues in the targeted pocket have favorable characteristics, such as shape, depth, and hydrophobicity, for drug-like small molecules.

The virtual screening results showed that compounds containing phenylquinoline and aminobenzamide were predicted to have negative binding free energy to the "super-open" pocket on ZIKV NS3pro, indicating thermodynamically favorable ligand–protein complex formation. The binding pose of RI07, the top compound in this series according to the docking study and its IC_{50} value, suggests that the 2-aminobenzamide group in the prospective molecules is particularly effective at interacting with the side chains of L76, W83, L85, V146, I147, G148, and L149 in the hydrophobic region. Its two amide oxygens also form hydrogen bonds with amine hydrogens of V155 and N152 at distances of 3.2 and 2.6 Å, respectively. The docking pose of other compounds, compared to RI07, shows similar binding poses (Figure S1) and comparable hydrogen bonding distances (Figures S2–S6), demonstrating the consistency of binding conformations in the identified allosteric pocket. On the other hand, the substituted phenylquinoline functional group, compared to the aminobenzamide group, interacts with a smaller number of residues, such as Q74, T118, D120, and I123, but its rigidity and unique structure fit exceptionally well in this narrow part of the pocket as shown in Figure 2B.

The estimated binding free energies of all six compounds in this series from the docking study correspond to the evaluated IC_{50} values ranging from 3.8 to 17.4 µM, as shown in Table 1, making them suitable inhibitor candidates against the novel allosteric pocket. It is evident that selected derivatives of RI07 did not improve the IC_{50} compared to RI07, and the IC_{50} values vary significantly with different substituent groups on the phenylquinoline functional group; this underscores the potential for further optimization to increase activity, especially with this functional group. In addition to the enzymatic activity assay, the predicted key chemical properties, such as water solubility, cell permeability, propensity for pan-assay interference, potassium channel blocking activity, polar surface area, and ICM ToxScore, are all within acceptable ranges for being drug-like molecules.

These findings confirmed our hypothesis that this novel allosteric pocket revealed transiently in the "super-open" conformation can be used as a target for a new type of antiviral for the Zika virus. In addition to inhibiting ZIKV protease, the identified compounds containing phenylquinoline and aminobenzamide in this series may be applied against other flaviviruses such as WNV, DENV, TBEV, and YFV by exploiting the presence of "super-open" pockets in their protease domains. This target is beneficial because of its conservation, hence, its lower propensity for treatment escaping mutations, and its favorable shape. The identified allosteric inhibitors targeting ZIKV NS2B-NS3 protease and preventing viral proliferation may be further optimized for improved efficacy, pharmacokinetics (PK), pharmacodynamics (PD), and reduced adverse side-effect profile.

5. Conclusions

Infections with ZIKV are current global health concerns due to the risk of microcephaly in newborns and Guillain-Barré syndrome in adults. Since there is no approved anti-ZIKV agent and vaccination strategies are limited due to antibody-dependent enhancement (ADE) effects observed in Dengue and other flaviviruses, The discovery of small-molecule orally available drugs as ZIKV infection treatments has been limited. We have reported a new strategy of targeting a "super-open" conformation of ZIKV NS2B-NS3 protease suggested by the analysis of existing X-ray crystal structures of WNV and ZIKV proteases. The newly identified pocket is a preferable target compared to an active site due to its shape, conservation, and hydrophobicity. We have identified six novel compounds containing phenylquinoline and aminobenzamide that inhibited ZIKV NS2B-NS3 protease at IC_{50} values ranging from 3.8 to 14.4 μM. The predicted pharmacokinetic properties of all compounds were also evaluated, showing promising drug-like properties. Targeting the identified transient pocket may prove to be a promising strategy for fighting other flaviviral infections as well due to the similar pattern of inactive-to-active state transitions of the viral NS2B-NS3 system.

Supplementary Materials: The following supporting information can be downloaded at https://www.mdpi.com/article/10.3390/v15051106/s1. Table S1. Code names, structures, and ICM binding scores of initial ten candidates from virtual screening against ZIKV NS2B-NS3 protease. Figure S1. Predicted binding conformations of RI07, RI22, RI23, RI24, RI27, and RI28 targeting the open conformation of NS2B-NS3 protease at the identified allosteric pocket. Figure S2. Two-dimensional interaction diagram for predicted pose of RI22 compound (L76, W83, L85, D86, A87, A88, T118, D120, I123, V146, I147, G148, L149, N152, G153, V154, and V155). Hydrogen bonding interactions are shown by grey dotted lines. Figure S3. Two-dimensional interaction diagram for predicted pose of RI23 compound (Q74, L76, W83, L85, D86, A87, I123, V146, I147, G148, L149, N152, G153, and V155). Hydrogen bonding interactions are shown by grey dotted lines. Figure S4. Two-dimensional interaction diagram for predicted pose of RI24 compound (Q74, L76, W83, L85, D86, A87, T118, I123, V146, I147, G148, L149, N152, G153, and V155). Hydrogen bonding interactions are shown by grey dotted lines. Figure S5. Two-dimensional interaction diagram for predicted pose of RI27 compound (Q74, L76, W83, L85, D86, A87, A88, I123, V146, I147, G148, L149, N152, G153, V154, and V155). Hydrogen bonding interactions are shown by grey dotted lines. Figure S6. Two-dimensional interaction diagram for predicted pose of RI28 compound (Q74, L76, W83, L85, D86, A87, A88, T118, I123, V146, I147, G148, L149, N152, G153, V154, and V155). Hydrogen bonding interactions are shown by grey dotted lines.

Author Contributions: Conceptualization, R.A. and A.V.T.; Methodology, S.A.S., C.-T.H., V.L., C.-H.C. and A.V.T.; Software, I.M. and R.A.; Validation, I.M., S.A.S., C.-T.H., V.L. and C.-H.C.; Formal analysis, I.M., S.A.S., C.-T.H., V.L. and C.-H.C.; Investigation, I.M., S.A.S., C.-T.H., V.L. and C.-H.C.; Resources I.M., S.A.S., C.-T.H., V.L., C.-H.C., A.V.T. and R.A.; Data curation, I.M. and S.A.S.; Writing—original draft, I.M., S.A.S., A.V.T. and R.A.; Writing—review and editing, I.M., J.K. and R.A.; Visualization, I.M. and S.A.S.; Supervision A.V.T. and R.A.; Project administrations A.V.T. and R.A.; Funding acquisition A.V.T. and R.A. All authors have read and agreed to the published version of the manuscript.

Funding: This work was supported in part by NIH R35GM131881 to R.A., and by NIH R01 NS105969-01 to A.V.T.

Institutional Review Board Statement: Not applicable.

Informed Consent Statement: Not applicable.

Data Availability Statement: All data generated or analyzed during this study are included in this published article.

Acknowledgments: We thank Alexander Aleshin for his help, discussions, and X-ray crystallography work which laid the foundation for our screen, and Conall Sauvey for useful discussions and assistance.

Conflicts of Interest: The authors declare no conflict of interest.

References

1. Campos, G.S.; Bandeira, A.C.; Sardi, S.I. Zika Virus Outbreak, Bahia, Brazil. *Emerg. Infect. Dis.* **2015**, *21*, 1885–1886. [CrossRef] [PubMed]
2. MacKenzie, J.S.; Gubler, D.J.; Petersen, L.R. Emerging flaviviruses: The spread and resurgence of Japanese encephalitis, West Nile and dengue viruses. *Nat. Med.* **2004**, *10*, S98–S109. [CrossRef] [PubMed]
3. Gulland, A. Zika virus is a global public health emergency, declares WHO. *BMJ* **2016**, *352*, i657. [CrossRef]
4. Bhargavi, B.S.; Moa, A. Global outbreaks of zika infection by epidemic observatory (EpiWATCH), 2016–2019. *Glob. Biosecurity* **2020**, *2*.
5. Musso, D.; Gubler, D.J. Zika Virus. *Clin. Microbiol. Rev.* **2016**, *29*, 487–524. [CrossRef]

6.	Hennessey, M.; Fischer, M.; Staples, J.E. Zika Virus Spreads to New Areas—Region of the Americas, May 2015–January 2016. *MMWR. Morb. Mortal. Wkly. Rep.* **2016**, *65*, 55–58. [CrossRef]

7.	Hazin, A.N.; Poretti, A.; Di Cavalcanti Souza Cruz, D.; Tenorio, M.; Van Der Linden, A.; Pena, L.J.; Brito, C.; Gil, L.H.V.; de Barros Miranda-Filho, D.; de Azevedo Marques, E.T.; et al. Computed Tomographic Findings in Microcephaly Associated with Zika Virus. *N. Engl. J. Med.* **2016**, *374*, 2193–2195. [CrossRef]

8.	Vogel, G. Evidence grows for Zika virus as pregnancy danger. *Science* **2016**, *351*, 1123–1124. [CrossRef]

9.	Solomon, I.H.; Milner, D.A.; Folkerth, R.D. Neuropathology of Zika Virus Infection. *J. Neuroinfect. Dis.* **2016**, *7*, 220. [CrossRef]

10.	Rasmussen, S.A.; Jamieson, D.J.; Honein, M.A.; Petersen, L.R. Zika Virus and Birth Defects—Reviewing the Evidence for Causality. *N. Engl. J. Med.* **2016**, *374*, 1981–1987. [CrossRef]

11.	Cuevas, E.L.; Tong, V.T.; Rozo, N.; Valencia, D.; Pacheco, O.; Gilboa, S.M.; Mercado, M.; Renquist, C.M.; González, M.; Ailes, E.C.; et al. Preliminary Report of Microcephaly Potentially Associated with Zika Virus Infection During Pregnancy—Colombia, January–November 2016. *MMWR. Morb. Mortal. Wkly. Rep.* **2016**, *65*, 1409–1413. [CrossRef] [PubMed]

12.	Enfissi, A.; Codrington, J.; Roosblad, J.; Kazanji, M.; Rousset, D. Zika virus genome from the Americas. *Lancet* **2016**, *387*, 227–228. [CrossRef] [PubMed]

13.	Cao-Lormeau, V.-M.; Blake, A.; Mons, S.; Lastère, S.; Roche, C.; Vanhomwegen, J.; Dub, T.; Baudouin, L.; Teissier, A.; Larre, P.; et al. Guillain-Barré Syndrome outbreak associated with Zika virus infection in French Polynesia: A case-control study. *Lancet* **2016**, *387*, 1531–1539. [CrossRef] [PubMed]

14.	Lucchese, G.; Kanduc, D. Zika virus and autoimmunity: From microcephaly to Guillain-Barré syndrome, and beyond. *Autoimmun. Rev.* **2016**, *15*, 801–808. [CrossRef] [PubMed]

15.	Parra, B.; Lizarazo, J.; Jiménez-Arango, J.A.; Zea-Vera, A.F.; González-Manrique, G.; Vargas, J.; Angarita, J.A.; Zuñiga, G.; Lopez-Gonzalez, R.; Beltran, C.L.; et al. Guillain–Barré Syndrome Associated with Zika Virus Infection in Colombia. *N. Engl. J. Med.* **2016**, *375*, 1513–1523. [CrossRef] [PubMed]

16.	Gorshkov, K.; Shiryaev, S.A.; Fertel, S.; Lin, Y.-W.; Huang, C.-T.; Pinto, A.; Farhy, C.; Strongin, A.Y.; Zheng, W.; Terskikh, A.V. Zika Virus: Origins, Pathological Action, and Treatment Strategies. *Front. Microbiol.* **2019**, *9*, 3252. [CrossRef] [PubMed]

17.	Deng, S.-Q.; Yang, X.; Wei, Y.; Chen, J.-T.; Wang, X.-J.; Peng, H.-J. A Review on Dengue Vaccine Development. *Vaccines* **2020**, *8*, 63. [CrossRef]

18.	Halstead, S.B. Dengvaxia sensitizes seronegatives to vaccine enhanced disease regardless of age. *Vaccine* **2017**, *35*, 6355–6358. [CrossRef]

19.	Li, H.; Clum, S.; You, S.; Ebner, K.E.; Padmanabhan, R. The Serine Protease and RNA-Stimulated Nucleoside Triphosphatase and RNA Helicase Functional Domains of Dengue Virus Type 2 NS3 Converge within a Region of 20 Amino Acids. *J. Virol.* **1999**, *73*, 3108–3116. [CrossRef]

20.	Gorbalenya, A.E.; Koonin, E.V.; Donchenko, A.P.; Blinov, V.M. Two related superfamilies of putative helicases involved in replication, recombination, repair and expression of DNA and RNA genomes. *Nucleic Acids Res.* **1989**, *17*, 4713–4730. [CrossRef]

21.	Wengler, G.; Wengler, G. The NS 3 Nonstructural Protein of Flaviviruses Contains an RNA Triphosphatase Activity. *Virology* **1993**, *197*, 265–273. [CrossRef] [PubMed]

22.	Mastrangelo, E.; Milani, M.; Bollati, M.; Selisko, B.; Peyrane, F.; Pandini, V.; Sorrentino, G.; Canard, B.; Konarev, P.V.; Svergun, D.I.; et al. Crystal Structure and Activity of Kunjin Virus NS3 Helicase; Protease and Helicase Domain Assembly in the Full Length NS3 Protein. *J. Mol. Biol.* **2007**, *372*, 444–455. [CrossRef] [PubMed]

23.	Luo, D.; Xu, T.; Watson, R.P.; Scherer-Becker, D.; Sampath, A.; Jahnke, W.; Yeong, S.S.; Wang, C.H.; Lim, S.P.; Strongin, A.; et al. Insights into RNA unwinding and ATP hydrolysis by the flavivirus NS3 protein. *EMBO J.* **2008**, *27*, 3209–3219. [CrossRef] [PubMed]

24.	Pierson, T.C.; Diamond, M.S. The continued threat of emerging flaviviruses. *Nat. Microbiol.* **2020**, *5*, 796–812. [CrossRef]

25.	Lee, H.; Ren, J.; Nocadello, S.; Rice, A.J.; Ojeda, I.; Light, S.; Minasov, G.; Vargas, J.; Nagarathnam, D.; Anderson, W.F.; et al. Identification of novel small molecule inhibitors against NS2B/NS3 serine protease from Zika virus. *Antivir. Res.* **2017**, *139*, 49–58. [CrossRef]

26.	Shiryaev, S.A.; Farhy, C.; Pinto, A.; Huang, C.-T.; Simonetti, N.; Ngono, A.E.; Dewing, A.; Shresta, S.; Pinkerton, A.B.; Cieplak, P.; et al. Characterization of the Zika virus two-component NS2B-NS3 protease and structure-assisted identification of allosteric small-molecule antagonists. *Antivir. Res.* **2017**, *143*, 218–229. [CrossRef]

27.	Shiryaev, S.A.; Strongin, A.Y. Structural and functional parameters of the flaviviral protease: A promising antiviral drug target. *Futur. Virol.* **2010**, *5*, 593–606. [CrossRef]

28.	Voss, S.; Nitsche, C. Inhibitors of the Zika virus protease NS2B-NS3. *Bioorganic Med. Chem. Lett.* **2020**, *30*, 126965. [CrossRef]

29.	Behnam, M.A.M.; Graf, D.; Bartenschlager, R.; Zlotos, D.P.; Klein, C.D. Discovery of Nanomolar Dengue and West Nile Virus Protease Inhibitors Containing a 4-Benzyloxyphenylglycine Residue. *J. Med. Chem.* **2015**, *58*, 9354–9370. [CrossRef]

30.	Nitsche, C. Strategies Towards Protease Inhibitors for Emerging Flaviviruses. *Dengue Zika Control. Antivir. Treat. Strateg.* **2018**, *1062*, 175–186. [CrossRef]

31.	Stiefelhagen, P.; Einecke, D. Therapy of chronic hepatitis C: Protease inhibitors significantly increase rate of healing. *MMW Fortschr. Med.* **2011**, *153*, 14–15. [CrossRef] [PubMed]

32.	De Leuw, P.; Stephan, C. Protease inhibitor therapy for hepatitis C virus-infection. *Expert Opin. Pharmacother.* **2018**, *19*, 577–587. [CrossRef] [PubMed]

33. Meewan, I.; Zhang, X.; Roy, S.; Ballatore, C.; O'donoghue, A.J.; Schooley, R.T.; Abagyan, R. Discovery of New Inhibitors of Hepatitis C Virus NS3/4A Protease and Its D168A Mutant. *ACS Omega* **2019**, *4*, 16999–17008. [CrossRef] [PubMed]
34. Aleshin, A.E.; Shiryaev, S.A.; Strongin, A.Y.; Liddington, R.C. Structural evidence for regulation and specificity of flaviviral proteases and evolution of the Flaviviridae fold. *Protein Sci.* **2007**, *16*, 795–806. [CrossRef] [PubMed]
35. Radichev, I.; Shiryaev, S.A.; Aleshin, A.E.; Ratnikov, B.I.; Smith, J.W.; Liddington, R.C.; Strongin, A.Y. Structure-based mutagenesis identifies important novel determinants of the NS2B cofactor of the West Nile virus two-component NS2B–NS3 proteinase. *J. Gen. Virol.* **2008**, *89*, 636–641. [CrossRef]
36. Lei, J.; Hansen, G.; Nitsche, C.; Klein, C.D.; Zhang, L.; Hilgenfeld, R. Crystal structure of Zika virus NS2B-NS3 protease in complex with a boronate inhibitor. *Science* **2016**, *353*, 503–505. [CrossRef]
37. Weinert, T.; Olieric, V.; Waltersperger, S.; Panepucci, E.; Chen, L.; Zhang, H.; Zhou, D.; Rose, J.; Ebihara, A.; Kuramitsu, S.; et al. Fast native-SAD phasing for routine macromolecular structure determination. *Nat. Methods* **2014**, *12*, 131–133. [CrossRef]
38. eMolecules: San Diego, "eMolecules". 2017. Available online: https://www.emolecules.com/ (accessed on 20 February 2023).
39. Abagyan, R.; Totrov, M.; Kuznetsov, D. ICM?A new method for protein modeling and design: Applications to docking and structure prediction from the distorted native conformation. *J. Comput. Chem.* **1994**, *15*, 488–506. [CrossRef]
40. Neves, M.A.C.; Totrov, M.; Abagyan, R. Docking and scoring with ICM: The benchmarking results and strategies for improvement. *J. Comput. Mol. Des.* **2012**, *26*, 675–686. [CrossRef]
41. Totrov, M. Atomic Property Fields: Generalized 3D Pharmacophoric Potential for Automated Ligand Superposition, Pharmacophore Elucidation and 3D QSAR. *Chem. Biol. Drug Des.* **2007**, *71*, 15–27. [CrossRef]
42. Li, Y.; Zhang, Z.; Phoo, W.W.; Loh, Y.R.; Wang, W.; Liu, S.; Chen, M.W.; Hung, A.W.; Keller, T.H.; Luo, D.; et al. Structural Dynamics of Zika Virus NS2B-NS3 Protease Binding to Dipeptide Inhibitors. *Structure* **2017**, *25*, 1242–1250.e3. [CrossRef] [PubMed]
43. Chan, J.F.-W.; Chik, K.K.-H.; Yuan, S.; Yip, C.C.-Y.; Zhu, Z.; Tee, K.-M.; Tsang, J.O.-L.; Chan, C.C.-S.; Poon, V.K.-M.; Lu, G.; et al. Novel antiviral activity and mechanism of bromocriptine as a Zika virus NS2B-NS3 protease inhibitor. *Antivir. Res.* **2017**, *141*, 29–37. [CrossRef] [PubMed]
44. Cregar-Hernandez, L.; Jiao, G.S.; Johnson, A.T.; Lehrer, A.T.; Wong, T.A.; Margosiak, S.A. Small Molecule Pan-Dengue and West Nile Virus NS3 Protease Inhibitors. *Antivir. Chem. Chemother.* **2011**, *21*, 209–217. [CrossRef] [PubMed]

 *viruses*

Article

Noncoding RNA of Zika Virus Affects Interplay between Wnt-Signaling and Pro-Apoptotic Pathways in the Developing Brain Tissue

Andrii Slonchak [1,2,*], Harman Chaggar [3], Julio Aguado [3], Ernst Wolvetang [3] and Alexander A. Khromykh [1,2,*]

1 School of Chemistry and Molecular Biosciences, The University of Queensland, Brisbane 4072, Australia
2 Australian Infectious Diseases Research Centre, Global Virus Network Centre of Excellence, Brisbane 4072, Australia
3 Australian Institute for Bioengineering and Nanotechnology, The University of Queensland, Brisbane 4072, Australia
* Correspondence: a.slonchak@uq.edu.au (A.S.); alexander.khromykh@uq.edu.au (A.A.K.)

Abstract: Zika virus (ZIKV) has a unique ability among flaviviruses to cross the placental barrier and infect the fetal brain causing severe abnormalities of neurodevelopment known collectively as congenital Zika syndrome. In our recent study, we demonstrated that the viral noncoding RNA (subgenomic flaviviral RNA, sfRNA) of the Zika virus induces apoptosis of neural progenitors and is required for ZIKV pathogenesis in the developing brain. Herein, we expanded on our initial findings and identified biological processes and signaling pathways affected by the production of ZIKV sfRNA in the developing brain tissue. We employed 3D brain organoids generated from induced human pluripotent stem cells (ihPSC) as an ex vivo model of viral infection in the developing brain and utilized wild type (WT) ZIKV (producing sfRNA) and mutant ZIKV (deficient in the production of sfRNA). Global transcriptome profiling by RNA-Seq revealed that the production of sfRNA affects the expression of >1000 genes. We uncovered that in addition to the activation of pro-apoptotic pathways, organoids infected with sfRNA-producing WT, but not sfRNA-deficient mutant ZIKV, which exhibited a strong down-regulation of genes involved in signaling pathways that control neuron differentiation and brain development, indicating the requirement of sfRNA for the suppression of neurodevelopment associated with the ZIKV infection. Using gene set enrichment analysis and gene network reconstruction, we demonstrated that the effect of sfRNA on pathways that control brain development occurs via crosstalk between Wnt-signaling and proapoptotic pathways.

Keywords: Zika virus; sfRNA; flavivirus; systems virology; transcriptomics; noncoding RNA; brain development; apoptosis; Wnt-signaling

Citation: Slonchak, A.; Chaggar, H.; Aguado, J.; Wolvetang, E.; Khromykh, A.A. Noncoding RNA of Zika Virus Affects Interplay between Wnt-Signaling and Pro-Apoptotic Pathways in the Developing Brain Tissue. *Viruses* **2023**, *15*, 1062. https://doi.org/10.3390/v15051062

Academic Editors: Yiping Li and Yuliang Liu

Received: 30 March 2023
Revised: 21 April 2023
Accepted: 23 April 2023
Published: 26 April 2023

1. Introduction

Zika Virus (ZIKV) is an emerging pathogen that belongs to the genus *Flavivirus* in the family *Flaviviridae* [1]. It is a small, enveloped virus with a single-stranded RNA genome of positive polarity. ZIKV is a significant public health concern with a recent outbreak resulting in almost 1,000,000 infections [2]. In urban and suburban environments, ZIKV can circulate between humans and mosquitoes [3]. Common symptoms of ZIKV infection include papular rash, arthritis, myalgia, non-purulent conjunctivitis, headache, fever, retro-orbital pain, edema, and vomiting [4]. In addition, a temporal and geographical association between Guillain-Barre syndrome and ZIKV outbreaks in the Pacific and the Americas has been described, suggesting the infections may lead to neurological complications, which is a common disease outcome for flaviviruses [4]. Moreover, ZIKV is the only known flavivirus to have teratogenic effects in humans [5]. Mothers with proven antecedent acute ZIKV infections had high rates of giving birth to infants with microcephaly, visual impairment, and global neurodevelopmental delay [6,7]. In 2016, the World Health Organization announced ZIKV outbreaks as a health emergency of international concern, stressing the importance of further research on this virus. To date, no specific antiviral and vaccines have been approved for use against the ZIKV infection and the pathogenesis of ZIKV in the developing brain is not completely understood.

One of the characteristics of flaviviral infection is the intracellular accumulation of a virus-derived, non-coding RNA, referred to as subgenomic flavivirus RNA or sfRNA [8]. It is produced from the viral 3′UTRs due to the unique ability of flavivirus RNA to resist degradation by the host 5′-3′ exoribonuclease XRN1 [8,9]. The 3′UTRs of mosquito-borne flaviviruses commonly contain at least two highly structured RNA elements that can stall XRN1 progression, resulting in the production of at least two sfRNA species of slightly different lengths [10,11]. These RNA elements are usually referred to as XRN1-resistant RNA elements (xrRNAs). The production of sfRNAs has been shown in all flaviviruses tested to date [11–13], including phylogenetically divergent dual-host flaviviruses and insect-specific flaviviruses [11]. They were demonstrated to promote viral replication in arthropod and vertebrate hosts and facilitate the transmission of mosquito-borne flaviviruses (reviewed in [12,14]). ZIKV was shown to produce two sfRNA species of different lengths due to the presence of two XRN1-resistant elements in the 3′UTR [15,16]. In the insect host, the production of sfRNA was demonstrated to facilitate ZIKV replication and secretion into saliva by inhibiting the apoptosis of infected cells in mosquito tissues [16]. In vertebrate hosts, ZIKV sfRNA was found to facilitate viral replication and promote apoptosis and cytopathic effects [17,18], which is opposite to its inhibitory effect on apoptosis in mosquitoes. It was also shown to be required for viral dissemination through the placental barrier into the fetal brain [17]. ZIKV sfRNA was shown to execute its functions by suppressing interferon signaling [17,18] and promoting apoptosis, acting in cooperation with the viral protein NS5 [17].

In a recent study, we employed human brain organoids as a model of the ZIKV infection in the developing brain and found that ZIKV sfRNA also promotes viral replication in neural tissues and is required for viral neuropathic effect as well as for virus-induced deaths of neural progenitor cells (NPCs) accompanied by the activation of Caspase-3 [17]. Brain organoids are stem-cell-derived 3D tissues that self-assemble into the structures of the developing brain, utilizing the intrinsic ability of human pluripotent stem cells (hP-SCs) to differentiate spontaneously into neural lineages [19]. They can be generated from embryonic (ehPSCs) or induced (ihPSCc) stem cells to resemble the entire brain (cerebral organoids) or be guided to become region-specific [19]. The advantage of organoids compared to cell monolayers is their ability to produce all relevant cell types and recapitulate the architecture of the developing brain. The human origin of organoids and their capacity to mount an innate immune response make them the most relevant accessible brain model for ZIKV research [20].

In this study, we employed ihPSCc-derived human brain organoids to further characterize the role of sfRNA in ZIKV pathogenesis in the developing brain. We performed transcriptome-wide gene expression profiling of organoids infected with WT ZIKV and the sfRNA-deficient mutant virus and identified genes that were differentially expressed compared to the mock. We then compared differentially expressed genes (DEGs) for each condition and identified those that were affected by the production of sfRNA. Using gene ontology enrichment analysis and network reconstruction we further confirmed our earlier findings regarding the ability of ZIKV sfRNA to promote apoptosis. In addition, we found that the production of sfRNA has a previously unknown inhibitory effect on signalling pathways that regulate brain development. We demonstrated that this effect likely results from the induction of pro-apoptotic signaling, which interconnects with pathways that control neuron differentiation and the development of the nervous system. Thus, we demonstrated that sfRNA is an important viral factor of ZIKV pathogenesis in the developing brain.

2. Materials and Methods

Generation of human brain organoids. Generations of human brain organoids have been described previously [21]. Human embryonic stem cell line GENEA022 (Genea Biocells, San Diego, CA, USA) was cultured on ECM Gel from Engelbreth-Holm-Swarm murine sarcoma (Sigma-Aldrich, St. Louis, MO, USA) in mTeSR medium (Stem Cell Technologies, Vancouver, BC, Canada). Brain organoids were generated using a protocol [19] in which patterned embryoid bodies were expanded for 4 days in N2 medium (DMEM/F12 (Gibco, Waltham, MA, USA) with 1% N-2 supplement (Gibco, USA), 2% B-27 supplement (Gibco, USA), 1% MEM Non-Essential Amino Acids (Gibco, USA), 100 U/mL Penicillin and 100 μg/mL Streptomycin (Gibco, USA), and 0.1% β-mercaptoethanol (Gibco, USA), with daily supplementation of 20 ng/mL bFGF (R&D, USA). Embryoid bodies were embedded in 15 μL of Matrigel (Stem Cell Technologies, Vancouver, BC, Canada) and media changed to a 1:1 mixture of Neurobasal medium (Invitrogen, Waltham, MA, USA) and DMEM/F12, containing 1:200 MEM-NEAA, 1:100 Glutamax (Gibco, Waltham, MA, USA), 1:100 N2 supplement, 1:50 B27 supplement, 1 s penicillin-streptomycin, 50 μM 2-mercaptoethanol, and 0.25% insulin solution (Sigma-Aldrich, St. Louis, MO, USA). Fresh media was replaced three times a week.

Viruses and infection. WT and sfRNA-deficient ZIKV were of the MR766 strain and have been described previously [16]. Brain organoids were utilized for a viral infection on day 15. A virus inoculum containing 10^5 FFU of each virus was added to a single organoid-containing well of a 24-well plate and was incubated at 37 °C O/N. Each ZIKV-infected organoid was then washed three times with culture media and transferred to a single well of a 24-well plate containing 500 μL of ND medium.

Foci-forming immunoassay. Ten-fold serial dilutions of culture fluids were prepared in DMEM media, supplemented with 2% FBS and 25 μL of each dilution, and were used to infect 10^4 Vero cells pre-seeded in 96 well plates. After 1 h of incubation at 37 °C, 180 μL overlay media (1:1 mixture of 2% carboxymethyl cellulose (Sigma-Aldrich, St. Louis, MO, USA) and 2× M199 medium supplemented with 5% FCS, 100 μg/mL streptomycin, 100 U/mL penicillin, and 2.2 g/L $NaHCO_3$) was added to the wells. At 3 dpi, the overlay medium was removed and the cells were fixed with 100 μL/well of ice-cold fixation solution (80% acetone, 20% PBS) for 30 min at −20 °C. After the fixative solution was removed, the monolayer was fully dried and plates were blocked with 150 μL/well of ClearMilk blocking solution (Pierce, Waltham, MA, USA) for 60 min. The plates were then incubated for 1 h with 50 μL/well of 4G2 mouse monoclonal antibody specific toZIKV envelope protein in 1:100 dillution, followed by 1 h incubation with 50 μL/well of 1:1000 dilution of anti-mouse IRDye 800 CW secondary antibody (LI-COR Biosciences, Lincoln, NE, USA). All antibodies were diluted with Clear Milk blocking buffer (Pierce, Waltham, MA, USA) and incubations were performed at 37 °C. After each incubation with the antibodies, the plates were washed 5 times with phosphate-buffered saline containing 0.05% Tween 20 (PBST). The plates were scanned using an LICOR Odyssey CLx Imaging System with the following settings: channel = 800 and 700, quality= medium, focus= 3.0 mm, intensity = auto, resolution= 42 μm.

RNA isolation. Eight individual organoids were combined in 1 mL of TRIreagent (Sigma, USA) and homogenized for 5 min at 30 Hz using a Tissue Lyser II (Qiagen, Hilden, Germany). RNA was then isolated according to the TRIreagent manufacturer's recommendations. The quantity and purity of the RNA were assessed using Nanodrop ND1000 (Thermo Fisher, Waltham, MA, USA).

Next-generation sequencing. RNA samples pooled from 8 brain organoids per sample, per condition, infected with WT or xrRNA2' mutant $ZIKV_{MR766}$ or mock-infected, were subjected to ribosomal RNA depletion using the Ribo-Zero Gold rRNA Removal Kit (Illumina, San Diego, CA, USA) followed by library preparation. Two pools of organoids were used in the experiment for each condition. Libraries were sequenced on an Illumina HiSeq4000 instrument generating single-end 150 bp reads. Image analysis was performed using the HiSeq Control Software (HCS) vHD 3.4.0.38 and Real-Time Analysis (RTA) v2.7.7

on the instrument computer. The Illumina bcl2fastq 2.20.0.422 pipeline was then used to produce the sequence data. Library preparation, sequencing, and data acquisition were performed by the University of Queensland Genomics Facility and the Australian Genomics Research Facility (AGRF). Quality control of sequencing data was performed using FastQC v.0.72. The data were then trimmed using TRIMMOMATIC v.0.36.4 and the following settings: ILLUMINACLIP: TruSeq3-PE:2:30:10 LEADING:32 TRAILING:32 SLIDINGWINDOW:4:20 MINLEN:25. Trimmed reads were mapped to the human genome assembly hg38 using HISAT2 v.2.1.0. A feature count was performed using HTSEQ v.0.9.1 in Union mode, with strand set to "Reverse" and end feature type set to "exon". RNA-Seq data generated in this study are deposited in the Gene Expression Omnibus database (accession number GSE230197).

Differential gene expression analysis. Differential gene expression analysis was performed using edgeR v.3.24.0. Low abundance reads (<0.5 cpm) were removed from the data set and the data were normalized using the trimmed mean of the M-values method (TMM). Normalized data were used to build the quasi-likelihood negative binomial generalized log-linear model. The test for differential expression relative to a threshold (glmTreat) was then applied to the contrasts WT-Mock, Mut-Mock, and (WT-Mock)-(Mut-Mock) to identify the genes that had significantly different expressions between the conditions by at least two-fold. Genes were considered differentially expressed if the FDR-corrected p-values were <0.05. Gene expression data were plotted using ggplot2 v.3.3.2.

Gene ontology and pathway enrichment analyses. Gene ontology (GO) and KEGG pathway enrichment analyses were performed using the Database for Annotation, Visualization, and Integrated Discovery (DAVID) v6.8. Enrichment data were exported from DAVID, combined with gene expression values, and z-scores were calculated using the R package GOplot v.1.0.2. Heatmaps were generated using heatmap.2 function of R-package gplots v3.1.1.

Network reconstruction. Gene interaction networks were reconstructed using Cytoscape v3.8.0. Individual networks were generated from DEGs associated with GO categories related to brain development and apoptosis using a local search with the GeneMANIA Cytoscape Plug-in. Interaction types considered in the network were "pathways", "genetic interactions", and "physical interactions". The networks were then merged using the "merge networks" function of Cytoscape, with the mode set to "Union" and the nod keys set to "gene name". Betweenness centrality values were calculated using the "Analyze Network" function of Cytoscape and represent the numbers of the shortest paths that pass through each nod in the network. A table of gene names with logFC values was then merged with the table of nods where logFCs were displayed as nod coloring.

Quantitative RT-PCR. Total organoids RNA (500 ng) was used to produce cDNA with qScript cDNA SuperMix (Quantabio, USA) according to the manufacturer's instructions. cDNA was diluted 1:10 and used as a template for qRT-PCR. PCR was performed in 20 μL of a reaction mix containing 1X SYBR Green PCR Master Mix (Applied Biosystems, Waltham, MA, USA), 10 pmoles of forward and reverse PCR primers, and 3 μL of diluted cDNA. Reactions were performed using the following cycling conditions: 95 °C for 5 min, 40 cycles of 95 °C for 5 s, and 60 °C for 20 s, followed by melting-curve analysis using QuantStudio 6 Flex Real Time PCR Instrument (Applied Biosystems, USA). Gene expression levels were normalized to *TBP*. Viral RNA levels were determined using the standard curve method by comparing the C_T values of the samples to the C_T values observed in the amplification of the serial dilutions (10^2–10^8 copies) of a PCR-amplified and purified ZIKV genomic fragment. For each experiment, RNA from 3 biological replicates was used and amplification of each cDNA sample was performed in technical triplicate. Negative controls were included for each set of primers. Primers for FOXG1, DLX5, and HIST1H1 are listed in Table S1. Primers for ISGs have been described previously [22].

Statistical analysis. Statistical analyses were performed using GraphPad Prism v.9.0 and R v4.1.2.

3. Results

3.1. Production of sfRNA Alters Transcriptome of ZIKV-Infected Human Brain Organoids

We recently showed that the production of sfRNA facilitates ZIKV replication in human brain organoids and is required for the viral neuropathic effect. Organoids infected with WT ZIKV exhibited signs of pathology starting from 6 days post-infection (dpi) and disintegrated by 15 dpi, while organoids infected with sfRNA-deficient ZIKV mutants survived the course of infection. We also found that organoids infected with the sfRNA-producing WT virus had stronger activation of Caspase-3 compared to those infected with the sfRNA-deficient mutant virus and demonstrated that the production of sfRNA induced the apoptosis of neural progenitor cells [17].

To further dissect how the production of sfRNA affects developing brain tissues during the ZIKV infection we infected human brain organoids at day-in-vitro 15 (DIV15) with WT ZIKV and sfRNA-deficient mutant viruses and isolated RNA for transcriptome profiling at 3 dpi. This time point was selected because we previously determined that at 3 dpi, viral replication in brain organoids was at its highest level, while the morphology and tissue structure of organoids remained intact [17]. We employed a previously designed xrRNA2' viral mutant that contains mutations in the XRN1-resistant structure xrRNA2 and produces neither sfRNA2 nor sfRNA1 in vertebrate cells [17]. RNA-Seq analysis demonstrated that infection with WT ZIKV led to the significant (FDR-corrected p-value < 0.05) up-regulation of 2334 genes and down-regulation of 1109 genes (Figure 1A and Table S2), while infection with sfRNA-deficient mutant virus significantly increased the expression of 971 genes and decreased the expression of only 3 genes (Figure 1B and Table S3). Consistent with previous observations [17,18], WT ZIKV induced the expression of interferon-stimulated genes (ISGs) (OAS1, IFIT2, IFITM1) and pro-inflammatory cytokines, e.g., CXCL10 (Figure 1A). Infection with the sfRNA-deficient ZIKV mutant also induced the expression of these genes, although to a lesser extent (Figure 2B). Notably, the down-regulated genes, that were specific to infection with the sfRNA-producing WT virus, included regulators of neuron differentiation, such as DLX6, DLX5, and FOXG1 (Figure 1A), which indicates that the production of sfRNA may contribute to the dysregulation of brain development in ZIKV infections.

Statistical comparisons of differentially expressed genes between brain organoids infected with WT ZIKV and sfRNA-deficient mutant ZIKV demonstrated that, in total, 1896 genes exhibited significantly different responses to infection with WT compared to the mutant virus (Figure 1C and Table S4). Amongst the genes with the most significant difference in expression between the WT and mutant virus infections were antiviral genes (e.g., IRF1, IFIT2, OASL) that exhibited a higher expression in the WT virus infection (Figure 2D). In addition, we found that other than the regulators of brain development, genes exclusively down-regulated in response to being infected with WT-virus included histones (e.g., HIST1H4L, HIST1H3E) and long noncoding RNAs (LINC00506, LINC01164, LINC00461, etc).

To validate the results of transcriptome profiling, we repeated the organoid infection and performed qRT-PCR for the mRNAs of several genes that exhibited significant differences in expression between organoids infected with WT and sfRNA-deficient ZIKV. We selected DLX5, HIST1H1, and FOXG1 as the representatives of downregulated genes and IFIT2, ISG15, MX1, OAS1, and TRIM25 as representatives of upregulated genes. The results of qRT-PCR were consistent with RNA-Seq data, showing significantly stronger down-regulation of FOXG1, HIST1H1, and DLX5 regarding WT virus infection compared to the mutant virus infection (Figure 2A). The expression of ISGs was also confirmed to be higher in WT ZIKV-infected organoids (Figure 2A). This finding was rather unexpected, as ZIKV sfRNA was previously shown by us and others to inhibit interferon (IFN) signaling and ISG expression in infected cells [17,18]. This discrepancy can be explained by the difference between the infections in the brain organoids used in this study and the infections in the cell monolayers utilized previously. As we recently showed, in brain organoids, ZIKV infects only a small proportion of cells [17], whereas, in a cell monolayer culture infected at a high multiplicity of infection, all cells are infected. As per the results, all cells grown in the monolayer culture should contain sfRNA and exhibit the same phenotype in regard to the antiviral response. In contrast, organoids have a large number of uninfected cells that do not contain sfRNA and also respond to IFNs produced by infected cells. As IFN production itself is not affected by sfRNA, we should expect that organoids infected with the WT virus, which exhibits higher replication efficiency, would produce more IFN, triggering a stronger antiviral response in uninfected bystander cells. In this case, the expression of ISGs should be proportional to the amount of viral RNA present in the organoids. To test this hypothesis, we determined the viral copy number in organoids infected with WT and sfRNA-deficient ZIKV (Figure 2B) and normalized gene expression to the viral RNA (Figure 2C). We found that the expression of ISGs per copy of viral RNA was significantly higher in organoids infected with the mutant virus compared to those with the WT infection, which is consistent with previous findings on the inhibitory effect of sfRNA on IFN signaling. Notably, the difference in the expression of FOXG1, HIST1H1, and DLX5 between organoids infected with WT and sfRNA-deficient viruses was still preserved after normalization (Figure 2C), which indicates that the unique effect of sfRNA on the suppression of these genes in infected cells cannot be attributed to the higher virus replication efficiency.

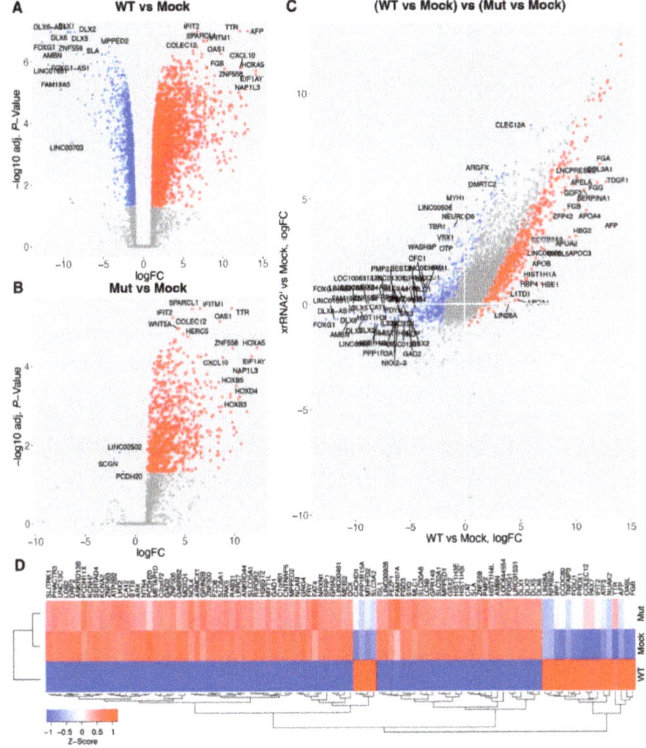

Figure 1. Differential host gene expression in iPSC-derived human brain organoids infected with WT and sfRNA-deficient ZIKV. Human brain organoids at DIV15 were infected with WT or sfRNA-deficient ZIKV mutant at a dose of 10^5 FFU. Total RNA was isolated at 3 dpi and analyzed by RNA-Seq. (**A**) Volcano plot showing differentially expressed genes in human brain organoids infected with WT ZIKV. (**B**) Volcano plot showing differentially expressed genes in human brain organoids

infected with sfRNA-deficient ZIKV mutant. In (**A,B**) genes with significantly different expression levels compared to the mock (FDR-corrected *p*-value < 0.05 and logFC > 1 or <−1) are shown in the color red for upregulated genes and blue for downregulated genes. The top most significant differentially expressed genes (DEGs) are labeled. (**C**) Comparison of gene expression levels between human brain organoids infected with WT and sfRNA-deficient (Mut) ZIKV. Genes with significantly different expressions (FDR-corrected *p*-value < 0.05 and logFC > 1 or <−1) are shown in the color red for genes with higher expression in WT infection compared to Mut virus infection and blue for genes with lower expression in WT infection compared to Mut virus infection. (**D**) Heat map representation of the expression levels of the top 100 most significant genes identified in (**C**).

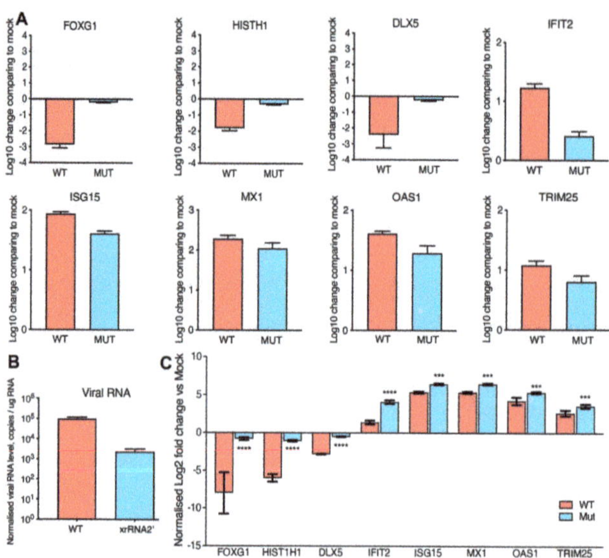

Figure 2. qRT-PCR validation of differentially expressed host genes identified by global transcriptome profiling. RNA isolated from human brain organoids infected with WT or sfRNA-deficient mutant viruses was subjected to cDNA synthesis and qRT-PCR. RNA from mock-infected organoids was used as a control. (**A**) Expression of the genes involved in brain development and ISGs determined using the $\Delta\Delta C_T$ method relative to the mock, with normalization to TBP mRNA level. (**B**) Viral RNA levels in infected organoids determined as copy numbers per ug of input RNA using the standard curve method. (**C**) Expression of host genes involved in brain development and ISGs normalized to viral RNA levels. The values are the means of three replicates ± SD. Statistical analysis was performed using Student's *t*-test: **** *p* < 0.0001, *** *p* < 0.001.

In summary, we demonstrated that the production of sfRNA by ZIKV results in the specific down-regulation of multiple genes in the developing brain tissue. Some of these genes are known to be involved in the regulation of brain development [23].

3.2. Production of sfRNA in ZIKV-Infected Human Brain Organoids Affects Activity of Multiple Pathways Related to Brain Development

To identify molecular processes and signaling pathways affected by the production of sfRNA in ZIKV-infected human brain organoids, the list of the most significantly affected genes (FDR-corrected *p*-value < 0.01 in Table S4) was subjected to gene ontology (GO) and KEGG pathway enrichment analyses. The information about the association between individual DEGs and enriched GO or KEGG categories was then combined with information about their logFC in WT vs mutant virus comparison to calculate z-scores that show how significantly each pathway or process was affected and in which direction the cumulative expression of the associated genes was shifted.

This analysis demonstrated that the production of sfRNA led to the inhibition of biological processes "nervous system development", "neuron differentiation", "axon guidance", "brain development", and "activation of Wnt-signaling", which counteracts neuron differentiation (Figure 3). Biological processes ("regulation of apoptotic processes", "inflammatory response", and "response to virus") exhibited higher activation in organoids infected with sfRNA-producing WT ZIKV compared to those infected with the sfRNA-deficient mutant (Figure 3). In molecular functions ontology, the enriched categories were related to DNA binding and transcription factor activity, and frizzled binding (Figure 3). As frizzled is a family of G protein-coupled receptors that mediate Wnt signaling [23], this further supports the conclusion regarding the effect of ZIKV sfRNA production on this pathway. The enrichment in cellular components ontology demonstrated that the production of sfRNA was associated with decreased expression of nucleosome components (Figure 3), which is consistent with reduced nucleosome assembly identified by biological process analysis. In addition, sfRNA-producing ZIKV was found to reduce the expression of genes that encode neuronal proteins, as indicated by negative z-scores observed for the categories "axon", "postsynaptic membrane", and "neuronal cell body". This provides additional evidence for the inhibitory effect of sfRNA on neuron differentiation and brain development.

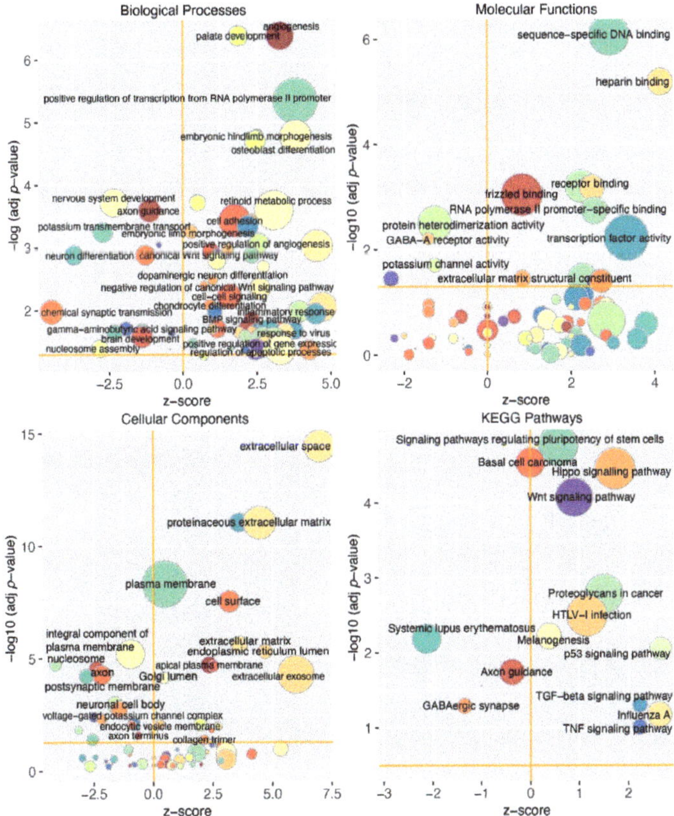

Figure 3. Functional classification of differentially expressed host genes that responded differently to infection with WT compared to sfRNA-deficient ZIKV in iPSC-derived human brain organoids. Size of bubbles on the plots is proportional to the number of associated genes; significantly enriched categories are labeled.

KEGG pathway enrichment analysis (Figure 3) demonstrated that the production of sfRNA was associated with increased activity of the Wnt pathway, Hippo signaling, inflammation (TNF-signaling), and p53-induced apoptosis, while the axon guidance pathway was suppressed by the sfRNA-producing virus. These results are generally consistent with GO enrichment analysis and provide additional evidence for the inhibition of brain development and activation of apoptosis resulting from sfRNA production. In addition, they identify the effect of sfRNA on Hippo signaling; however, the role of this pathway in the mammalian nervous system is not entirely clear, with different evidence suggesting its role in neuron differentiation as well as neuroinflammation and neuron apoptosis.

In summary, gene ontology and KEGG pathway enrichment analyses confirmed the previously identified pro-apoptotic function of ZIKV sfRNA in the developing neural tissue. In addition, it showed that the production of sfRNA by ZIKV is required for the inhibition of brain development and neuron differentiation.

3.3. ZIKV sfRNA Affects Brain Development via a Link between Pro-Apoptotic Signaling and Wnt Pathway

To obtain an understanding of why the production of sfRNA results in the suppression of brain development, we looked at the associations between individual DEGs and the most affected biological processes and signaling pathways. We found that GO categories related to neural development had many shared genes and seemed in part synonymous (Figure 4A,B). We also noticed that these categories shared several genes with pathways related to apoptosis (e.g., WNT7A, WNT5A, SFRP1, Figure 4A,B). Therefore, we hypothesized that affected pathways may interconnect and, in some way, affect each other.

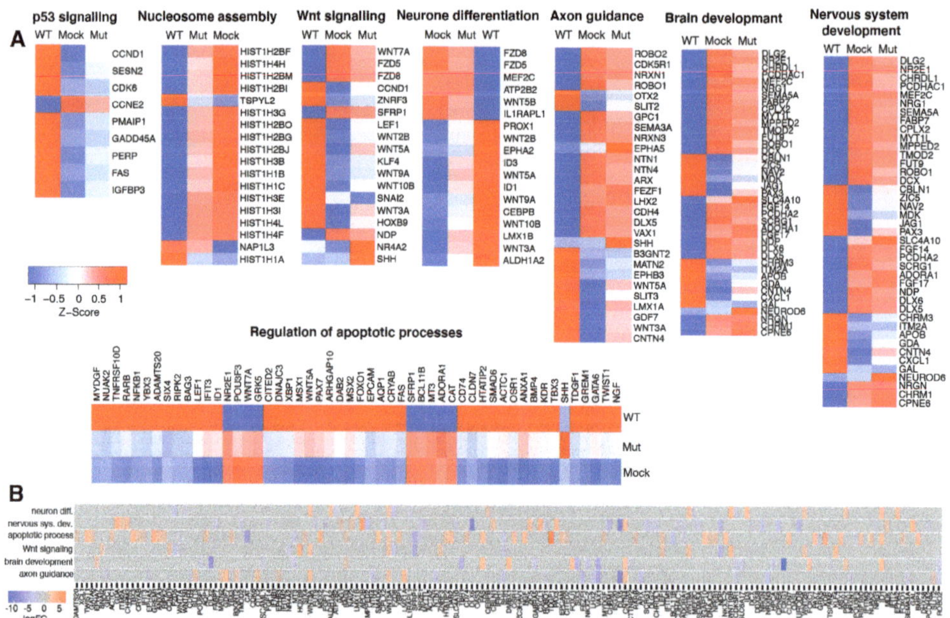

Figure 4. Expression of the host genes associated with brain development and antiviral response in iPSC-derived human brain organoids infected with WT and sfRNA-deficient ZIKV. (**A**) Expression of host genes associated with biological processes and signaling pathways differentially affected by WT and sfRNA-deficient ZIKV. (**B**) Heat map showing association with multiple biological processes of individual differentially expressed host genes.

To test this hypothesis, we reconstructed gene interaction networks for each biological pathway and GO category shown in Figure 4A (Figure 5), merged them, and explored the topology of the resulting network. The merged network consisted of five clusters, four of which were connected, which indicates cross-talk between all of the affected processes, except nucleosome assembly (Figure 6). Network analysis also demonstrated that networks reconstructed from genes associated with brain development and nervous system development were virtually identical (Figure 6), which indicates that these GO terms are highly synonymous. In the merged network, they formed a common cluster with axon guidance genes. Another major cluster in the resulting network consisted of combined genes related to Wnt signaling and neuron differentiation, which indicates that these processes are tightly interconnected (Figure 6). As expected, this cluster was connected to the "brain and nervous system developments, axon guidance" subnetwork (Figure 6). In addition, it appeared to be highly connected to the subnetwork reconstructed from apoptosis-related genes (Figure 6). Wnt signaling/neuron differentiation and apoptosis subnetworks contained multiple shared nods, including WNT7A, SFRP1, LEF1, and others (Figure 6). The connections between different functional categories of genes affected by the production of sfRNA in ZIKV-infected human brain organoids identified by network analysis suggest that the production of sfRNA likely affects brain development indirectly, due to connections between apoptosis pathways and Wnt-signaling, which in turn regulates neuron differentiation.

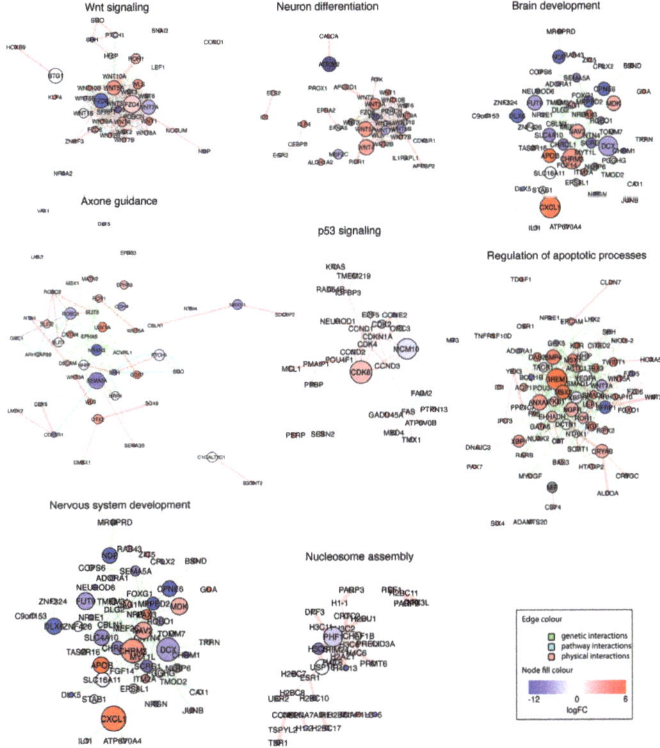

Figure 5. Networks of interactions between the genes affected by the production of sfRNA in ZIKV infection of human brain organoids. Genes associated with each enriched functional category shown in Figure 5A were used to reconstruct the networks of genetic, physical, and pathway interactions. Nod sizes indicate betweenness centrality, the logFC values are for ((WT−Mock) − (Mut−Mock)) contrast.

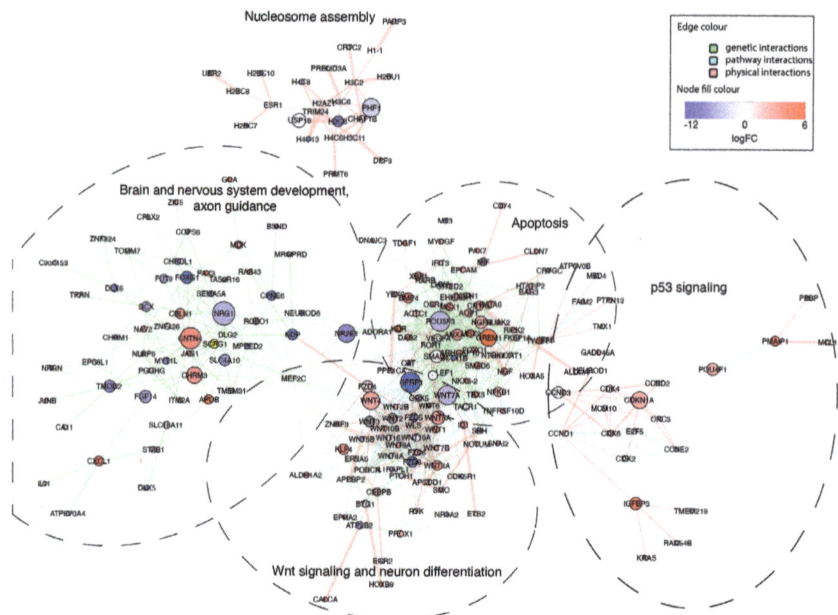

Figure 6. Interconnection between sfRNA-affected biological processes in ZIKV-infected human brain organoids. Individual networks were reconstructed from sfRNA-affected genes related to the most enriched pathways and GO BP terms: Wnt signaling pathway, nervous system development, neuron differentiation, axon guidance, brain development, regulation of apoptotic processes, p53 signaling pathway, and nucleosome assembly. Networks were then merged using the union method and gene names as query keys. The resulting network was analyzed to determine betweenness centrality values for the nods (visualized as nod size). The logFC values are for ((WT−Mock) − (Mut−Mock)) comparison.

Therefore, we propose the following model to explain the inhibitory effect of sfRNA production on brain development. In ZIKV-infected neural progenitors, sfRNA triggers apoptosis. The activation of pro-apoptotic pathways also activates Wnt-signaling. The Wnt signaling pathway inhibits neuron differentiation, which leads to disruptions in brain development.

4. Discussion

The distinctive feature of ZIKV pathogenesis is the ability of the virus to pass through the placental barrier and establish infection in the fetal brain, causing neurodevelopment abnormalities, such as microcephaly [24]. To date, the molecular mechanisms responsible for this effect of the Zika infection are not fully understood. Our recent results demonstrated that sfRNA acts as one of the viral factors of ZIKV pathogenesis in the developing brain [17]. Production of sfRNA was also shown to be a requirement for trans-placental dissemination of the virus and infection of the fetal brain in the mouse pregnancy model [17,25]. In addition, in human brain organoids and cultured neural progenitors, WT virus-producing sfRNA induced the apoptosis of infected cells, while viruses deficient in sfRNA production were much less potent in inducing apoptosis and were unable to cause the death of the infected brain organoids [17].

Herein, we used the RNA-seq of ZIKV-infected human brain organoids to identify the entire spectrum of biological processes and signaling pathways affected by ZIKV sfRNA. Consistent with previous studies [26–28], we found that infection with WT ZIKV inhibits the expression of multiple genes involved in brain development, neuron differentiation, and nucleosome assembly, and it activates the expression of genes involved in the antiviral response, apoptosis, and inflammation. However, down-regulation of the genes involved in brain development and apoptosis in organoids infected with the sfRNA-deficient ZIKV mutant was not observed (Figures 1 and 3). Notably, the activation of antiviral and pro-inflammatory genes was evident in organoids infected with the mutant virus, as well as in wild type infection, although it was less profound due to the lower viral loads. The comparable activation of antiviral genes by both viruses indicates that differences in replication between WT and mutant ZIKV only cause subtle quantitative differences in gene expression and are not responsible for the virtually non-existing effect of the sfRNA-deficient ZIKV infection on the expression of genes involved in brain development (Figure 1). Among the genes highly suppressed by WT, but not by sfRNA-deficient ZIKV, were FOXG1 and LHX2 (Figure 1A,B). FOXG1 is a transcription factor that has been previously linked to a variety of congenital brain disorders, including postnatal microcephaly [29]. LHX2 is another neural transcription factor that regulates neural differentiation by suppressing the Wnt signaling [30]. Decreased expression of these genes was previously identified as a risk factor for the development of ZIKV-induced microcephaly in a human discordant twins study [31]. In addition, ZIKV infection was shown to inhibit the expression of FOXG1 in primary human fetal neural progenitors [32]. Here, we demonstrate that this activity of ZIKV requires sfRNA.

One of the pathways highly affected by the production of ZIKV sfRNA during neuro-infection was Wnt-signaling. This pathway is generally involved in the regulation of development, and in neural development it controls the differentiation of neural progenitor cells [23]. Wnt signaling is initiated by the binding of secreted Wnt ligands to plasma membrane receptors of the Frizzled family and co-receptor of the low-density lipoprotein receptor-related protein family. The resulting receptor-ligand complex recruits Dishev-elled protein (DVL), which in turn recruits the Axin and glycogen synthase kinase 3β (GSK3β) complex that are the components of the β-catenin destruction complex. This causes dissociation of the β-catenin destruction complex and leads to the stabilization and accumulation of β-catenin, which then translocates to the nucleus, forms a complex with TCF/LEF transcription factors, and regulates the expression of the target genes [33]. Dysregulation of the Wnt pathway during infection was previously proposed as a potential cause of ZIKV-associated microcephaly [34]. Brain development relies on the proliferation and differentiation of neural progenitor cells (NPC). NPCs can divide symmetrically, producing two NPCs or asymmetrically producing one NPC and one neuron. The path taken by the NPCs is controlled by the Wnt pathway, which inhibits differentiation and favors symmetric division. Wnt is in turn controlled by a transcription factor FOXG1, which inhibits Wnt pathway activity and promotes differentiation. According to our data, the production of sfRNA is associated with the downregulation of FOXG1 and up-regulation of multiple genes involved in the Wnt pathway (Figures 1 and 4). Therefore, it should inhibit the asymmetric division of NPCs, thus preventing the generation of mature neurons and the formation of the brain.

Collectively, our data demonstrate that the production of ZIKV sfRNA is required for the inhibition of brain development via the activation of Wnt-signaling. However, it is unclear why sfRNA targets pathways involved in brain development if their suppression is unlikely to be beneficial for virus replication/survival. Therefore, we hypothesized that the effect of sfRNA on Wnt signaling can be indirect and result from crosstalk between Wnt-signaling and another pathway, which is the primary target of sfRNA relevant to virus-host interactions. This hypothesis was supported by gene interaction network analysis. It demonstrated that Wnt-signaling tightly interconnects with a network of pro-apoptotic genes activated by sfRNA (Figure 6). We recently demonstrated the requirement of sfRNA for the induction of apoptosis [17], and here we demonstrate that the production of sfRNA is associated with the activation of 8 genes that belong to p53 signaling and 42 genes related to other pro-apoptotic pathways, including Fas and TNF signaling (Figures 4 and 5). Notably, some of the apoptosis-related genes (e.g., SFRP1, GRK5, WNT5A, WNT7A, and LEF1) were also components of the Wnt-signaling pathway (Figures 4B and 6). Interconnection between Wnt-signaling and apoptosis was reported previously [35,36]. In particular, Wnt7a was shown to inhibit apoptosis [37,38]. Herein, we observed the inhibitory effect of ZIKV sfRNA on the expression of Wnt7a, which is consistent with the activation of apoptosis by sfRNA. In contrast, Wnt5a induces apoptosis [39] and it was up-regulated in brain organoids infected with WT, but not sfRNA-deficient ZIKV. Therefore, we propose that apoptosis is likely to be the primary target of the sfRNA-dependent modulation of Wnt-signaling, while brain development is affected indirectly due to the shared role of Wnt-signaling in both processes.

Another category of genes whose expression was affected by the production of ZIKV sfRNA was related to nucleosome assembly and was primarily represented by a different member of the histone 1H family. These genes were significantly more down-regulated by WT ZIKV compared to the sfRNA-deficient virus, except for HIST1H1A, which was up-regulated. In addition, the gene encoding for nucleosome assembly protein TSPYL2 was up-regulated only by the sfRNA-producing virus. The inhibitory effect of ZIKV infection on the expression of histones was shown previously [32]; however, the functional significance of this effect for viral infection is currently unclear. The H1 family is currently the most understudied group of histones, and their biological functions are not fully understood. In addition, our network analysis did not identify any connections between this group of genes and other pathways affected by sfRNA. Therefore, it is unclear why sfRNA causes suppression of the expression of the H1 family of histones. TSPYL2 is known as an inhibitor of cell-cycle progression [40] and activation of its expression may be related to cell-cycle arrest associated with ZIKV infection [41]. However, further research is required to understand the effect of the ZIKV infection on chromatin and the role of sfRNA in the modulation of nucleosome assembly.

In summary, our study for the first time identifies the effects of ZIKV sfRNA production on gene expression in the developing brain tissue, using human brain organoids as an ex vivo model of infection. It reveals the requirement of sfRNA for the inhibitory effect of ZIKV infection on neural development and identifies Wnt-signaling as the sfRNA-affected pathway responsible for it. Our comprehensive pathway enrichment and network reconstruction analysis establishes the link between the Wnt-pathway and pro-apoptotic activity of sfRNA and suggests that the inhibition of brain development by ZIKV sfRNA is an indirect effect of its function on the activation of apoptosis.

Supplementary Materials: The following supporting information can be downloaded at: https://www.mdpi.com/article/10.3390/v15051062/s1, Table S1: sequences of the oligonucleotides used in the study; Table S2: differentially expressed genes in human brain organoids infected with WT ZIKV; Table S3: Differentially expressed genes in human brain organoids infected with sfRNA-deficient ZIKV mutant; Table S4: brain organoid genes with expression significantly affected by the production of sfRNA: ((WT-Mock)-(Mut-Mock)) contrast.

Author Contributions: Conceptualization, A.S. and A.A.K.; methodology, A.S., H.C. and J.A.; software, A.S.; formal analysis, A.S.; investigation, A.S., H.C. and J.A.; writing—original draft preparation, A.S. and A.A.K.; writing—review and editing, A.S. and A.A.K.; visualization, A.S.; supervision, A.A.K. and E.W.; project administration, A.A.K. and E.W.; funding acquisition, A.A.K., A.S., J.A. and E.W. All authors have read and agreed to the published version of the manuscript.

Funding: This research was funded by National Health and Medical Research Council (NHMRC) Project grant number APP1127916 and Ideas grant number 2012809 to A.A.K., NHMRC Ideas grant number GNT2021272 to A.S., NHMRC Ideas grants number APP1138795, APP1127976, APP1144806, and APP1130168 to E.W.; Australian Research Council (ARC) Discovery grant number DP210103401 to E.W.; NHMRC Ideas grant number APP2001408 to J.A.; The University Of Queensland Early Career Researcher Grant number UQECR2058457 to J.A.; and a Jérôme Lejeune Postdoctoral Fellowship and Brisbane Children's Hospital Foundation grant number 50308 to J.A. E.W. was supported by the BrAshA-T Foundation and the Perry Cross Spinal Research Foundation.

Institutional Review Board Statement: The study was conducted in accordance with the Declaration of Helsinki, and approved by the University of Queensland Human Research Ethics Committee (protocol code 2019/HE000159, 20 October 2020).

Informed Consent Statement: Informed consent was obtained from all subjects involved in the study.

Data Availability Statement: All data is available within the manuscript and Supplementary Materials.

Conflicts of Interest: The authors declare no conflict of interest.

References

1. Pierson, T.C.; Diamond, M.S. The Emergence of Zika Virus and Its New Clinical Syndromes. *Nature* **2018**, *560*, 573–581. [CrossRef] [PubMed]
2. Baronti, C.; Lieutaud, P.; Bardsley, M.; de Lamballerie, X.; Resman Rus, K.; Korva, M.; Petrovec, M.; Avsic-Zupanc, T.; Matusali, G.; Meschi, S.; et al. The Importance of Biobanking for Response to Pandemics Caused by Emerging Viruses: The European Virus Archive As an Observatory of the Global Response to the Zika Virus and COVID-19 Crisis. *Biopreservation Biobanking* **2020**, *18*, 561–569. [CrossRef]
3. Valentine, M.J.; Murdock, C.C.; Kelly, P.J. Sylvatic Cycles of Arboviruses in Non-Human Primates. *Parasites Vectors* **2019**, *12*, 463. [CrossRef] [PubMed]
4. Falcao, M.B.; Cimerman, S.; Luz, K.G.; Chebabo, A.; Brigido, H.A.; Lobo, I.M.; Timerman, A.; Angerami, R.N.; Cunha, C.A.; Bacha, H.A.; et al. Management of Infection by the Zika Virus. *Ann. Clin. Microbiol. Antimicrob.* **2016**, *15*, 57. [CrossRef]
5. Mlakar, J.; Korva, M.; Tul, N.; Popović, M.; Poljšak-Prijatelj, M.; Mraz, J.; Kolenc, M.; Rus, K.R.; Vipotnik, T.V.; Vodušek, V.F.; et al. Zika Virus Associated with Microcephaly. *N. Engl. J. Med.* **2016**, *374*, 951–958. [CrossRef]
6. Calvet, G.; Aguiar, R.S.; Melo, A.S.O.; Sampaio, S.A.; de Filippis, I.; Fabri, A.; Araujo, E.S.M.; de Sequeira, P.C.; de Mendonça, M.C.L.; de Oliveira, L.; et al. Detection and Sequencing of Zika Virus from Amniotic Fluid of Fetuses with Microcephaly in Brazil: A Case Study. *Lancet Infect. Dis.* **2016**, *16*, 653–660. [CrossRef]
7. Rice, M.E.; Galang, R.R.; Roth, N.M.; Ellington, S.R.; Moore, C.A.; Valencia-Prado, M.; Ellis, E.M.; Tufa, A.J.; Taulung, L.A.; Alfred, J.M.; et al. Vital Signs: Zika-Associated Birth Defects and Neurodevelopmental Abnormalities Possibly Associated with Congenital Zika Virus Infection—U.S. Territories and Freely Associated States, 2018. *Morb. Mortal. Wkly. Rep.* **2018**, *67*, 858–867. [CrossRef]
8. Pijlman, G.P.; Funk, A.; Kondratieva, N.; Leung, J.; Torres, S.; van der Aa, L.; Liu, W.J.; Palmenberg, A.C.; Shi, P.-Y.Y.; Hall, R.A.; et al. A Highly Structured, Nuclease-Resistant, Noncoding RNA Produced by Flaviviruses Is Required for Pathogenicity. *Cell Host Microbe* **2008**, *4*, 579–591. [CrossRef]
9. Funk, A.; Truong, K.; Nagasaki, T.; Torres, S.; Floden, N.; Balmori Melian, E.; Edmonds, J.; Dong, H.; Shi, P.-Y.; Khromykh, A. a RNA Structures Required for Production of Subgenomic Flavivirus RNA. *J. Virol.* **2010**, *84*, 11407–11417. [CrossRef]
10. Villordo, S.M.; Carballeda, J.M.; Filomatori, C.V.; Gamarnik, A.V. RNA Structure Duplications and Flavivirus Host Adaptation. *Trends Microbiol.* **2016**, *24*, 270–283. [CrossRef]
11. Slonchak, A.; Parry, R.; Pullinger, B.; Sng, J.D.J.; Wang, X.; Buck, T.F.; Torres, F.J.; Harrison, J.J.; Colmant, A.M.G.; Hobson-Peters, J.; et al. Structural Analysis of 3′UTRs in Insect Flaviviruses Reveals Novel Determinants of SfRNA Biogenesis and Provides New Insights into Flavivirus Evolution. *Nat. Commun.* **2022**, *13*, 1279. [CrossRef]
12. Slonchak, A.; Khromykh, A.A. Subgenomic Flaviviral RNAs: What Do We Know after the First Decade of Research. *Antivir. Res.* **2018**, *159*, 13–25. [CrossRef] [PubMed]
13. Setoh, Y.X.; Amarilla, A.A.; Peng, N.Y.; Slonchak, A.; Periasamy, P.; Figueiredo, L.T.M.; Aquino, V.H.; Khromykh, A.A. Full Genome Sequence of Rocio Virus Reveal Substantial Variations from the Prototype Rocio Virus SPH 34675 Sequence. *Arch. Virol.* **2018**, *163*, 255–258. [CrossRef] [PubMed]
14. Clarke, B.D.; Roby, J.A.; Slonchak, A.; Khromykh, A.A. Functional Non-Coding RNAs Derived from the Flavivirus 3′ Untranslated Region. *Virus Res.* **2015**, *206*, 53–61. [CrossRef]
15. Akiyama, B.M.; Laurence, H.M.; Massey, A.R.; Costantino, D.A.; Xie, X.; Yang, Y.; Shi, P.-Y.Y.; Nix, J.C.; Beckham, J.D.; Kieft, J.S. Zika Virus Produces Noncoding RNAs Using a Multi-Pseudoknot Structure That Confounds a Cellular Exonuclease. *Science* **2016**, *354*, 1148–1152. [CrossRef] [PubMed]
16. Slonchak, A.; Hugo, L.E.; Freney, M.E.; Hall-Mendelin, S.; Amarilla, A.A.; Torres, F.J.; Setoh, Y.X.; Peng, N.Y.G.; Sng, J.D.J.; Hall, R.A.; et al. Zika Virus Noncoding RNA Suppresses Apoptosis and Is Required for Virus Transmission by Mosquitoes. *Nat. Commun.* **2020**, *11*, 2205. [CrossRef]
17. Slonchak, A.; Wang, X.; Aguado, J.; Sng, J.D.J.; Chaggar, H.; Freney, M.E.; Yan, K.; Torres, F.J.; Amarilla, A.A.; Balea, R.; et al. Zika Virus Noncoding RNA Cooperates with the Viral Protein NS5 to Inhibit STAT1 Phosphorylation and Facilitate Viral Pathogenesis. *Sci. Adv.* **2022**, *8*, eadd8095. [CrossRef]
18. Pallarés, H.M.; Soledad, G.; Navarro, C.; Villordo, S.M.; Merwaiss, F.; De Borba, L.; Gonzalez, M.M.; Ledesma, L.; Ojeda, D.S.; Henrion-Lacritick, A.; et al. Zika Virus Subgenomic Flavivirus RNA Generation Requires Cooperativity between Duplicated RNA Structures That Are Essential for Productive Infection in Human Cells. *J. Virol.* **2020**, *94*, 343–363. [CrossRef]

19. Lancaster, M.A.; Renner, M.; Martin, C.A.; Wenzel, D.; Bicknell, L.S.; Hurles, M.E.; Homfray, T.; Penninger, J.M.; Jackson, A.P.; Knoblich, J.A. Cerebral Organoids Model Human Brain Development and Microcephaly. *Nature* **2013**, *501*, 373–379. [CrossRef]
20. Qian, X.; Nguyen, H.N.; Jacob, F.; Song, H.; Ming, G. Using Brain Organoids to Understand Zika Virus-Induced Microcephaly. *Development* **2017**, *144*, 952–957. [CrossRef]
21. Setoh, Y.X.; Amarilla, A.A.; Peng, N.Y.G.; Griffiths, R.E.; Carrera, J.; Freney, M.E.; Nakayama, E.; Ogawa, S.; Watterson, D.; Modhiran, N.; et al. Determinants of Zika Virus Host Tropism Uncovered by Deep Mutational Scanning. *Nat. Microbiol.* **2019**, *4*, 876–887. [CrossRef] [PubMed]
22. Slonchak, A.; Clarke, B.; Mackenzie, J.; Amarilla, A.A.; Setoh, Y.X.; Khromykh, A.A. West Nile Virus Infection and Interferon Alpha Treatment Alter the Spectrum and the Levels of Coding and Noncoding Host RNAs Secreted in Extracellular Vesicles. *BMC Genom.* **2019**, *20*, 474. [CrossRef] [PubMed]
23. Clevers, H.; Nusse, R. Wnt/β-Catenin Signaling and Disease. *Cell* **2012**, *149*, 1192–1205. [CrossRef]
24. Honein, M.A.; Dawson, A.L.; Petersen, E.E.; Jones, A.M.; Lee, E.H.; Yazdy, M.M.; Ahmad, N.; Macdonald, J.; Evert, N.; Bingham, A.; et al. Birth Defects among Fetuses and Infants of US Women with Evidence of Possible Zika Virus Infection during Pregnancy. *JAMA* **2017**, *317*, 59–68. [CrossRef]
25. Sparks, H.; Monogue, B.; Akiyama, B.; Kieft, J.; Beckham, J.D. Disruption of Zika Virus XrRNA1-Dependent SfRNA1 Production Results in Tissue-Specific Attenuated Viral Replication. *Viruses* **2020**, *12*, 1177. [CrossRef]
26. Tiwari, S.K.; Dang, J.; Qin, Y.; Lichinchi, G.; Bansal, V.; Rana, T.M. Zika Virus Infection Reprograms Global Transcription of Host Cells to Allow Sustained Infection. *Emerge Microbes Infect.* **2017**, *5*, 50. [CrossRef]
27. Zhang, F.; Hammack, C.; Ogden, S.C.; Cheng, Y.; Lee, E.M.; Wen, Z.; Qian, X.; Nguyen, H.N.; Li, Y.; Yao, B.; et al. Molecular Signatures Associated with ZIKV Exposure in Human Cortical Neural Progenitors. *Nucleic Acids Res.* **2016**, *44*, 8610–8620. [CrossRef] [PubMed]
28. Sun, X.; Hua, S.; Chen, H.-R.R.; Ouyang, Z.; Einkauf, K.; Tse, S.; Ard, K.; Ciaranello, A.; Yawetz, S.; Sax, P.; et al. Transcriptional Changes during Naturally Acquired Zika Virus Infection Render Dendritic Cells Highly Conducive to Viral Replication. *Cell Rep.* **2017**, *21*, 3471–3482. [CrossRef]
29. Kumamoto, T.; Hanashima, C. Evolutionary Conservation and Conversion of Foxg1 Function in Brain Development. *Dev. Growth Differ.* **2017**, *59*, 258–269. [CrossRef]
30. Hou, P.-S.; Chuang, C.-Y.; Kao, C.-F.; Chou, S.-J.; Stone, L.; Ho, H.-N.; Chien, C.-L.; Kuo, H.-C. LHX2 Regulates the Neural Differentiation of Human Embryonic Stem Cells via Transcriptional Modulation of PAX6 and CER1. *Nucleic Acids Res.* **2013**, *41*, 7753–7770. [CrossRef]
31. Caires-Júnior, L.C.; Goulart, E.; Melo, U.S.; Araujo, B.S.H.; Alvizi, L.; Soares-Schanoski, A.; De Oliveira, D.F.; Kobayashi, G.S.; Griesi-Oliveira, K.; Musso, C.M.; et al. Discordant Congenital Zika Syndrome Twins Show Differential In Vitro Viral Susceptibility of Neural Progenitor Cells. *Nat. Commun.* **2018**, *9*, 475. [CrossRef]
32. Jiang, X.; Dong, X.; Li, S.H.; Zhou, Y.P.; Rayner, S.; Xia, H.M.; Gao, G.F.; Yuan, H.; Tang, Y.P.; Luo, M.H. Proteomic Analysis of Zika Virus Infected Primary Human Fetal Neural Progenitors Suggests a Role for Doublecortin in the Pathological Consequences of Infection in the Cortex. *Front. Microbiol.* **2018**, *9*, 1067. [CrossRef] [PubMed]
33. Lee, E.; Salic, A.; Krüger, R.; Heinrich, R.; Kirschner, M.W. The Roles of APC and Axin Derived from Experimental and Theoretical Analysis of the Wnt Pathway. *PLoS Biol.* **2003**, *1*, e10. [CrossRef]
34. Lin, J.-J.; Chin, T.-Y.; Chen, C.-P.; Chan, H.-L.; Wu, T.-Y. Zika Virus: An Emerging Challenge for Obstetrics and Gynecology. *Taiwan J. Obs. Gynecol.* **2017**, *56*, 585–592. [CrossRef] [PubMed]
35. Zimmerman, Z.F.; Kulikauskas, R.M.; Bomsztyk, K.; Moon, R.T.; Chien, A.J. Activation of Wnt/β-Catenin Signaling Increases Apoptosis in Melanoma Cells Treated with Trail. *PLoS ONE* **2013**, *8*, e69593. [CrossRef]
36. Pai, S.G.; Carneiro, B.A.; Mota, J.M.; Costa, R.; Leite, C.A.; Barroso-Sousa, R.; Kaplan, J.B.; Chae, Y.K.; Giles, F.J. Wnt/Beta-Catenin Pathway: Modulating Anticancer Immune Response. *J. Hematol. Oncol.* **2017**, *10*, 101. [CrossRef] [PubMed]
37. Ai, P.; Xu, X.; Xu, S.; Wei, Z.; Tan, S.; Li, J. Overexpression of Wnt7a Enhances Radiosensitivity of Non-Small-Cell Lung Cancer via the Wnt/JNK Pathway. *Biol. Open* **2020**, *9*, bio050575. [CrossRef]
38. Carta, L.; Sassoon, D. Wnt7a Is a Suppressor of Cell Death in the Female Reproductive Tract and Is Required for Postnatal and Estrogen-Mediated Growth1. *Biol Reprod* **2004**, *71*, 444–454. [CrossRef]
39. Peng, S.; Zhang, J.; Chen, J.; Wang, H. Effects of Wnt5a Protein on Proliferation and Apoptosis in JAR Choriocarcinoma Cells. *Mol. Med. Rep.* **2011**, *4*, 99–104. [CrossRef]
40. Magni, M.; Buscemi, G.; Maita, L.; Peng, L.; Chan, S.Y.; Montecucco, A.; Delia, D.; Zannini, L. TSPYL2 Is a Novel Regulator of SIRT1 and P300 Activity in Response to DNA Damage. *Cell Death Differ.* **2019**, *26*, 918–931. [CrossRef]
41. Zhang, N.; Zhang, N.; Qin, C.F.; Liu, X.; Shi, L.; Xu, Z. Zika Virus Disrupts Neural Progenitor Development and Leads to Microcephaly in Mice. *Cell Stem Cell* **2016**, *19*, 120–126. [CrossRef]

Article

Development and Characterization of Efficient Cell Culture Systems for Genotype 1 Hepatitis E Virus and Its Infectious cDNA Clone

Putu Prathiwi Primadharsini [1], Shigeo Nagashima [1], Toshinori Tanaka [1], Suljid Jirintai [1,2], Masaharu Takahashi [1], Kazumoto Murata [1] and Hiroaki Okamoto [1,*]

[1] Division of Virology, Department of Infection and Immunity, Jichi Medical University School of Medicine, Shimotsuke, Tochigi 329-0414, Japan

[2] Division of Pathology, Department of Basic Veterinary Medicine, Inner Mongolia Agricultural University College of Veterinary Medicine, Hohhot 010018, China

* Correspondence: hokamoto@jichi.ac.jp; Tel.: +81-285-58-7404

Abstract: Hepatitis E virus (HEV) is a major cause of acute viral hepatitis globally. Genotype 1 HEV (HEV-1) is responsible for multiple outbreaks in developing countries, causing high mortality rates in pregnant women. However, studies on HEV-1 have been hindered by its poor replication in cultured cells. The JE04-1601S strain recovered from a Japanese patient with fulminant hepatitis E who contracted HEV-1 while traveling to India was serially passaged 12 times in human cell lines. The cell-culture-generated viruses (passage 12; p12) grew efficiently in human cell lines, but the replication was not fully supported in porcine cells. A full-length cDNA clone was constructed using JE04-1601S_p12 as a template. It was able to produce an infectious virus, and viral protein expression was detectable in the transfected PLC/PRF/5 cells and culture supernatants. Consistently, HEV-1 growth was also not fully supported in the cell culture of cDNA-derived JE04-1601S_p12 progenies, potentially recapitulating the narrow tropism of HEV-1 observed in vivo. The availability of an efficient cell culture system for HEV-1 and its infectious cDNA clone will be useful for studying HEV species tropism and mechanisms underlying severe hepatitis in HEV-1-infected pregnant women as well as for discovering and developing safer treatment options for this condition.

Keywords: hepatitis E virus; genotype 1; serial passages; cell culture system; infectious cDNA clone; replication efficiency; species tropism

Citation: Primadharsini, P.P.; Nagashima, S.; Tanaka, T.; Jirintai, S.; Takahashi, M.; Murata, K.; Okamoto, H. Development and Characterization of Efficient Cell Culture Systems for Genotype 1 Hepatitis E Virus and Its Infectious cDNA Clone. *Viruses* **2023**, *15*, 845. https://doi.org/10.3390/v15040845

Academic Editors: Yiping Li and Yuliang Liu

Received: 1 March 2023
Revised: 21 March 2023
Accepted: 25 March 2023
Published: 26 March 2023

1. Introduction

Hepatitis E virus (HEV) is a single-stranded positive-sense RNA virus with an approximately 7.2 kb genome. It is a member of the family *Hepeviridae*, subfamily *Orthohepevirinae*. This subfamily comprises four genera, including the genus *Paslahepevirus*. Most human infections involve the species *Paslahepevirus balayani* genotypes 1, 2, 3, and 4 and less frequently 7 [1].

The genome contains a short 5′-untranslated region (UTR) with a 7-methylguanosine cap, three open reading frames (ORFs) and a short 3′-UTR terminated by the poly(A) tract [2,3]. ORF1 is translated from genomic RNA, while ORF2 and ORF3 are translated from a subgenomic RNA strand [4,5]. ORF1 encodes a non-structural polyprotein containing multiple functional domains involved in viral replication: methyltransferase (MeT), Y domain, papain-like cysteine protease (PCP), hypervariable region (HVR), X or macro domain, helicase (Hel), and RNA-dependent RNA polymerase (RdRp) [6–9]. ORF2 encodes the capsid protein, which plays a crucial role during virion assembly and viral attachment to the host cell and is the major target for neutralizing antibodies [10,11]. ORF3 encodes a multifunctional phosphoprotein required for virion egress [12–14] and is a functional ion channel acting as a viroporin [15]. ORF4—exclusively present in genotype 1 HEV

(HEV-1)—encodes a novel protein identified in the coding sequence of ORF1, is synthesized only under conditions of endoplasmic reticulum (ER) stress, and is short-lived, as it is degraded quickly by the host proteasome [16,17].

HEV infection is distributed globally. Among the four major genotypes infecting humans, HEV-1 and genotype 2 HEV (HEV-2) are restricted to humans and have been responsible for multiple outbreaks in developing countries where the virus is transmitted through drinking contaminated water [18]. Although HEV-1 typically affects people in developing countries, such as South Asia and most countries in sub-Saharan Africa [19], imported infections have been reported in industrialized countries, including Japan [20–22]. HEV-1 infection in pregnant women frequently leads to infant mortality or premature delivery. Particularly in the third trimester, HEV-1 infection has been linked to a poor prognosis, where the fatality rate can reach up to 30% [23,24]. Genotype 3 HEV (HEV-3) and genotype 4 HEV (HEV-4), conversely, primarily affect populations in industrialized countries and are mainly transmitted through zoonotic foodborne routes, with less frequent routes including through solid organ transplantation or transfusion of blood products [25].

Most HEV infections are self-limiting but carry a risk of progression to chronic infection in those with an immunocompromised status [26]. Ribavirin has been used to treat certain instances of clinical HEV infection, such as chronic cases (caused by HEV-3, HEV-4, and genotype 7 HEV (HEV-7)) or acute fulminant cases (caused by HEV-1, HEV-3, and HEV-4) [27,28]. However, it has a number of major side effects, such as anemia, and although the latest study does not suggest a clear signal of human teratogenicity for ribavirin, it should be used with caution in pregnant women, a major risk group [29].

The development of cell culture systems for HEV-3 and HEV-4 strains has progressed significantly in recent decades [30,31]. However, HEV-1 replicates poorly in cultured cells due to a lack of an efficient cell culture system [16,32–36], which has hindered research on its life cycle, determinants of its species tropism to humans, and mechanisms underlying its severity in pregnant women, as well as the discovery and development of safer treatment options, particularly for pregnant women with fulminant hepatitis caused by HEV-1 infection.

To overcome this obstacle, we attempted to develop an efficient cell culture system for HEV-1 and to construct an infectious cDNA clone of HEV-1 using the JE04-1601S strain, which was recovered from a Japanese patient with fulminant hepatitis E who contracted HEV infection while traveling to India [37] and has been serially passaged 12 times, as a template. Currently, there are seven subtypes of HEV-1 (1a to 1g) [38]. The strain used in the present study is subtype 1f, which is one of the most common circulating human HEV subtypes in South Asia, including India [39] and Bangladesh [40].

2. Materials and Methods

2.1. Cell Culture

PLC/PRF/5 (ATCC No. CRL-8024; American Type Culture Collection, Manassas, VA, USA), HepG2/C3A (ATCC No. HB-8065), and A549_1-1H8 (a subclone of A549, No. RCB0098; RIKEN BRC Cell Bank, Tsukuba, Japan) cells were grown in growth medium which consists of Dulbecco's modified Eagle's medium (DMEM; Thermo Fisher Scientific, Waltham, MA, USA), supplemented with 10% heat-inactivated fetal bovine serum (FBS) (Thermo Fisher Scientific), at 37 °C in a humidified 5% CO_2 atmosphere, as previously described [41]. LLC-PK1 (ATCC No. CL-101) cells were grown in medium 199 with Earle's Salt (Thermo Fisher Scientific) and 2.2 g/l $NaHCO_3$, supplemented with 3% heat-inactivated FBS. PK15 (ATCC No. CCL-33) cells were grown in DMEM, supplemented with 10% heat-inactivated FBS and 0.1 mM non-essential amino acid (NEAA) (Thermo Fisher Scientific). IBRS-2 (Cellosaurus, CVCL_4528; Swiss Institute of Bioinformatics, Lausanne, Switzerland) cells were grown in DMEM supplemented with 10% heat-inactivated FBS. For A549_1-1H8, following virus inoculation, growth medium was replaced with maintenance medium (50% DMEM, and 50% medium 199 with Earle's Salt and 2.2 g/l $NaHCO_3$) containing 2% heat-inactivated FBS and 30 mM $MgCl_2$. All types of culture media contain

100 U/mL penicillin G, 100 µg/mL streptomycin, and 2.5 µg/mL amphotericin B, except for the medium used in Section 2.11.

2.2. Viruses

A serum sample of an HEV-1f strain recovered from a Japanese patient with fulminant hepatitis E who contracted the infection while traveling to India (JE04-1601S, 2.8×10^6 copies/mL; referred to as S5 in [37]) was filtrated through a 0.22 µm microfilter (Millex-GV; Merck Millipore, Darmstadt, Germany), aliquoted as virus stocks, and then stored at -80 °C. Filtrated culture medium containing the cell-culture-adapted HEV-3b strain (JE03-1760F passage 26, JE03-1760F_p26; 1.5×10^8 copies/mL) [42] and filtrated culture medium containing the cell-culture-adapted HEV-4c strain (HE-JF5/15F passage 24, HE-JF5/15F_p24; 2.0×10^8 copies/mL) [31,43] were also utilized in this study.

2.3. Virus Inoculation and Serial Passages

Passage 0 (p0) was carried out in a monolayer of PLC/PRF/5 cells using the filtrated JE04-1601S with a viral load of 1.5×10^6 copies/well in a six-well plate (Iwaki, Shizuoka, Japan). Culture medium of p0 was filtrated and then inoculated onto a monolayer of A549_1-1H8 cells in a six-well plate with a viral load of 1.1×10^4 copies/well and was regarded as p1. Serial passages were performed until p12 with an inoculum titer of 1.0×10^5 copies/well (except for p2 and p4 with 4.8×10^4 and 6.8×10^4 copies/well, respectively) (Table 1). The volume of inoculum was 200 µL/well. The general protocol for cell culture is described in Section 2.1.

Table 1. Sources and HEV RNA titers of inocula used for serial inoculations onto A549_1-1H8 cells.

Passage	Inoculum	Cells	Viral Load of HEV Inoculated in Each Well (Copies Per Well) [a]
0	Serum (JE04-1601S_wt)	PLC/PRF/5	1.5×10^6
1	Culture supernatant (26th day after the first inoculation)	A549_1-1H8	1.1×10^4
2	Culture supernatant (62nd day after the second inoculation)	A549_1-1H8	4.8×10^4
3	Culture supernatant (42nd day after the third inoculation)	A549_1-1H8	1.0×10^5
4	Culture supernatant (42nd day after the fourth inoculation)	A549_1-1H8	6.8×10^4
5	Culture supernatant (70th day after the fifth inoculation)	A549_1-1H8	1.0×10^5
6	Culture supernatant (54th day after the sixth inoculation)	A549_1-1H8	1.0×10^5
7	Culture supernatant (64th day after the seventh inoculation)	A549_1-1H8	1.0×10^5
8	Culture supernatant (30th day after the eighth inoculation)	A549_1-1H8	1.0×10^5
9	Culture supernatant (36th day after the ninth inoculation)	A549_1-1H8	1.0×10^5
10	Culture supernatant (26th day after the tenth inoculation)	A549_1-1H8	1.0×10^5
11	Culture supernatant (54th day after the eleventh inoculation)	A549_1-1H8	1.0×10^5
12	Culture supernatant (38th day after the twelfth inoculation)	A549_1-1H8	1.0×10^5

[a] Quantification of HEV RNA was performed after filtration of the serum sample or the culture supernatant through a 0.22 µm microfilter.

Inoculation of culture supernatants (JE03-1760F_p26, HE-JF5/15F_p24, or JE04-1601S_p12) was performed in a monolayer of PLC/PRF/5, A549_1-1H8, HepG2/C3A, PK15, IBRS-2, or LLC-PK1 cells in a six-well plate with a titer of 1.0×10^5 copies/well or 1.0×10^6 copies/well, unless otherwise stated. After incubation at room temperature for 1 h, the cells were washed five times with phosphate-buffered saline without Mg^{2+} and Ca^{2+} (PBS[−]), and 2 mL of the respective medium (as described in Section 2.1) was added to each well. The cells were then incubated at 35.5 °C in a humidified 5% CO_2 atmosphere, as previously described [41]. Every other day, half of the culture medium (1 mL) was replaced with fresh respective medium (as described in Section 2.1).

Inoculum was filtrated through a microfilter with a pore size of 0.22 µm before being inoculated. The collected culture medium was centrifuged at $1300 \times g$ at room temperature for 2 min, and then the supernatant was stored at -80 °C until use.

2.4. Quantification of HEV RNA

Total RNA was extracted from culture supernatants of inoculated or transfected cells using TRIzol-LS reagent (Thermo Fisher Scientific) or from cultured cells using TRIzol reagent (Thermo Fisher Scientific). The quantification of HEV RNA was performed by real-time reverse transcription (RT)-polymerase chain reaction (PCR) using a LightCycler apparatus (Roche Diagnostics KK, Tokyo, Japan) with a QuantiTect Probe RT-PCR kit (Qiagen, Tokyo, Japan), a primer set, and a probe targeting the overlapping region of ORF2 and ORF3, according to the previously described method [44]. The limit of detection by RT-PCR used in the current study is 2.0×10^1 RNA copies/mL.

2.5. Immunocapture RT-PCR Assay

The immunocapture RT-PCR assay was performed as described previously with some modifications [44]. Briefly, anti-ORF2 monoclonal antibody (MAb) (H6225) [44] (2 µg/mL) was mixed with protein G magnetic beads (Bio-Rad, Hercules, CA, USA) in PBS(-) containing 0.1% bovine serum albumin (BSA), and then the mixture was rotated at room temperature for 2 h. Membrane-associated particles in the culture supernatants of HEV-1-infected cells were pre-treated with 0.1% sodium deoxycholate (DOC-Na) and 0.1% trypsin at 37 °C for 2 h. After washing the magnetic beads twice with PBS(-) containing 0.1% Tween 20, the treated HEV particles and magnetic beads were incubated in PBS(-) containing 0.1% BSA by rotating at room temperature for 2 h. The supernatants were collected and the magnetic beads were washed three times with PBS(-) containing 0.1% Tween 20. Total RNA in the supernatant and magnetic beads was extracted with TRIzol-LS reagent and TRIzol reagent, respectively, and then subjected to quantification of HEV RNA as described in Section 2.4.

2.6. The Determination and Analysis of Full-Length and Partial Genome Sequences of JE04-1601S Strains

The full-length genomic sequences of the JE04-1601S strains (wild-type, p10, and p12) were determined according to the method described previously [42]. In brief, total RNA extracted from a serum sample (for wild-type isolate) or culture medium (for p10 and p12 isolates) was subjected to cDNA synthesis followed by nested PCR of eight overlapping regions including the extreme 5′- and 3′-terminal regions. The amplified regions excluding the primer sequences were nucleotide (nt) 1–132 (132 base pairs (bp)), nt 19–1296 (1278 bp), nt 1058–2094 (1037 bp), nt 2024–3129 (1106 bp), nt 2938–4695 (1758 bp), nt 4598–6376 (1779 bp), nt 6297–7122 (826 bp), and nt 7050–7192 (143 bp); the nt positions are numbered in accordance with the JE04-1601S genome obtained in the present study. The extreme 5′-end sequence (nt 1–132) was determined using the First Choice RLM-RACE kit (Ambion, Austin, TX, USA) [45]. Amplification of the 3′-end sequence (nt 7050–7192 (143 bp): poly(A) tail excluded) was performed in accordance with the previously described method [45]. The amplification product was sequenced on both strands directly or after cloning into the plasmid vector (T-vector pMD20; TaKaRa Bio, Shiga, Japan), using the BigDye Terminator v3.1 Cycle Sequencing Kit on an ABI PRISM 3130*xl* Genetic Analyzer (Thermo Fisher Scientific). The sequence analysis was performed using the Genetyx Mac ver. 22 (Genetyx, Tokyo, Japan).

A portion of the ORF1 region (nt 2651–3200, 550 bp) of the cell-culture-generated variants of JE04-1601S (p0 to p12) was determined. In brief, total RNA extracted from culture medium was reverse-transcribed with SuperScript IV (Thermo Fisher Scientific) and then subjected to PCR in the presence of ExTaq (TaKaRa Bio), using primers HE570 (sense: TGCTTATCGGGAGACTTGC, nt 2632–2650) and HE571 (anti-sense: GTGCTCAAAGTC-GATGGCTG, nt 3201–3220). The thermal cycler conditions were 94 °C for 2 min, 35 cycles (94 °C, 30 s; 55 °C, 30 s; 72 °C, 75 s), and then 72 °C for 7 min.

A phylogenetic tree was constructed using the neighbor-joining tree of Jukes–Cantor distances based on the entire genomic sequences of three JE04-1601S strains, all known genotype 1 strains (1a, n = 13; 1b, n = 10; 1c, n = 2; 1d, n = 1; 1e, n = 1; 1f, n = 32; 1g, n = 18; and unclassified subtypes, n = 2) [38], and each one of prototype strains of genotypes 2–8 [38]. Multiple alignments were generated using the MUSCLE software program, version 3.5 [46].

2.7. Construction of Full-Length Infectious cDNA Clones

To generate a full-length infectious cDNA clone of JE04-1601S_p12, RNA was extracted from culture medium containing JE04-1601S_p12 using TRIzol-LS. cDNA was synthesized using SuperScript IV (Thermo Fisher Scientific) with primer SSP-T (Table 2). Using the synthesized cDNA as template, three fragments covering the entire JE04-1601S_p12 genome were amplified by PCR with Platinum SuperFi II DNA Polymerase (Thermo Fisher Scientific) (see Figure 4). Fragment 1-1 (F1-1) was amplified with primers 1601S-1 and 1601S-7, fragment 1-2 (F1-2) was amplified with primers 1601S-8 and 1601S-2, and fragment 2 (F2) was amplified with primers 1601S-3 and 1601S-4 (Table 2). The three overlapping amplified fragments were designed to share a 15-nt homologous sequence.

Table 2. The primers used to construct the pJE04-1601S_p12 and pJE04-1601S_p12-GAA clones.

Name	Polarity	Sequence (5′ to 3′)	Note
SSP-T		AAGGATCCGTCGACATCGATAATACGTTTTTTTTTTTTTTTT	cDNA synthesis
1601S-1	+	GCTTAATACGACTCACTATAGCAGACCACATATGTGGTCGATGCC	T7 promoter (underlined) and positive-strand sequence (nt 1–25 [a])
1601S-2	−	TTGCATCGGAGATGCCCACCTCG	Negative-strand sequence (nt 3598–3620)
1601S-3	+	GGTGGGCATCTCCGATGCAATCG	Positive-strand sequence (nt 3601–3623)
1601S-4	−	GCCCCAAGGGGTTATGCTAGTTTTTTTTTTTTTTTTT TTTTTTTTTTTTTTTCAGGGAGCGCGAAACGCAGAAAAGAG	pUC19 vector, poly(A), and negative-strand sequence (nt 7167–7192)
1601S-5	+	CTAGCATAACCCCTTGGGGCCTC	pUC19 vector
1601S-6	−	TATAGTGAGTCGTATTAAGCTTGGCG	T7 promoter (underlined) and pUC19 vector
1601S-7	−	ACAGAATGGATTGGCCGACTCCC	Negative-strand sequence (nt 2088–2110)
1601S-8	+	AGTCGGCCAATCCATTCTGTGGC	Positive-strand sequence (nt 2091–2113)
1601S-9	+	GGTGGGCATCTCCGATGCAACTCTAAACGGGTCTTGAGGGG	Positive-strand sequence (nt 3601–3620) and pUC19 vector
1601S-*Spe*I-F	+	TTGGGCAGAAACTAGTGTTCACCC	Positive-strand sequence (nt 3384–3407), *Spe*I site (underlined)
1601S-GAA-R	−	ATCGAGGCGGCACCTTTGAAGGCAGCCACCTGC	Negative-strand sequence (nt 4654–4686), mutated nucleotide (underlined)
1601S-GAA-F	+	AGGTGCCGCCTCGATAGTGCTTTGCAGTGAGTACC	Positive-strand sequence (nt 4672–4706), mutated nucleotide (underlined)
1601S-*Spe*I-R	−	CGCCATTAGTACTAGTAAAATAAAGATCC	Negative-strand sequence (nt 6197–6225), *Spe*I site (underlined)

[a] The nucleotide positions are numbered in accordance with the JE04-1601S strain obtained in the present study.

The amplified fragments were purified using a FastGene Gel Extraction Kit (Nippon Genetics Europe, Tokyo, Japan). First, F1-1 and F1-2 were subcloned into pUC19 vector, which harbors the T7 promoter. A cDNA clone of pJE03-1760F [47] used as vector for F1-1 and F1-2 was linearized using inverse PCR in accordance with the previously described method [48] with the pJE03-1760F clone as a template, a high-fidelity DNA polymerase (KOD Plus ver. 2; Toyobo, Osaka, Japan), and primers 1601S-9 and 1601S-6 (Table 2). The amplicons were then purified. F1-1 and F1-2 were fused to generate pJE04-1601S_p12 F1 using the In-Fusion Snap Assembly (TaKaRa Bio), according to the protocols provided by the manufacturer. In brief, an In-Fusion reaction was performed in a total volume of 10 μL, containing 2 μL of 5× In-Fusion Snap Assembly Master Mix, 100 ng of each fragment (F1-1, F1-2, and pUC19 vector), and dH$_2$O. The reaction mix was incubated at 50 °C for 15 min and then placed on ice, where 2.5 μL of each mixture was transformed into *Escherichia coli* Stellar Competent Cells (TaKaRa Bio). The plasmids were then extracted, and the F1-1 and F1-2 regions of pJE04-1601S_p12 were sequenced using Sanger's method described in Section 2.6. Next, the F1 fragment was amplified by PCR, with the pJE04-1601S_p12 F1 as a template, a high-fidelity DNA polymerase KOD Plus ver. 2, and primers 1601S-1 and 1601S-2 (Table 2), followed by purification of the amplicons.

To construct the full-length cDNA clone of pJE04-1601S_p12 under the T7 promoter, purified F1 and F2 were fused using the In-Fusion Snap Assembly. The pUC19 with T7 promoter and poly(A) tract used as a vector was linearized by inverse PCR with the cDNA clone of pJE03-1760F as a template, a high-fidelity DNA polymerase KOD Plus ver. 2, and primers 1601S-5 and 1601S-6 (Table 2). The amplicons were then purified and used for the In-Fusion reaction. The reaction was performed using 200 ng of each fragment (F1, F2, and pUC19 vector), as described above. The plasmids were then extracted, and the sequences of T7 promoter, full-genome, and poly(A) tract were confirmed by Sanger's method, as described in Section 2.6.

In addition, as a negative control, pJE04-1601S_p12-GAA was generated by mutating the conserved RNA replication motif GDD to GAA (Asp1551Ala [nt A4677C, nt T4678C] and Asp1552Ala [nt A4680C, nt T4681C]). The pJE04-1601S_p12 cDNA clone was used as the template. In brief, two fragments covering the entire JE04-1601S_p12 genome were amplified by PCR with KOD Plus ver. 2. F1 GAA (nt 3384–4686; starting at *Spe*I site to the mutation) was amplified with primers 1601S-*Spe*I-F and 1601S-GAA-R, while F2 GAA (nt 4672–6225; starting at the mutation to *Spe*I site) was amplified with primers 1601S-GAA-F and 1601S-*Spe*I-R. The vector was generated by digesting the pJE04-1601S_p12 clone with *Spe*I-HF (New England Biolabs, Tokyo, Japan) at nt 3395 and nt 6210. The digested cDNA as well as the amplicons (F1 GAA and F2 GAA) were then purified. To generate the full-length cDNA clone of pJE04-1601S_p12-GAA, the three fragments were fused using the In-Fusion Snap Assembly. The reaction was performed using 100 ng of F1 GAA and F2 GAA each and 50 ng of vector, as described above. The sequence between two *Spe*I sites of pJE04-1601S_p12-GAA was confirmed by Sanger's method, as described in Section 2.6.

2.8. In Vitro Transcription and Transfection of RNA Transcripts to PLC/PRF/5 Cells

The full-length cDNA clone and its replication-defective mutant (pJE04-1601S_p12 and pJE04-1601S_p12-GAA, respectively) were each linearized with *Nhe*I (New England Biolabs), and the RNA transcripts were synthesized with T7 RNA polymerase using AmpliScribe™ T7-*Flash*™ Transcription Kit (Epicentre Biotechnologies, Madison, WI, USA). After in vitro transcription, RNA transcripts of the cDNA clones were capped using a ScriptCap m7G Capping System (Epicentre Biotechnologies). The integrity and yield of the synthesized RNAs were determined by agarose gel electrophoresis. An aliquot (2.5 µg) of the capped RNA was transfected into confluent PLC/PRF/5 cells in a well of a six-well plate using the TransIT-mRNA transfection kit (Mirus Bio, Madison, WI, USA) in accordance with the manufacturer's recommendations. Following incubation at 37 °C for two days, the cells were washed with PBS(-), and then the culture medium was replaced with 2 mL of growth medium, and the cells were incubated at 35.5 °C. Every other day, half of the culture medium (1 mL) was replaced with fresh growth medium. The collected culture medium was centrifuged at $1300\times g$ at room temperature for 2 min, and the supernatants were stored at −80 °C until use.

2.9. Western Blotting

To detect the expression of ORF2 and ORF3 proteins in the cells transfected with RNA transcripts of pJE04-1601S_p12 and pJE04-1601S_p12-GAA, the proteins in the culture supernatants were separated by sodium dodecyl sulfate-polyacrylamide gel electrophoresis (SDS-PAGE) and blotted onto polyvinylidene difluoride (PVDF) membranes (0.45 µm) (Merck-Millipore), immunodetected with an anti-HEV ORF2 MAb (H6253) [44], or anti-ORF3 MAb (TA0529) [49] and enhanced chemiluminescence HRP-conjugated anti-mouse IgM from goat (Santa Cruz Biotechnology, Santa Cruz, CA, USA), and then visualized by a chemiluminescence assay using SuperSignal West Atto Chemiluminescent Substrate (Thermo Fisher Scientific) with an ImageQuant LAS 500 (GE Healthcare, Turnpike Fairfield, CT, USA), as described previously [47].

2.10. Immunofluorescence Assays

HEV-1-infected PLC/PRF/5 cells seeded into eight-well chamber slides (Watson, Tokyo, Japan) were subjected to immunofluorescence staining according to the previously described method [47]. The primary antibody used was anti-HEV ORF2 MAb (H6253) [44] or anti-ORF3 MAb (TA0529) [49], and the secondary antibody was Alexa-Fluor 488-conjugated anti-mouse IgM (Thermo Fisher Scientific). Nuclei were counterstained with 4′,6-diamidino-2-phenylindole dihydrochloride (DAPI, Thermo Fisher Scientific). Slide glasses were mounted with Fluoromount/Plus medium (Diagnostic BioSystems, Pleasanton, CA, USA) and then viewed under an FV1000 confocal laser microscope (Olympus, Tokyo, Japan).

2.11. Sensitivity of HEV-1 to Ribavirin in a Cell Culture System

Monolayers of PLC/PRF/5 cells in a 24-well plate were inoculated with 1.0×10^5 copies of cDNA-derived JE04-1601S_p12/well in growth medium without FBS containing 40 or 160 µM ribavirin (Fujifilm Wako, Osaka, Japan) in DMSO (final concentration, 1%) and then subsequently incubated at 37 °C for 2 h. After incubation, the cells were washed five times with PBS(-), and 0.5 mL of growth medium containing 40 or 160 µM ribavirin in DMSO (final concentration, 1%) was added to each well, followed by incubation at 35.5 °C. Every other day, half of the culture medium was replaced with fresh growth medium containing 40 or 160 µM ribavirin in DMSO (final concentration, 1%). The collected culture supernatants were centrifuged at $1300 \times g$ at room temperature for 2 min, and the supernatants were stored at −80 °C until use. The concentrations of ribavirin used in the current study were determined according to our previous report on the evaluation of drug effect on HEV growth in cultured cells—both concentrations were considered to not cause any significant toxicity—and dose-dependent inhibition on HEV growth was demonstrated [50,51].

2.12. Lactate Dehydrogenase (LDH) Cytotoxicity Assay

The cytotoxicity of the drug treatment was quantified by measuring lactate dehydrogenase (LDH) activity released into the culture medium using an LDH cytotoxicity assay kit (Nacalai Tesque, Kyoto, Japan) according to the manufacturer's protocol. In brief, 100 µL culture supernatants in a 96-well plate were added with 100 µL substrate solution. The plate was protected from light and incubated for 20 min at room temperature. Following the addition of 50 µL stop solution, absorbance was measured at 490 nm using an iMark microplate reader. Measured values were normalized to the value of vehicle control.

2.13. Nucleotide Sequence Accession Numbers

The nucleotide sequences of HEV isolates determined in the present study have been deposited in the GenBank/EMBL/DDBJ databases under the following numbers: LC753635 (JE04-1601S), LC753636 (JE04-1601S_p10), LC753637 (JE04-1601S_p12), LC753638 (pJE04-1601S_p12), and LC753639 (pJE04-1601S_p12-GAA).

3. Results

3.1. Serial Passages of the JE04-1601S Strain

As the initial step in the serial passages of JE04-1601S (p0), a serum sample containing the HEV-1f strain (JE04-1601S_wild-type (wt)) was inoculated onto PLC/PRF/5 cells (Figure 1, Table 1). HEV RNA became detectable in the culture supernatant at 1.3×10^3 copies/mL at 4 days postinoculation (dpi), and its load increased to 1.3×10^5 copies/mL at 26 dpi. The first passage (p1) on A549_1-1H8 cells (Figure 1) was carried out using the culture medium from 26 dpi. The HEV RNA became detectable at 4.6×10^2 copies/mL at 4 dpi and continued to increase, peaking at 6.9×10^5 copies/mL at 62 dpi. Twelve consecutive passages were carried out in A549_1-1H8 cells (Figure 1, Table 1). The time required for the HEV RNA to be detectable in the culture medium shortened during serial passages; up to p8, it started to appear at 4 dpi, whereas from p9 onward, it started to become detectable at 2 dpi. The interval between the inoculation of the cultures and the peak virus titer was shortened as well. During passages, the HEV RNA continued to increase to higher titers in the culture medium, finally peaking at approximately 10^8 copies/mL from p10 onward. In addition, the yield of the virus also increased during the passages. Collectively, these results suggested that the virus adapted to growth in cell culture. There was no cytopathic effect observed in either PLC/PRF/5 or A549_1-1H8 cells during the serial passages.

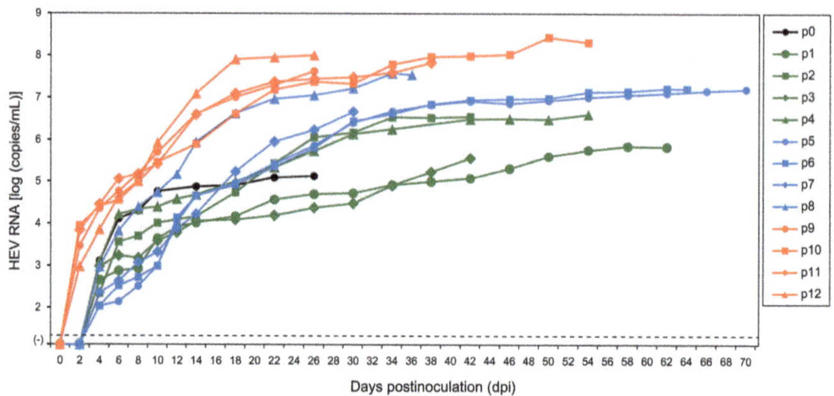

Figure 1. Quantification of HEV RNA in culture supernatants of PLC/PRF/5 (for passage 0, p0) inoculated with serum sample of JE04-1601S/wild-type and in culture supernatants of A549_1-1H8 cells inoculated with culture supernatants of p0, p1, p2, p3, p4, p5, p6, p7, p8, p9, p10, or p11 that were harvested on the final day of each passage (see Table 1). The harvested culture supernatant of each passage was purified by passing through a microfilter with a pore size of 0.22 μm (see Materials and Methods) and then inoculated onto A549_1-1H8 cells. The dotted horizontal line represents the limit of detection by real-time RT-PCR used in the current study, at 2.0×10^1 RNA copies/mL. Each passage was performed for three wells and one representative well showing median viral titer at 10 and 18 days postinoculation was selected, and culture media collected serially from the selected well were subjected to quantification of HEV RNA.

A phylogenetic tree was constructed based on the entire genomic sequences of three JE04-1601S strains obtained in the present study (JE04-1601S_wt, JE04-1601S_p10, and JE04-1601S_p12), with all known genotype 1 strains [38] and each one of the prototype strains of genotypes 2–8 [38] (Figure 2). The phylogenetic tree confirmed that the p10 and p12 isolates segregated into a subcluster within subtype 1f together with the wild-type parent.

3.2. Mutational Characteristics in Serial Passages of the JE04-1601S Strain

To identify the molecular mechanisms underlying the adaptation of JE04-1601S to growth in cell culture, we determined the complete nucleotide sequence of p10 and p12 as indicated in Figure 2 and compared the sequences of these variants to the sequence of its wild-type parent. Table 3 compares the mutations over the entire genome and the amino acid differences within the three ORFs between the wild-type JE04-1601S and its cell-culture-produced variants. Nucleotide mutations were restricted to ORF1 and ORF2. The p10 isolate had 14 nucleotide mutations, including three mutations at nt 3667, 5090, and 5528 with a mixed nucleotide population of T as well as C that the wild-type virus possessed. A mutation occurred at nt 2988, resulting in an amino acid alteration from Ala to Gly in the helicase region of ORF1. Compared to the p10 isolate, there are nine additional nucleotide mutations in the p12 isolate, where mutations at nt 477 and nt 6613 resulted in amino acid changes from Ser to Phe in the methyltransferase region of ORF1 and from Gly to Ala in ORF2, respectively. Furthermore, one mutation at nt 3667 found in the p10 isolate (mixed nucleotide population of T and C) had a backward mutation to C in the p12 isolate.

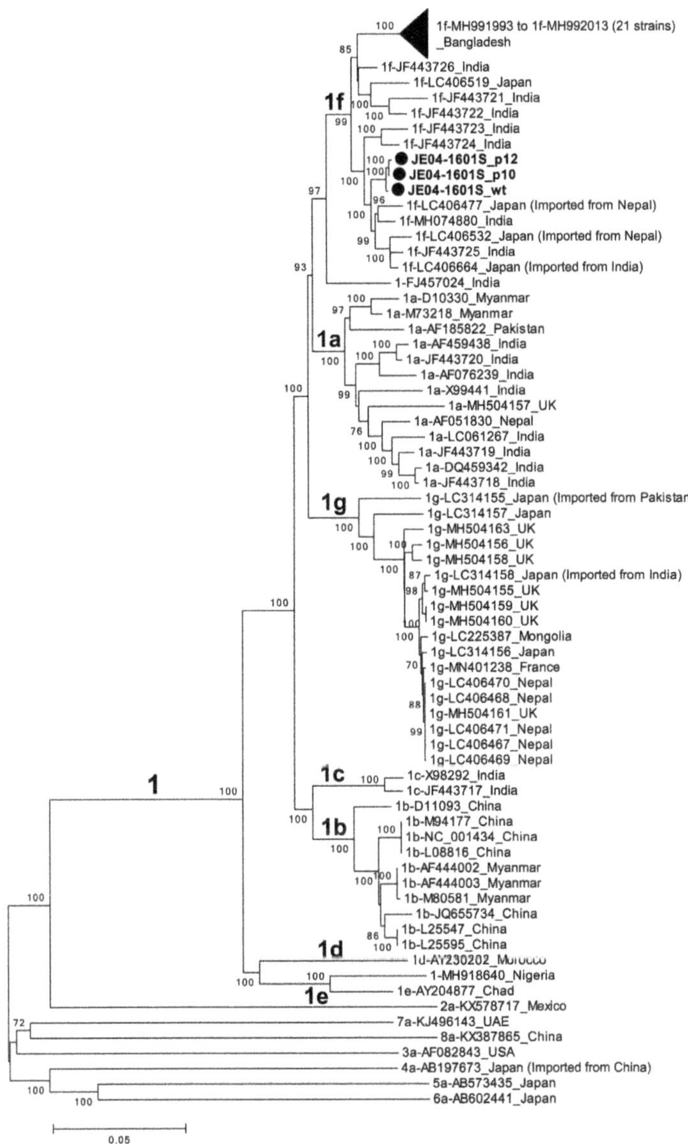

Figure 2. A phylogenetic tree constructed using the neighbor-joining tree of Jukes–Cantor distances based on the entire genomic sequences of three JE04-1601S strains obtained in the present study (JE04-1601S_wild-type (wt), JE04-1601S_p10, and JE04-1601S_p12), all known genotype 1 strains (1a, n = 13; 1b, n = 10; 1c, n = 2; 1d, n = 1; 1e, n = 1; 1f, n = 32; 1g, n = 18; and unclassified subtypes, n = 2), and each one of the prototype strains of genotypes 2–8. The three JE04-1601S strains obtained in the present study are highlighted with closed circles for clarity. Each reference sequence is shown with the genotype/subtype, followed by the accession number and the name of the country in which it was detected. The bootstrap values ($\geq 70\%$) of the nodes are indicated as a percentage of data obtained from 1000 resamplings. Tips are collapsed for 21 Bangladeshi 1f strains with similar sequences. The scale bar (0.05) represents the number of nucleotide substitutions per site.

Table 3. A comparison of the sequence of the original JE04-1601S strain in serum and its cell-culture-produced variants over the entire genome.

Nucleotide Position	Region (Domain) [a]	Nucleotide [b]			Amino Acid	
		JE04-1601S_wt	Passage 10 (p10)	Passage 12 (p12)	Position	Substitution
244	ORF1 (MeT)	C	T	T	73	-
433	ORF1 (MeT)	C	C	T	136	-
477	ORF1 (MeT)	C	C	T	151	Ser to Phe
617	ORF1 (MeT)	C	T	T	198	-
895	ORF1 (Y)	A	A	G	290	-
1270	ORF1 (Y)	C	T	T	415	-
1738	ORF1 (PCP)	C	T	T	571	-
1870	ORF1 (PCP)	T	C	C	615	-
2311	ORF1 (HVR)	C	T	T	762	-
2356	ORF1 (X)	T	C	C	777	-
2728	ORF1 (X)	C	T	T	901	-
2988	ORF1 (Hel)	C	G	G	988	Ala to Gly
3112	ORF1 (Hel)	T	C	C	1029	-
3298	ORF1 (Hel)	T	T	C	1091	-
3667	ORF1 (RdRp)	C	Y[b]	C	1214	-
4417	ORF1 (RdRp)	T	C	C	1464	-
4603	ORF1 (RdRp)	T	T	C	1526	-
5090	ORF1 (RdRp)	C	Y	T	1689	-
5528	ORF2	C	Y	T	128	-
6613	ORF2	G	G	C	490	Gly to Ala
6770	ORF2	C	C	T	542	-

[a] MeT, methyltransferase; Y, Y domain; PCP, papain-like cysteine protease; HVR, hypervariable region; X, X or macro domain; Hel, helicase; RdRp, RNA-dependent RNA polymerase. [b] Y = U/C.

The first non-synonymous mutation (Ala to Gly) was observed in the helicase region of ORF1 (nt 2988, aa 988) in the p10 isolate. To determine when it started to occur, the sequences of wild-type JE04-1601S and its cell-culture-generated variants (p0 to p12) within nt 2651–3200 were compared (Table 4). The non-synonymous mutation in the serial passages emerged during p4 and was maintained until p12.

Table 4. A comparison of the sequences of wild-type JE04-1601S and its cell-culture-generated variants (p0 to p12) within nt 2651–3200.

Wild-Type or Variants of JE04-1601S	Nucleotide Position		
	2728	2988	3112
Wild-type (serum)	C	C	T
p0 (culture supernatant, 26 dpi [a])	C	C	C
p1 (culture supernatant, 62 dpi)	T	C	C
p2 (culture supernatant, 42 dpi)	T	C	C
p3 (culture supernatant, 60 dpi)	T	C	C
p4 (culture supernatant, 70 dpi)	T	S [b]	C
p5 (culture supernatant, 54 dpi)	T	G	C
p6 (culture supernatant, 64 dpi)	T	G	C
p7 (culture supernatant, 30 dpi)	T	G	C
p8 (culture supernatant, 36 dpi)	T	G	C
p9 (culture supernatant, 26 dpi)	T	G	C
p10 (culture supernatant, 54 dpi)	T	G	C
p11 (culture supernatant, 38 dpi)	T	G	C
p12 (culture supernatant, 26 dpi)	T	G	C

[a] dpi, days postinoculation. [b] S = G/C.

3.3. Replication Ability of JE04-1601S_p12 in Various Cell Lines

To examine the replication ability of the cell-culture-generated JE04-1601S_p12 in various cell lines, the culture medium of p12 at 26 dpi (peak HEV RNA titer) was filtered through a 0.22 μm microfilter and then inoculated onto three cell lines of human origin (lung-adenocarcinoma-derived A549_1-1H8 cells and hepatocellular-carcinoma-derived PLC/PRF/5 and HepG2/C3A cells) and three porcine-kidney-derived cell lines (PK15, IBRS-2, and LLC-PK1 cells) at 1×10^5 and 1×10^6 copies/well (Figure 3). The virus growth was observed for 60 days. In addition to the HEV-1 inoculum, inoculation was also carried out using HEV-3 (JE03-1760F_p26) and HEV-4 (HE-JF5/15F_p24) strains for comparison.

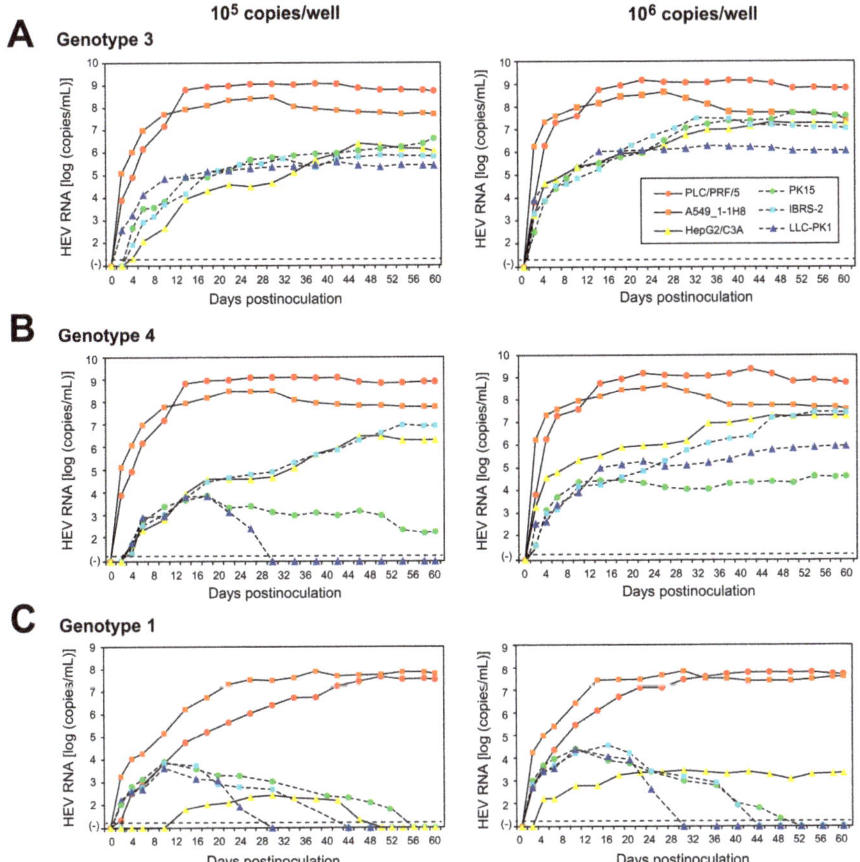

Figure 3. Quantification of HEV RNA in culture supernatants of human-derived cell lines (PLC/PRF/5, A549_1-1H8, and HepG2/C3A cells; indicated with continuous lines) and in culture supernatants of porcine-kidney-derived cell lines (PK15, IBRS-2, and LLC-PK1 cells; indicated with dotted lines) inoculated with the HEV-3 (JE03-1760F_p26) strain (**A**), the HEV-4 (HE-JF5/15F_p24) strain (**B**), or the HEV-1 (JE04-1601S_p12) strain (**C**) at a titer of 1.0×10^5 copies/well (left panels) or 1.0×10^6 copies/well (right panels) in six-well plates. The virus growth was observed for 60 days. The dotted horizontal line represents the limit of detection by real-time RT-PCR used in the current study, at 2.0×10^1 RNA copies/mL. Each inoculation was performed for three wells, one representative well showing median viral titer at 20 and 60 days postinoculation was selected, and culture media collected serially from the selected well were subjected to quantification of HEV RNA.

The HEV growth was supported in both human- and porcine-derived cell lines inoculated with HEV-3 or HEV-4 at 1×10^5 copies/well (Figure 3A,B, left panels) and 1×10^6 copies/well (Figure 3A,B, right panels). High replication efficiency was observed in both PLC/PRF/5 and A549_1 1H8 cells, where the virus titer in the culture medium peaked at 10^9 copies/mL for PLC/PRF/5 cells and at 10^8 copies/mL for A549_1-1H8 cells. Although the virus replication was less efficient in the remaining cell lines, the HEV RNA titer in the culture medium gradually increased, peaking at 10^7, 10^7, 10^7, and 10^6 copies/mL in HEV-3-inoculated HepG2/C3A, PK15, IBRS-2, and LLC-PK1 cells, respectively. Meanwhile, the HEV RNA titer peaked at 10^7, 10^4, 10^7, and 10^5 copies/mL in HEV-4-inoculated HepG2/C3A, PK15, IBRS-2, and LLC-PK1, respectively. Only the inoculation of HEV-4 at 1×10^5 copies/well to LLC-PK1 cells was unable to maintain virus replication, as the virus titer in the culture medium started decreasing at 22 dpi and became undetectable from 30 dpi onward. This is possibly dose-dependent, as HEV-4 reached a peak titer of 10^5 copies/mL when LLC-PK1 cells were inoculated with a titer of 1×10^6 copies/well, and the virus titer in the culture medium was maintained until 60 dpi.

In contrast, the virus replication was only maintained in cell lines of human origin for inoculation with JE04-1601S_p12—albeit with a much lower efficiency in HepG2/C3A cells—where the HEV RNA titer in the culture medium peaked at nearly 10^8 copies/mL for both PLC/PRF/5 and A549_1-1H8 cells and at 10^3 copies/mL for HepG2/C3A cells (Figure 3C). Although the HEV RNA was initially detectable at low titer in the culture medium of all porcine-derived cell lines—peaking at approximately 10^4 copies/mL—it gradually decreased and became undetectable from 30, 44, and 52 dpi onward for LLC-PK1, IBRS-2, and PK15, respectively, despite the fact that the cell-culture-adapted virus was inoculated with a high titer (10^6 copies/well).

Supporting these results, following the removal of the membrane and ORF3 of HEV particles in the culture supernatants of the human-derived cell lines, nearly 100% of the HEV particles were captured by anti-ORF2 MAb at mid-cultivation (20 days postinoculation) and at the end of cultivation (60 days postinoculation) (Table 5). In contrast, only 10–20% of membrane-unassociated particles were captured by anti-ORF2 MAb at mid-cultivation, and HEV RNA was undetectable at the final cultivation day, in the culture supernatants of infected porcine-derived cell lines (Table 5). This suggested that although HEV-1 might replicate at the initial cultivation days in the porcine-derived cell lines, the replication was not fully supported in these cells.

Table 5. Reactivity of anti-ORF2 with membrane-unassociated JE04-1601S_p12 (HEV-1) particles as evaluated by immunoprecipitation and real-time RT-PCR.

Inoculated Cells	% of Captured HEV Particles [a] in the Total HEV-1 Per Tube	
	20 Days Postinoculation	60 Days Postinoculation
PLC/PRF/5	98.7	95.8
A549_1-1H8	98.9	98.5
HepG2/C3A	95.6	100.0
PK15	10.1	N/A [b]
IBRS-2	12.3	N/A
LLC-PK1	20.5	N/A

[a] Membrane-unassociated HEV particles generated from the culture supernatants of JE04-1601S_p12-infected cells following treatment with 0.1% sodium deoxycholate (DOC-Na) and 0.1% trypsin at 37 °C for 2 h were subjected to immunoprecipitation, and the precipitates were then subjected to real-time RT-PCR. [b] Not applicable (HEV RNA was undetectable in culture supernatants).

3.4. Construction of an Infectious cDNA Clone of JE04-1601S_p12 and Transfection of Its RNA Transcript to PLC/PRF/5 Cells

Given the lack of an infectious cDNA clone of HEV-1 with high replication efficiency, JE04-1601S from p12 was used as a template to construct one (Figure 4A). Three fragments (F1-1, F1-2, and F2) covering the whole genome of the JE04-1601S_p12 strain were generated by RT-PCR and then cloned into a pUC19 vector in a stepwise manner according to the In-Fusion cloning method (Figure 4B). Sequence analyses revealed that the resulting infectious cDNA clone had been constructed correctly.

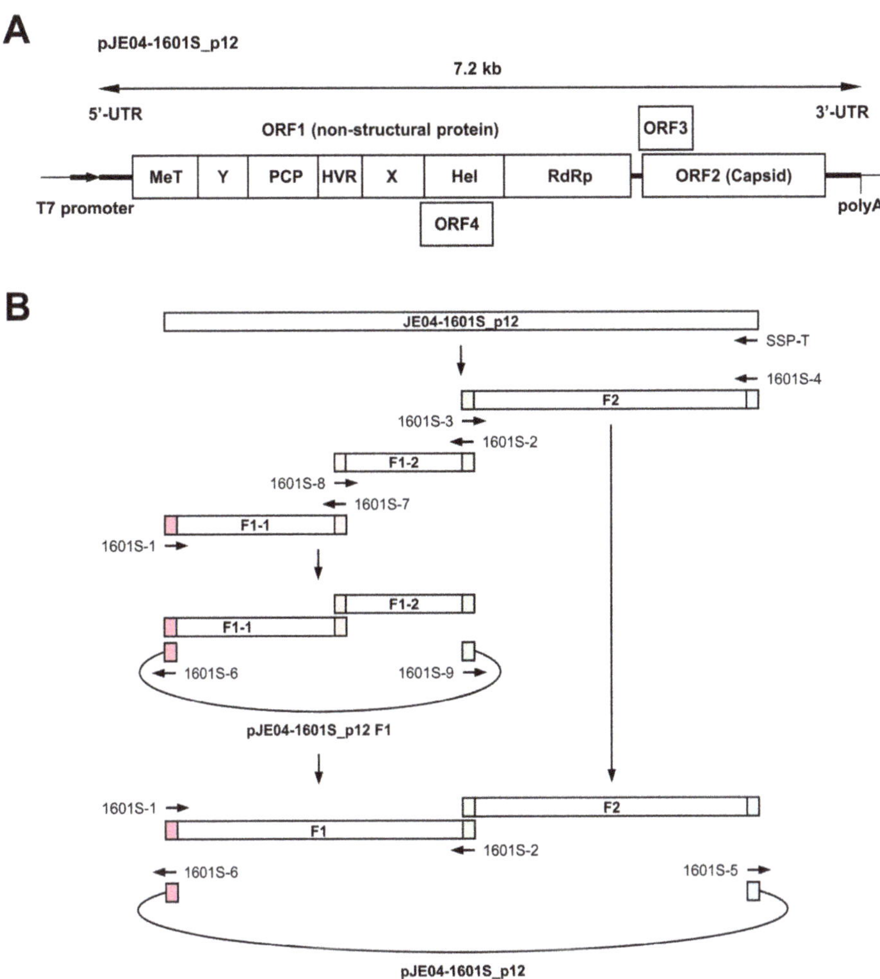

Figure 4. A schematic representation of the full-length genome of the JE04-1601S_p12 strain (**A**) and the strategy to construct its full-length cDNA clone (pJE04-1601S_p12) (**B**). Three fragments covering its whole genome were generated by reverse transcription-polymerase chain reaction (RT-PCR) and then cloned into the pUC19 vector in a stepwise manner using the In-Fusion cloning method. The 15-bp overlaps at their ends are highlighted with same colors. MeT, methyltransferase; Y, Y domain; PCP, papain-like cysteine protease; HVR, hypervariable region; X, X or macro domain; Hel, helicase; and RdRp, RNA-dependent RNA polymerase.

To examine the capability of the resulting JE04-1601S_p12 cDNA clone to produce a progeny virus, the RNA transcript of pJE04-1601S_p12 was transfected into PLC/PRF/5 cells. To monitor the virus production, the HEV RNA titer in the culture supernatants of the transfected cells was quantified (Figure 5A). The HEV RNA titer gradually increased until 20 days posttransfection (dpt), peaking at 2.9×10^8 copies/mL, and stayed at ~10^8 copies/mL thereafter. In contrast, a gradual decrease in the HEV RNA titer was observed in the culture supernatants of PLC/PRF/5 cells transfected with the RNA transcript of a replication-defective mutant (pJE04-1601S_p12-GAA), which expressed functionally disrupted RdRp.

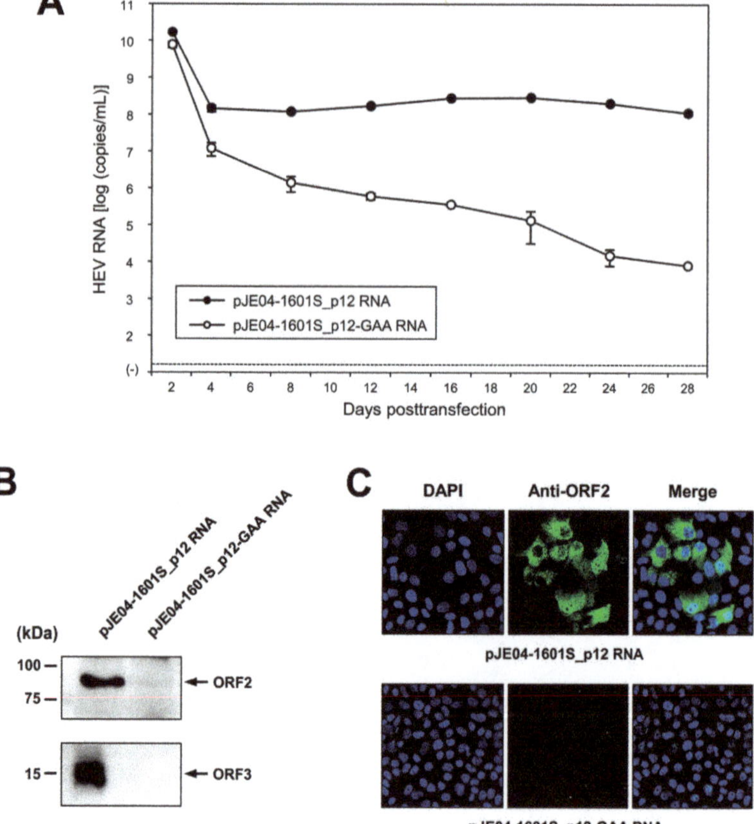

Figure 5. Capability of the cDNA clone of JE04-1760S_p12 to produce infectious progeny viruses. (**A**) Quantification of HEV RNA in culture supernatants. RNA transcript of pJE04-1601S_p12 was transfected to PLC/PRF/5 cells, along with RNA transcript of its replication-defective mutant (pJE04-1601S_p12-GAA), which served as a negative control. HEV growth was observed for 28 days. The data are presented as the mean ± standard deviation (SD) for two wells each. The dotted horizontal line represents the limit of detection by real-time RT-PCR used in the current study, at 2.0×10^1 RNA copies/mL. RNA transfection experiment was performed twice for two wells each, and representative result was shown. (**B**) A Western blot analysis of the culture supernatants transfected with RNA transcript of pJE04-1601S_p12 or that of pJE04-1601S_p12-GAA to examine the expression of HEV ORF2 (upper panel) and ORF3 proteins (lower panel) at day 28 posttransfection. (**C**) Immunofluorescence staining of the cells transfected with the RNA transcript of pJE04-1601S_p12 (upper panel) or that of pJE04-1601S_p12-GAA (lower panel) to examine the HEV ORF2 protein expression at day 28 posttransfection. For Western blotting and immunofluorescence assay, results representative of two experiments are shown.

Culture supernatants from 28 dpt were then subjected to Western blotting to examine the expression of the viral proteins. The specific bands of ORF2 (Figure 5B, upper panel) and ORF3 (Figure 5B, lower panel) proteins were only detected in the culture supernatants of cells transfected with the RNA transcript of pJE04-1601S_p12 and were undetectable in the culture supernatants of pJE04-1601S_p12-GAA RNA-transfected cells. To examine the intracellular expression of the ORF2 protein, the transfected PLC/PRF/5 cells at 28 dpt were subjected to an immunofluorescence assay (IFA). ORF2 protein was expressed abundantly in the cells transfected with the RNA transcript of pJE04-1601S_p12 (Figure 5C, upper panel), in contrast to the pJE04-1601S_p12-GAA RNA-transfected cells (Figure 5C, lower panel), in which the expression of ORF2 protein was undetectable. Taken together, these results indicate that the HEV-1 cDNA clone is capable of producing an infectious virus.

3.5. Characterization of cDNA-Derived JE04-1601S_p12 Progenies in the Cell Culture System

To characterize the cDNA-derived JE04-1601S_p12 progenies, they were inoculated onto PLC/PRF/5 (Figure 6A) and A549_1-1H8 (Figure 6B) cells at a titer of 1×10^5 and 1×10^6 copies/well in a six-well plate. During 28 days of observation, the HEV RNA titer in the culture medium of both inoculated cell lines increased gradually and dose-dependently, peaking at 9.7×10^7 copies/mL in PLC/PRF/5 cells and 8.6×10^6 copies/mL in A549_1-1H8 cells, suggesting that the cDNA-derived JE04-1601S_p12 progenies were infectious.

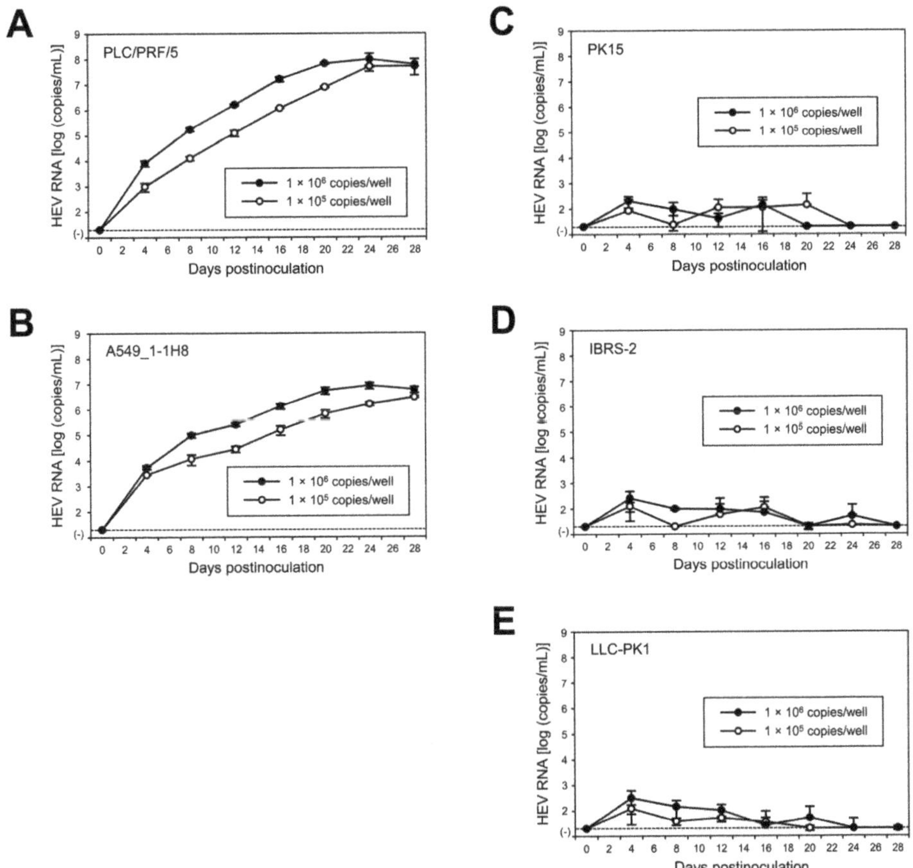

Figure 6. Species tropism of HEV-1 to humans in cell culture of the cDNA-derived JE04-1601S_p12 progeny viruses. Quantification of HEV RNA in culture supernatants of the human-derived cell lines (left panels) PLC/PRF/5 (**A**) and A549_1-1H8 (**B**) cells as well as the porcine-kidney-derived cell lines (right panels) PK15 (**C**), IBRS-2 (**D**), and LLC-PK1 (**E**) cells inoculated with cDNA-derived JE04-1601S_p12 progeny viruses. Inoculum titers were 1.0×10^5 copies/well or 1.0×10^6 copies/well in six-well plates. HEV growth was observed for 28 days. The data are presented as the mean ± SD for three wells each. The dotted horizontal line represents the limit of detection by real-time RT-PCR used in the current study, at 2.0×10^1 RNA copies/mL. Each inoculation was of single experiment for three wells.

To further examine the species tropism of HEV-1 to humans in the cell culture of the cDNA-derived JE04-1601S_p12 progenies, the progenies were inoculated to the non-human-derived (porcine kidney) cell lines of PK15 (Figure 6C), IBRS-2 (Figure 6D), and LLC-PK1 (Figure 6E). Despite being inoculated with the same titers as in cell lines of human origin (1×10^5 and 1×10^6 copies/well), the genomic RNA of HEV-1 in porcine-derived cell lines was initially detectable at low titer, peaking at approximately 10^2 copies/mL, then continued to decrease. These results further support the results described in Section 3.3 that the replication of HEV-1 is not fully supported in porcine-derived cell lines.

Ribavirin is currently used to treat certain cases of clinical HEV infections, such as cases of chronic or acute fulminant hepatitis. To examine the sensitivity of the cDNA-derived JE04-1601S_p12 progenies to ribavirin, the cDNA-derived JE04-1601S_p12 progenies were inoculated to PLC/PRF/5 cells (1×10^5 copies/well of 24-well plate) in the presence of 40 or 160 µM ribavirin in DMSO (final concentration, 1%). The virus kinetics were then observed for 28 days. The HEV RNA titer in the culture supernatant decreased in a dose-dependent manner. Ribavirin at 40 µM decreased the HEV RNA titer to 9.5×10^1 copies/mL on the final day of observation (28 dpi), and HEV RNA became undetectable from 16 dpi onward in the culture supernatants of the cells treated with 160 µM ribavirin (Figure 7A). In addition, at 28 dpi, HEV RNA was undetectable in the cells treated with 160 µM ribavirin. Supporting this result, at 28 dpi, ORF2 protein expression was undetectable in PLC/PRF/5 cells treated with 160 µM ribavirin (Figure 7B). These results indicated that the sensitivity of HEV-1 to ribavirin was reproducible in the cell culture of cDNA-derived progenies and that the inhibition effect of ribavirin on HEV growth could be monitored long-term in this cell culture system. LDH cytotoxicity assay carried out using culture supernatants of the PLC/PRF/5 cells inoculated with JE04-1601S_p12 progenies and treated with 160 µM ribavirin, from the final day of cultivation (28 dpi) and mid-cultivation (12 dpi)—where the HEV RNA in culture supernatants are still detectable—suggested no significant toxicity caused by the treatment with 160 µM ribavirin at least until 28 days of treatment (Table 6). These results are supported by the IFA image (Figure 7B, right panel) in which no discernible morphological changes are observed on the nuclei.

Table 6. Lactate dehydrogenase (LDH) released into the culture supernatants of PLC/PRF/5 cells inoculated with JE04-1601S_p12 progenies and treated with ribavirin.

Treatment	LDH Release (Mean ± SD) [a]	
	12 Days Postinoculation	**28 Days Postinoculation**
No drug treatment	2.4% ± 0.3%	2.4% ± 0.3%
40 µM Ribavirin	2.1% ± 0.4%	2.4% ± 0.3%
160 µM Ribavirin	2.8% ± 0.3%	3.7% ± 0.3%

[a] LDH release was determined in the culture supernatants from the final day of cultivation (28 days postinoculation) in comparison to that at 12 days postinoculation (mid-cultivation) to examine the cytotoxicity caused by the ribavirin treatment. Data represent the mean ± SD of triplicate wells.

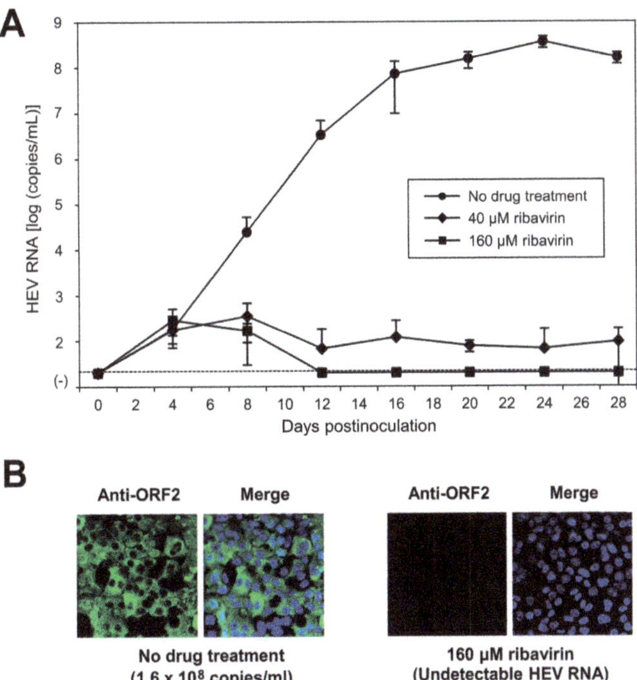

Figure 7. Sensitivity of HEV-1 to ribavirin in the cell culture system. (**A**) Quantification of HEV RNA in culture supernatants of PLC/PRF/5 cells inoculated with cDNA-derived JE04-1601S_p12 progeny viruses (1.0×10^5 copies/well) in the presence of 40 or 160 µM ribavirin in DMSO (final concentration, 1%). HEV kinetics were observed for 28 days. The data are presented as the mean $\pm$ SD for three wells each. The dotted horizontal line represents the limit of detection by real-time RT-PCR used in the current study, at 2.0×10^1 RNA copies/mL. The inoculation was of single experiment for three wells. (**B**) Immunofluorescence staining of PLC/PRF/5 cells inoculated with cDNA-derived JE04-1601S_p12 progenies in the presence of 160 µM ribavirin (right panel) to examine the HEV ORF2 protein expression at day 28 postinoculation in comparison to the ORF2 protein expression in untreated control cells (left panel). Results representative of two experiments are shown.

4. Discussion

Infection with HEV-1 is a major public health concern in South Asian countries and African countries [19], with particularly high fatality rates seen in pregnant women [23,24]. In addition to the tendency for HEV-1 infection to cause severe hepatitis in pregnant women, unlike infections caused by HEV-3, HEV-4, HEV-7, or the rat HEV—which are zoonotic—infection with HEV-1 is characterized by its species tropism to humans [52]. Although the exact determinants and mechanisms behind the severity of HEV-1 infection in pregnant women as well as its tropism to humans remain elusive, studies on these subjects—in addition to efforts to search for and develop safer treatment options for pregnant women—have been hampered by the lack of an efficient cell culture system for HEV-1 [16,33]. This obstacle prompted us to attempt to develop such a cell culture system and an infectious cDNA clone of HEV-1 with high replication efficiency.

The HEV-1 strain (JE04-1601S) used in the present study was of subtype 1f (Figure 2) and was recovered from a Japanese patient with fulminant hepatitis E who contracted the infection while traveling to India [37]. Previously, we have reported that two cell lines, PLC/PRF/5 and A549 cells, supported efficient propagation of HEV-3 [41] and HEV-4 [43], and therefore, we selected these two cell lines for the cultivation of HEV-1 as well. Similarly, as shown in Figure 3 in the current study, in the first ten days or so, the HEV RNA titer in the culture supernatants of A549_1-1H8 cells was higher than that in PLC/PRF/5 cells; in addition, the HEV RNA reached the peak level faster, and thus, in order to shorten the time required for consecutive passages, we carried out the passages in A549_1-1H8 cells. Following the initial propagation in PLC/PRF/5 cells, the culture medium containing JE04-1601S_p0 was serially passaged up to 12 times in A549_1-1H8 cells, resulting in markedly increased replication efficiency with no cytopathic effect (Figure 1) and indicating that the virus had adapted to growth in cell culture. These results were consistent with our previous reports on the serial passages of both HEV-3 (JE03-1760F) [42] and HEV-4 (HE-JF5/15F) [43] strains, in which adaptation to growth in cell culture resulted in the shortening of the interval between the inoculation of cultures and maximum virus yield, as well as an increase in the virus yield.

Increased replication efficiency in vitro might be the result of non-synonymous and synonymous mutations that emerged during the serial passages [42,53]. A comparison of the full-genome sequences of JE04-1601S_wt and its variants obtained from cell culture (p10 and p12, Table 2) demonstrated that both non-synonymous and synonymous mutations occurred during serial passages. The first non-synonymous mutation emerged during p4 and was maintained until p12 (Table 3), whereas one of the three mixed nucleotide populations in the p10 strain had a backward mutation in the p12 strain. Although mutations can occur frequently over the entire HEV genome during serial passages for adaptation to growth in cell culture [31,42,53], random mutations might emerge during the passages and only the selected mutations important for the virus are consistently maintained, which can result in viral fitness and thus heightened replication efficiency in vitro [9]. Further studies are warranted to determine which mutations are responsible for the heightened replication by site-directed mutagenesis using a reverse genetic system.

The inoculation of a culture supernatant containing JE04-1601S_p12 to various cell lines demonstrated that while HEV-3 and HEV-4 growth is supported in both human- and porcine-derived cell lines, HEV-1 growth was only fully supported in cell lines of human origin, in which HEV-1 replicated efficiently to reach an HEV RNA titer beyond 10^8 copies/mL in the culture supernatants of PLC/PRF/5 and A549_1-1H8-inoculated cells, whereas it was not fully supported in porcine-derived cell lines as the replication was only demonstrated at the initial cultivation days (Figure 3 and Table 5), potentially recapitulating the narrow tropism of HEV-1 observed in vivo.

Although HEV-1 can replicate in various cell lines of human origin, the virus titer in the culture supernatant of the infected cells generally peaked at 10^4 to 10^5 copies/mL in previous reports by other study groups [16,33–36,54,55]. However, in the present study, the peak HEV RNA titer reached approximately 10^8 copies/mL during serial passages (Figure 1), indicating an efficient replication capacity in our cell culture system. In addition, the robust cell culture system for HEV-1 in the present study is also capable of monitoring virus kinetics over a relatively long period, which will be useful for various studies on this genotype, including the evaluation of the efficacy of novel anti-HEV candidates.

Several factors might play role in the efficient replication achieved by the JE04-1601S strain during the passages. In addition to a possible role of adaptive nucleotide mutations, we cannot rule out the possible contribution of the clinical background of the patient from whom the sample originated, as he was suffering from fulminant hepatitis E with a high viral RNA titer in the serum (2.8×10^6 copies/mL). Our previously reported efficient cell culture system for HEV-4 (HE-JF5/15F) [43] was also based on a clinical sample from a patient with fulminant hepatitis E. The contribution of the clinical background to an efficient cell culture system has also been reported for the widely used hepatitis C virus JFH-1 [56], which was isolated from a patient with fulminant hepatitis C.

The potential contribution of differences in the multiplication efficiency of diverse HEV strains in cell culture should also be considered. Recently, it was reported that two mutations in the ORF1 of HEV-1 (Ala317Thr and Val1120Ile) resulted in enhanced virus replication in vitro [33], which may be associated with fulminant hepatitis. The HEV-1 strain used in the present study (JE04-1601S_wt) originally possessed Thr317 and Ile1120. This may also have influenced its higher infective capability than other HEV-1 strains, which might have contributed in part to the efficient replication achieved during serial passages.

Another important factor influencing the efficient replication might be ORF4, which is specific to HEV-1. It is short-lived due to its quick degradation by the host proteasome and is synthesized only under conditions of ER stress [16,17]. ER stress might be an ideal cellular condition for the optimal replication of HEV-1, as can be seen in pregnant women, and can partly explain the tendency for this high-risk group to progress to fulminant hepatitis. Other conditions that induce ER stress can also predispose individuals to develop fulminant-hepatitis-associated HEV-1 infection. In a limited analysis of HEV-1 strains, five out of seven sequences suggested to harbor a proteasome-resistant ORF4 were obtained from fulminant hepatitis patients [16]. Given that the HEV-1 strain used in the present study was obtained from a fulminant hepatitis patient, the expression of ORF4 might have been high, and it might have been proteasome-resistant, thus conferring the ability to replicate efficiently in cell culture. Further analyses using our efficient cell culture system for HEV-1 to confirm this hypothesis are warranted.

Given the absence of an infectious full-length HEV-1 clone with high replication efficiency, we used the cell-culture-adapted JE04-1601S from p12 as the template to construct a full-length cDNA clone using the In-Fusion cloning method (Figure 4). Results from the transfection of its RNA transcript to PLC/PRF/5 cells demonstrated that it successfully produced an infectious virus where the HEV RNA titer in the culture supernatant exceeded 10^8 copies/mL (Figure 5A), which is around the same as the highest RNA titer observed in the serial passages (Figure 1, p12). The detectable expression of viral proteins by Western blotting (ORF2 and ORF3) in the culture supernatant of the transfected PLC/PRF/5 cells (Figure 5B) and the detectable expression of ORF2 intracellularly by IFA (Figure 5C) further support the notion that the cDNA clone of JE04-1601S_p12 is capable of producing an infectious virus.

The characterization of the cDNA-derived JE04-1601S_p12 progenies demonstrated that the HEV RNA titer in the culture medium of inoculated PLC/PRF/5 (Figure 6A) and A549_1-1H8 (Figure 6B) cells increased gradually and dose-dependently, peaking at nearly 10^8 copies/mL in PLC/PRF/5 cells and indicating that the cDNA-derived JE04-1601S_p12 progenies were indeed infectious in the cell lines of human origin. In contrast, despite being inoculated with the same titers as in human-derived cell lines, the HEV-1 growth in non-human-derived (porcine kidney) cell lines was only temporary with the HEV RNA titer in the culture medium being initially detectable at low titer, peaking at approximately 10^2 copies/mL, then continued to decrease thereafter (Figure 6C–E). These results are consistent with the results demonstrated in Figure 3 and Table 5, suggesting that HEV-1 replication is not fully supported in porcine-derived cell lines, further recapitulating the narrow tropism of HEV-1 observed in vivo. Of note, it is likely that the cDNA-derived JE04-1601S_p12 progenies would replicate even more efficiently following serial passages, as evidenced by the serial passages of its wild-type strain.

In terms of the species tropism of HEV-1 to humans as further confirmed in the present study, multiple viral and host factors might play a role in the restricted tropism. Genotype-specific codon usage bias in HEV-1 is generally stronger than that of HEV-3 and HEV-4. HEV-3 and HEV-4 strains derived from either human or swine have more diverse codon usage patterns in their ORFs than HEV-1 [57]. Furthermore, in a correspondence analysis based on relative synonymous codon usage data, HEV genotypes appeared to cluster into HEV-1, and HEV-3 and HEV-4; based on ORF1, HEV-1 is clearly separated from other groups, partially reflecting the fact that HEV-1 infection is restricted to human hosts, while HEV-3 and HEV-4 strains were found in various animal species and were capable of cross-species transmission [58]. Previous attempts have been made to establish HEV-1 infection in non-human hosts (swine) in vitro and in vivo using intergenotypic chimeras [59–62]. In this strategy, HEV-1 is used as the genomic backbone, and various genomic regions are replaced with the corresponding regions of HEV-3 and HEV-4. However, the intergenotypic chimeras were unable to infect swine either in vitro or in vivo, suggesting that swine cells might lack essential host factors required to establish infection in pigs. In addition, it might reflect the functional importance of species-specific protein–protein interactions during HEV replication [52]. It will be interesting to further elucidate this topic in future studies while taking advantage of the availability of an efficient cell culture system of HEV-1 as well as the cDNA clone with high replication efficiency established in the present study.

Ribavirin is currently used to treat certain cases of clinical HEV infection, such as chronic or acute fulminant cases [27,28]. The further characterization of the cDNA-derived JE04-1601S_p12 progenies revealed that treatment with ribavirin strongly inhibited HEV-1 growth in our cell culture system, with the inhibition maintained over a long period of time (Figure 7). However, although a clear signal of human teratogenicity for ribavirin treatment for pregnant women—where HEV-1 infection leads to mortality in 30% of cases [23]—was not suggested, it has to be administered with caution [29]. The HEV-1 cDNA clone with its high replication efficiency and the robust cell culture system for HEV-1 developed in the present study will be valuable tools for the discovery and development of safer treatment options for this major risk group.

In the present study, we performed only the quantification of HEV RNA in culture supernatants to demonstrate virus growth kinetics. However, since the secreted form of ORF2 protein (ORF2s) is detectable in the culture supernatants of infected cells [63,64], the detection of ORF2s over time would be valuable to monitor virus growth kinetics during propagation at the protein levels. In our lab, we have been using PLC/PRF/5 cells for various HEV experiments, such as to monitor virus growth kinetics during propagation, as well as during the evaluation of drug effect, where we are able to obtain data on HEV replication ability [41–43,50,51,53]. Although we consider that the data on PLC/PRF/5 cells should be sufficient enough to draw the present conclusion, it is important to show that the adapted strain also replicates in other hepatoma cell lines which are more commonly used, such as Huh7 or S10-3 cells, in our future study.

In conclusion, we established a robust cell culture system for HEV-1 and an infectious cDNA clone of HEV-1 with high replication efficiency; the successful development of these tools might be the result of the adaptation of the virus to growth in cell culture, which may be attributed to selected mutations emerging during serial passages. In addition, the fact that the serum sample used for the initial passage was obtained from a fulminant hepatitis E patient with a high HEV RNA load (and thus the possible presence of proteasome-resistant ORF4) and that it originally had Thr317 and Ile1120 in the ORF1 region might partially explain its higher infective capability than other HEV-1 strains. Since the strain used here is of subtype 1f and harbors Thr317 and Ile1120—which are two among multiple factors possibly contributing to the high replication efficiency demonstrated in the current study—future work using other HEV-1 subtypes that do not harbor these amino acids shall provide the explanation for this efficiency. JE04-1601S_p12 and its cDNA-derived progenies replicated efficiently in cell lines of human origin but the replication was not fully supported in porcine-derived cell lines, potentially recapitulating the narrow species tropism in vivo. The efficient cell culture system for HEV-1 and its infectious cDNA clone with high replication efficiency established in the present study will be useful for the further elucidation of the determinants of HEV species tropism, the mechanism underlying the development of severe hepatitis in pregnant women infected with HEV-1, and the discovery and development of safer treatment options, particularly for pregnant women with fulminant hepatitis caused by HEV-1 infection.

Author Contributions: Conceptualization, P.P.P., T.T. and H.O.; methodology, P.P.P., S.N., T.T., S.J. and M.T.; formal analysis, P.P.P., S.N., T.T., S.J. and H.O.; data curation, P.P.P. and H.O.; writing—original draft preparation, P.P.P.; writing—review and editing, P.P.P. and H.O.; supervision, K.M. and H.O.; funding acquisition, H.O. All authors have read and agreed to the published version of the manuscript.

Funding: This work was supported in part by the Research Program on Hepatitis from the Japan Agency for Medical Research and Development, AMED (to H.O., JP22fk0210075).

Institutional Review Board Statement: This study was conducted in accordance with the guidelines of the Declaration of Helsinki and was approved by the Institutional Ethics Committee of the Jichi Medical University School of Medicine, Tochigi, Japan, under code eki14-82 (21 January 2015).

Informed Consent Statement: Informed consent was obtained from human subjects involved in the study, as declared in our previous study [37].

Data Availability Statement: All data are presented in the manuscript.

Acknowledgments: We would like to thank the native English-speaking scientists of Japan Medical Communication (https://www.japan-mc.co.jp/, accessed on 26 March 2023) for the expert linguistic services provided.

Conflicts of Interest: The authors declare no conflict of interest.

References

1. Purdy, M.A.; Drexler, J.F.; Meng, X.J.; Norder, H.; Okamoto, H.; Van der Poel, W.H.M.; Reuter, G.; de Souza, W.M.; Ulrich, R.G.; Smith, D.B. ICTV virus taxonomy profile: *Hepeviridae* 2022. *J. Gen. Virol.* **2022**, *103*, 9. [CrossRef] [PubMed]
2. Tam, A.W.; Smith, M.M.; Guerra, M.E.; Huang, C.C.; Bradley, D.W.; Fry, K.E.; Reyes, G.R. Hepatitis E virus (HEV): Molecular cloning and sequencing of the full-length viral genome. *Virology* **1991**, *185*, 120–131. [CrossRef]
3. Kabrane-Lazizi, Y.; Meng, X.J.; Purcell, R.H.; Emerson, S.U. Evidence that the genomic RNA of hepatitis E virus is capped. *J. Virol.* **1999**, *73*, 8848–8850. [CrossRef]
4. Graff, J.; Torian, U.; Nguyen, H.; Emerson, S.U. A bicistronic subgenomic mRNA encodes both the ORF2 and ORF3 proteins of hepatitis E virus. *J. Virol.* **2006**, *80*, 5919–5926. [CrossRef]
5. Ichiyama, K.; Yamada, K.; Tanaka, T.; Nagashima, S.; Jirintai; Takahashi, M.; Okamoto, H. Determination of the 5′-terminal sequence of subgenomic RNA of hepatitis E virus strains in cultured cells. *Arch. Virol.* **2009**, *154*, 1945–1951. [CrossRef]
6. Koonin, E.V.; Gorbalenya, A.E.; Purdy, M.A.; Rozanov, M.N.; Reyes, G.R.; Bradley, D.W. Computer-assisted assignment of functional domains in the nonstructural polyprotein of hepatitis E virus: Delineation of an additional group of positive-strand RNA plant and animal viruses. *Proc. Natl. Acad. Sci. USA* **1992**, *89*, 8259–8263. [CrossRef]
7. Panda, S.K.; Varma, S.P. Hepatitis E: Molecular virology and pathogenesis. *J. Clin. Exp. Hepatol.* **2013**, *3*, 114–124. [CrossRef] [PubMed]
8. Kenney, S.P.; Meng, X.J. Hepatitis E virus genome structure and replication strategy. *Cold Spring Harb. Perspect. Med.* **2019**, *9*, a031724. [CrossRef] [PubMed]
9. Primadharsini, P.P.; Nagashima, S.; Okamoto, H. Genetic variability and evolution of hepatitis E virus. *Viruses* **2019**, *11*, 456. [CrossRef]
10. Kalia, M.; Chandra, V.; Rahman, S.A.; Sehgal, D.; Jameel, S. Heparan sulfate proteoglycans are required for cellular binding of the hepatitis E virus ORF2 capsid protein and for viral infection. *J. Virol.* **2009**, *83*, 12714–12724. [CrossRef]
11. Xing, L.; Wang, J.C.; Li, T.C.; Yasutomi, Y.; Lara, J.; Khudyakov, Y.; Schofield, D.; Emerson, S.U.; Purcell, R.H.; Takeda, N.; et al. Spatial configuration of hepatitis E virus antigenic domain. *J. Virol.* **2011**, *85*, 1117–1124. [CrossRef]
12. Yamada, K.; Takahashi, M.; Hoshino, Y.; Takahashi, H.; Ichiyama, K.; Nagashima, S.; Tanaka, T.; Okamoto, H. ORF3 protein of hepatitis E virus is essential for virion release from infected cells. *J. Gen. Virol.* **2009**, *90*, 1880–1891. [CrossRef] [PubMed]
13. Emerson, S.U.; Nguyen, H.T.; Torian, U.; Burke, D.; Engle, R.; Purcell, R.H. Release of genotype 1 hepatitis E virus from cultured hepatoma and polarized intestinal cells depends on open reading frame 3 protein and requires an intact PXXP motif. *J. Virol.* **2010**, *84*, 9059–9069. [CrossRef]
14. Nagashima, S.; Takahashi, M.; Jirintai; Tanaka, T.; Yamada, K.; Nishizawa, T.; Okamoto, H. A PSAP motif in the ORF3 protein of hepatitis E virus is necessary for virion release from infected cells. *J. Gen. Virol.* **2011**, *92*, 269–278. [CrossRef] [PubMed]
15. Ding, Q.; Heller, B.; Capuccino, J.M.; Song, B.; Nimgaonkar, I.; Hrebikova, G.; Contreras, J.E.; Ploss, A. Hepatitis E virus ORF3 is a functional ion channel required for release of infectious particles. *Proc. Natl. Acad. Sci. USA* **2017**, *114*, 1147–1152. [CrossRef] [PubMed]
16. Nair, V.P.; Anang, S.; Subramani, C.; Madhvi, A.; Bakshi, K.; Srivastava, A.; Shalimar; Nayak, B.; Ranjith Kumar, C.T.; Surjit, M. Endoplasmic reticulum stress induced synthesis of a novel viral factor mediates efficient replication of genotype-1 hepatitis E virus. *PLoS Pathog.* **2016**, *12*, e1005521. [CrossRef]
17. Yadav, K.K.; Boley, P.A.; Fritts, Z.; Kenney, S.P. Ectopic expression of genotype 1 hepatitis E virus ORF4 increases genotype 3 HEV viral replication in cell culture. *Viruses* **2021**, *13*, 75. [CrossRef]
18. Aggarwal, R.; Goel, A. Natural history, clinical manifestations, and pathogenesis of hepatitis E virus genotype 1 and 2 infections. *Cold Spring Harb. Perspect. Med.* **2019**, *9*, a032136. [CrossRef]
19. Nelson, K.E.; Labrique, A.B.; Kmush, B.L. Epidemiology of genotype 1 and 2 hepatitis E virus infections. *Cold Spring Harb. Perspect. Med.* **2019**, *9*, a031732. [CrossRef]
20. Miyahara, K.; Miyake, Y.; Yasunaka, T.; Ikeda, F.; Takaki, A.; Iwasaki, Y.; Kobashi, H.; Kang, J.H.; Takahashi, K.; Arai, M.; et al. Acute hepatitis due to hepatitis E virus genotype 1 as an imported infectious disease in Japan. *Intern. Med.* **2010**, *49*, 2613–2616. [CrossRef]

21. Nishizawa, T.; Primadharsini, P.P.; Namikawa, M.; Yamazaki, Y.; Uraki, S.; Okano, H.; Horiike, S.; Nakano, T.; Takaki, S.; Kawakami, M.; et al. Full-length genomic sequences of new subtype 1g hepatitis E virus strains obtained from four patients with imported or autochthonous acute hepatitis E in Japan. *Infect. Genet. Evol.* **2017**, *55*, 343–349. [CrossRef] [PubMed]

22. Kobayashi, S.; Mori, A.; Sugiyama, R.; Li, T.C.; Fujii, Y.; Yato, K.; Matsuda, M.; Shiota, T.; Katsumata, M.; Iwamoto, T.; et al. Isolation and genome sequencing of hepatitis E virus genotype 1 imported from India to Japan. *Jpn. J. Infect. Dis.* **2022**, *75*, 604–607. [CrossRef] [PubMed]

23. Perez-Gracia, M.T.; Suay-Garcia, B.; Mateos-Lindemann, M.L. Hepatitis E and pregnancy: Current state. *Rev. Med. Virol.* **2017**, *27*, e1929. [CrossRef]

24. Khuroo, M.S. Hepatitis E and pregnancy: An unholy alliance unmasked from Kashmir, India. *Viruses* **2021**, *13*, 1329. [CrossRef]

25. Kamar, N.; Pischke, S. Acute and persistent hepatitis E virus genotype 3 and 4 infection: Clinical features, pathogenesis, and treatment. *Cold Spring Harb. Perspect. Med.* **2019**, *9*, a031872. [CrossRef]

26. Ma, Z.; de Man, R.A.; Kamar, N.; Pan, Q. Chronic hepatitis E: Advancing research and patient care. *J. Hepatol.* **2022**, *77*, 1109–1123. [CrossRef] [PubMed]

27. Dalton, H.R.; Kamar, N. Treatment of hepatitis E virus. *Curr. Opin. Infect. Dis.* **2016**, *29*, 639–644. [CrossRef]

28. European Association for the Study of the Liver. EASL clinical practice guidelines on hepatitis E virus infection. *J. Hepatol.* **2018**, *68*, 1256–1271. [CrossRef]

29. Sinclair, S.M.; Jones, J.K.; Miller, R.K.; Greene, M.F.; Kwo, P.Y.; Maddrey, W.C. Final results from the ribavirin pregnancy registry, 2004–2020. *Birth. Defects Res.* **2022**, *114*, 1376–1391. [CrossRef]

30. Meister, T.L.; Bruening, J.; Todt, D.; Steinmann, E. Cell culture systems for the study of hepatitis E virus. *Antiviral. Res.* **2019**, *163*, 34–49. [CrossRef]

31. Okamoto, H. Hepatitis E virus cell culture models. *Virus Res.* **2011**, *161*, 65–77. [CrossRef]

32. Okamoto, H. Culture systems for hepatitis E virus. *J. Gastroenterol.* **2013**, *48*, 147–158. [CrossRef]

33. Wang, B.; Tian, D.; Sooryanarain, H.; Mahsoub, H.M.; Heffron, C.L.; Hassebroek, A.M.; Meng, X.J. Two mutations in the ORF1 of genotype 1 hepatitis E virus enhance virus replication and may associate with fulminant hepatic failure. *Proc. Natl. Acad. Sci. USA* **2022**, *119*, e2207503119. [CrossRef] [PubMed]

34. Wu, X.; Dao Thi, V.L.; Liu, P.; Takacs, C.N.; Xiang, K.; Andrus, L.; Gouttenoire, J.; Moradpour, D.; Rice, C.M. Pan-genotype hepatitis E virus replication in stem cell-derived hepatocellular systems. *Gastroenterology* **2018**, *154*, 663–674.e7. [CrossRef] [PubMed]

35. Knegendorf, L.; Drave, S.A.; Dao Thi, V.L.; Debing, Y.; Brown, R.J.P.; Vondran, F.W.R.; Resner, K.; Friesland, M.; Khera, T.; Engelmann, M.; et al. Hepatitis E virus replication and interferon responses in human placental cells. *Hepatol. Commun.* **2018**, *2*, 173–187. [CrossRef]

36. Capelli, N.; Dubois, M.; Pucelle, M.; Da Silva, I.; Lhomme, S.; Abravanel, F.; Chapuy-Regaud, S.; Izopet, J. Optimized hepatitis E virus (HEV) culture and its application to measurements of HEV infectivity. *Viruses* **2020**, *12*, 139. [CrossRef]

37. Takahashi, M.; Tanaka, T.; Takahashi, H.; Hoshino, Y.; Nagashima, S.; Jirintai; Mizuo, H.; Yazaki, Y.; Takagi, T.; Azuma, M.; et al. Hepatitis E virus (HEV) strains in serum samples can replicate efficiently in cultured cells despite the coexistence of HEV antibodies: Characterization of HEV virions in blood circulation. *J. Clin. Microbiol.* **2010**, *48*, 1112–1125. [CrossRef]

38. Smith, D.B.; Izopet, J.; Nicot, F.; Simmonds, P.; Jameel, S.; Meng, X.J.; Norder, H.; Okamoto, H.; van der Poel, W.H.M.; Reuter, G.; et al. Update: Proposed reference sequences for subtypes of hepatitis E virus (species *Orthohepevirus A*). *J. Gen. Virol.* **2020**, *101*, 692–698. [CrossRef]

39. Pathak, R.; Barde, P.V. Detection of genotype 1a and 1f of hepatitis E virus in patients treated at tertiary care hospitals in Central India. *Intervirology* **2017**, *60*, 201–206. [CrossRef]

40. Baki, A.A.; Haque, W.; Giti, S.; Khan, A.A.; Rahman, M.M.; Jubaida, N.; Rahman, M. Hepatitis E virus genotype 1f outbreak in Bangladesh, 2018. *J. Med. Virol.* **2021**, *93*, 5177–5181. [CrossRef] [PubMed]

41. Tanaka, T.; Takahashi, M.; Kusano, E.; Okamoto, H. Development and evaluation of an efficient cell-culture system for hepatitis E virus. *J. Gen. Virol.* **2007**, *88*, 903–911. [CrossRef]

42. Lorenzo, F.R.; Tanaka, T.; Takahashi, H.; Ichiyama, K.; Hoshino, Y.; Yamada, K.; Inoue, J.; Takahashi, M.; Okamoto, H. Mutational events during the primary propagation and consecutive passages of hepatitis E virus strain JE03-1760F in cell culture. *Virus Res.* **2008**, *137*, 86–96. [CrossRef]

43. Tanaka, T.; Takahashi, M.; Takahashi, H.; Ichiyama, K.; Hoshino, Y.; Nagashima, S.; Mizuo, H.; Okamoto, H. Development and characterization of a genotype 4 hepatitis E virus cell culture system using a HE-JF5/15F strain recovered from a fulminant hepatitis patient. *J. Clin. Microbiol.* **2009**, *47*, 1906–1910. [CrossRef] [PubMed]

44. Takahashi, M.; Hoshino, Y.; Tanaka, T.; Takahashi, H.; Nishizawa, T.; Okamoto, H. Production of monoclonal antibodies against hepatitis E virus capsid protein and evaluation of their neutralizing activity in a cell culture system. *Arch. Virol.* **2008**, *153*, 657–666. [CrossRef] [PubMed]

45. Okamoto, H.; Takahashi, M.; Nishizawa, T.; Fukai, K.; Muramatsu, U.; Yoshikawa, A. Analysis of the complete genome of indigenous swine hepatitis E virus isolated in Japan. *Biochem. Biophys. Res. Commun.* **2001**, *289*, 929–936. [CrossRef]

46. Edgar, R.C. MUSCLE: A multiple sequence alignment method with reduced time and space complexity. *BMC Bioinform.* **2004**, *5*, 113. [CrossRef] [PubMed]

47. Yamada, K.; Takahashi, M.; Hoshino, Y.; Takahashi, H.; Ichiyama, K.; Tanaka, T.; Okamoto, H. Construction of an infectious cDNA clone of hepatitis E virus strain JE03-1760F that can propagate efficiently in cultured cells. *J. Gen. Virol.* **2009**, *90*, 457–462. [CrossRef]
48. Sasaki, J.; Kusuhara, Y.; Maeno, Y.; Kobayashi, N.; Yamashita, T.; Sakae, K.; Takeda, N.; Taniguchi, K. Construction of an infectious cDNA clone of Aichi virus (a new member of the family *Picornaviridae*) and mutational analysis of a stemloop structure at the 5' end of the genome. *J. Virol.* **2001**, *75*, 8021–8030. [CrossRef]
49. Takahashi, M.; Yamada, K.; Hoshino, Y.; Takahashi, H.; Ichiyama, K.; Tanaka, T.; Okamoto, H. Monoclonal antibodies raised against the ORF3 protein of hepatitis E virus (HEV) can capture HEV particles in culture supernatant and serum but not those in feces. *Arch. Virol.* **2008**, *153*, 1703–1713. [CrossRef]
50. Primadharsini, P.P.; Nagashima, S.; Nishiyama, T.; Takahashi, M.; Murata, K.; Okamoto, H. Development of recombinant infectious hepatitis E virus harboring the nanoKAZ gene and its application in drug screening. *J. Virol.* **2022**, *96*, e0190621. [CrossRef]
51. Primadharsini, P.P.; Nagashima, S.; Takahashi, M.; Murata, K.; Okamoto, H. Ritonavir blocks hepatitis E virus internalization and clears hepatitis E virus in vitro with ribavirin. *Viruses* **2022**, *14*, 2440. [CrossRef]
52. Primadharsini, P.P.; Nagashima, S.; Okamoto, H. Mechanism of cross-species transmission, adaptive evolution and pathogenesis of hepatitis E virus. *Viruses* **2021**, *13*, 909. [CrossRef]
53. Nagashima, S.; Kobayashi, T.; Tanaka, T.; Tanggis; Jirintai, S.; Takahashi, M.; Nishizawa, T.; Okamoto, H. Analysis of adaptive mutations selected during the consecutive passages of hepatitis E virus produced from an infectious cDNA clone. *Virus Res.* **2016**, *223*, 170–180. [CrossRef] [PubMed]
54. Sayed, I.M.; Seddik, M.I.; Gaber, M.A.; Saber, S.H.; Mandour, S.A.; El-Mokhtar, M.A. Replication of hepatitis E virus (HEV) in primary human-derived monocytes and macrophages in vitro. *Vaccines* **2020**, *8*, 239. [CrossRef] [PubMed]
55. El-Mokhtar, M.A.; Othman, E.R.; Khashbah, M.Y.; Ismael, A.; Ghaliony, M.A.; Seddik, M.I.; Sayed, I.M. Evidence of the extrahepatic replication of hepatitis E virus in human endometrial stromal cells. *Pathogens* **2020**, *9*, 295. [CrossRef] [PubMed]
56. Wakita, T.; Pietschmann, T.; Kato, T.; Date, T.; Miyamoto, M.; Zhao, Z.; Murthy, K.; Habermann, A.; Krausslich, H.G.; Mizokami, M.; et al. Production of infectious hepatitis C virus in tissue culture from a cloned viral genome. *Nat. Med.* **2005**, *11*, 791–796. [CrossRef]
57. Zhou, J.H.; Li, X.R.; Lan, X.; Han, S.Y.; Wang, Y.N.; Hu, Y.; Pan, Q. The genetic divergences of codon usage shed new lights on transmission of hepatitis E virus from swine to human. *Infect. Genet. Evol.* **2019**, *68*, 23–29. [CrossRef]
58. Sun, J.; Ren, C.; Huang, Y.; Chao, W.; Xie, F. The effects of synonymous codon usages on genotypic formation of open reading frames in hepatitis E virus. *Infect. Genet. Evol.* **2020**, *85*, 104450. [CrossRef]
59. Feagins, A.R.; Cordoba, L.; Sanford, D.J.; Dryman, B.A.; Huang, Y.W.; LeRoith, T.; Emerson, S.U.; Meng, X.J. Intergenotypic chimeric hepatitis E viruses (HEVs) with the genotype 4 human HEV capsid gene in the backbone of genotype 3 swine HEV are infectious in pigs. *Virus Res.* **2011**, *156*, 141–146. [CrossRef]
60. Chatterjee, S.N.; Devhare, P.B.; Pingle, S.Y.; Paingankar, M.S.; Arankalle, V.A.; Lole, K.S. Hepatitis E virus (HEV)-1 harbouring HEV-4 non-structural protein (ORF1) replicates in transfected porcine kidney cells. *J. Gen. Virol.* **2016**, *97*, 1829–1840. [CrossRef]
61. Cordoba, L.; Feagins, A.R.; Opriessnig, T.; Cossaboom, C.M.; Dryman, B.A.; Huang, Y.W.; Meng, X.J. Rescue of a genotype 4 human hepatitis E virus from cloned cDNA and characterization of intergenotypic chimeric viruses in cultured human liver cells and in pigs. *J. Gen. Virol.* **2012**, *93*, 2183–2194. [CrossRef]
62. Tian, D.; Yugo, D.M.; Kenney, S.P.; Lynn Heffron, C.; Opriessnig, T.; Karuppannan, A.K.; Bayne, J.; Halbur, P.G.; Meng, X.J. Dissecting the potential role of hepatitis E virus ORF1 nonstructural gene in cross-species infection by using intergenotypic chimeric viruses. *J. Med. Virol.* **2020**, *92*, 3563–3571. [CrossRef] [PubMed]
63. Montpellier, C.; Wychowski, C.; Sayed, I.M.; Meunier, J.C.; Saliou, J.M.; Ankavay, M.; Bull, A.; Pillez, A.; Abravanel, F.; Helle, F.; et al. Hepatitis E Virus Lifecycle and Identification of 3 Forms of the ORF2 Capsid Protein. *Gastroenterology* **2018**, *154*, 211–223.e8. [CrossRef] [PubMed]
64. Yin, X.; Ying, D.; Lhomme, S.; Tang, Z.; Walker, C.M.; Xia, N.; Zheng, Z.; Feng, Z. Origin, antigenicity, and function of a secreted form of ORF2 in hepatitis E virus infection. *Proc. Natl. Acad. Sci. USA* **2018**, *115*, 4773–4778. [CrossRef] [PubMed]

Article

Host Desmin Interacts with RABV Matrix Protein and Facilitates Virus Propagation

Wen Zhang [1,2], Yuming Liu [1,2], Mengru Li [1,2], Jian Zhu [1,2], Xiaoning Li [1,2,*], Ting Rong Luo [1,2,*] and Jingjing Liang [1,2,3,*]

1 State Key Laboratory for Conservation and Utilization of Subtropical Agro-Bioresources, Guangxi University, Nanning 530004, China
2 Laboratory of Animal Infectious Diseases, College of Animal Sciences and Veterinary Medicine, Guangxi University, Nanning 530004, China
3 Department of Pathobiology, School of Veterinary Medicine, University of Pennsylvania, Philadelphia, PA 19104, USA
* Correspondence: xiaoningli@gxu.edu.cn (X.L.); tingrongluo@gxu.edu.cn (T.R.L.); jiliang@vet.upenn.edu (J.L.)

Abstract: Microfilaments and microtubules, two crucial structures of cytoskeletal networks, are usurped by various viruses for their entry, egress, and/or intracellular trafficking, including the Rabies virus (RABV). Intermediate filaments (IFs) are the third major component of cytoskeletal filaments; however, little is known about the role of IFs during the RABV infection. Here, we identified the IF protein desmin as a novel host interactor with the RABV matrix protein, and we show that this physical interaction has a functional impact on the virus lifecycle. We found that the overexpression of desmin facilitates the RABV infection by increasing the progeny virus yield, and the suppression of endogenous desmin inhibits virus replication. Furthermore, we used confocal microscopy to observe that the RABV-M co-localizes with desmin in IF bundles in the BHK-21 cells. Lastly, we found that mice challenged with RABV displayed an enhanced expression of desmin in the brains of infected animals. These findings reveal a desmin/RABV-M interaction that positively regulates the virus infection and suggests that the RABV may utilize cellular IFs as tracks for the intracellular transport of viral components and efficient budding.

Keywords: Rabies virus; matrix protein; desmin; intermediate filaments; viral-host protein interaction

Citation: Zhang, W.; Liu, Y.; Li, M.; Zhu, J.; Li, X.; Luo, T.R.; Liang, J. Host Desmin Interacts with RABV Matrix Protein and Facilitates Virus Propagation. *Viruses* **2023**, *15*, 434. https://doi.org/10.3390/v15020434

Academic Editors: Yiping Li and Yuliang Liu

Received: 30 December 2022
Revised: 28 January 2023
Accepted: 31 January 2023
Published: 4 February 2023

1. Introduction

The Rabies virus (RABV) is a negative-stranded RNA virus of the *Rhabdoviridae* family and is the causative agent of rabies. In the past few decades, efforts to defeat rabies have had substantial success in North America and Western Europe. However, rabies is still a neglected tropical disease in developing countries in Africa and Asia, resulting in a high annual global death count [1,2]. A more comprehensive understanding of the molecular aspects of RABV infection and transmission is warranted for the development of novel and effective post-exposure therapies [3,4].

The RABV encodes five viral proteins, including the nucleoprotein (N), phosphoprotein (P), matrix protein (M), glycoprotein (G), and viral polymerase protein (L), all of which are components of the mature virion. RABV-M is a versatile structural protein that plays multiple roles during virus infection. The central function of the M protein is coordinating virion assembly and budding [5]. To this end, the late (L) budding domain of the M protein hijacks select host proteins (e.g., E3 ubiquitin ligase Nedd4) to facilitate the late step in virus-cell separation [6,7]. The M protein serves as a bridge between the virion envelope and the nucleocapsid core, and also plays a role in viral ribonucleoprotein (RNP) complex condensation, which modulates the balance of viral genome transcription and replication [1,8–10]. In addition, the M protein interacts with RelAp43 to modulate the NF-κB pathway and, moreover, cooperates with the P protein to modulate the Jak-Stat pathway, thus contributing to viral immune evasion [11–13]. Lastly, the M protein is essential to induce host cell apoptosis and is thought to be one of the determinants of virus pathogenesis [14–16]. It should be noted that most of the diverse functions of the M protein are achieved by its interaction and recruitment of specific host factors. Thus, further investigations of the M-host interactome may help to reveal and elucidate the key steps required for RABV infection and transmission.

Towards this end, we utilized a GST pull-down assay coupled with LC-MS/MS analysis to identify the full complement of M-host interactors that may contribute to the virus lifecycle and/or subvert the cellular pathway. Here, we describe the identification of the host intermediate filament protein desmin as a novel interactor with RABV-M. Desmin is one of the members of the type III intermediate filaments (IFs) protein family. As the major structural components of IFs in skeletal, cardiac, and some smooth muscle cells, desmin is important for the maintenance and integrity of the cytoskeleton [17]. Our identification of desmin as a novel host interactor with RABV-M is particularly intriguing since RABV exploits the cytoskeleton network to facilitate different stages of its lifecycle and the development of pathological processes [18–20].

In this study, we first demonstrated the association of RABV-M and desmin in vitro, and further confirmed that RABV-M interacts with endogenously-expressed desmin in Baby Hamster Kidney cells (BHK-21). In addition, we found that desmin functions as a positive regulator during the RABV infection; while the overexpression of desmin increased the virus yield, the suppression of endogenous desmin inhibited the virus release. Interestingly, we observed co-localization of RABV-M with the desmin-formed IF bundles in the BHK-21 cells via confocal microscopy. Importantly, our in vivo investigation showed that RABV challenge enhanced the expression level of desmin in a mouse model. In sum, our findings suggest that RABV-M interacts with and hijacks desmin to facilitate the virus infection.

2. Materials and Methods

2.1. Cells, Plasmids, and Virus

The HEK293T, BHK-21, and BSR cells were maintained in Dulbecco's Modified Eagle Medium (Corning, NY, USA) supplemented with 10% fetal bovine serum (Gibco, MT, USA), penicillin (100 U/mL)/streptomycin (100 μg/mL) (Invitrogen, MA, USA), and the cells were grown at 37 °C in a humidified 5% CO_2 incubator. The M genes from RABV strains RC-HL, CVS, and GX01 were cloned into pGEX-4T-1vector for prokaryotic expression. The DNA sequence encoding myc-tagged RABV-M (RC-HL strain) and flag-tagged desmin (Mus musculus) were cloned into the pcDNA 3.0 vector for eukaryotic expression. RABV (RC-HL strain) was grown and titrated, as described previously [21]. For virus titration, the virus in the cell culture supernatant was serially diluted and then incubated with BSR cells that overlay with DMEM, containing 1% methylcellulose and 2% FBS for 48 h. The cells were fixed with methanol and then immunostained with RABV-N antibody (kindly provided by Dr. Minamoto, Gifu University, Japan), and subsequently visualized via Alexa Fluor 488-anti-mouse-IgG secondary antibody (Beyotime, Shanghai, China). The fluorescent focus unit (FFU) assay was used to calculate the virus titer.

2.2. GST-Pulldown Assay

The GST alone or GST-tagged RABV-M fusion proteins were expressed in the BL-21 cells and subsequently purified and conjugated into glutathione (GSH) beads (Beyotime, Shanghai, China). The cell extracts from the BHK-21 cells were incubated with the GSH beads described above at 4 °C for 6 h with continuous rotating. The protein complexes were pulled down with beads and subjected to Western blot analysis. The rabbit anti-desmin polyclonal antibody (Proteintech, IL, USA, 16520-1-AP, 1:10,000 dilution) was used to detect desmin, the rabbit anti-RABV-M antiserum (LSBio, WA, USA, LS-C369074, 1:2000 dilution) was used to detect the M fusion proteins, and the mouse anti-GST antibody (Abmart, NJ, USA, M20007, 1:2000 dilution) was used to detect the control GST protein.

2.3. Immunoprecipitation Assay

The HEK293T or BHK-21 cells seeded in 6 well plates were transfected with the indicated plasmid combinations using Lipofectamine 2000 reagent (Invitrogen, MA, USA). At 24 h post transfection, the cells were harvested and lysed, and the cell extracts were subjected to Western blot analysis and co-immunoprecipitation. Briefly, the cell extracts were incubated with either mouse IgG or anti-myc (ABclonal, MA, USA, AE010, 1:100 dilution) antibody overnight at 4 °C with continuous rotation, and then the protein A/G agarose beads (Santa Cruz, CA, USA) were added to the mixtures and incubated for 6 h with continuous rotation. After incubation, the beads were collected via centrifugation and washed 5 times. The input cell extracts and immunoprecipitates were then detected by Western blotting with rabbit anti-desmin (Proteintech, IL, USA, 16520-1-AP, 1:10,000 dilution), rabbit anti-RABV-M (LSBio, WA, USA, LS-C369074, 1:2000 dilution) antisera, mouse anti-flag (ABclonal, MA, USA, AE004, 1:2000 dilution), or mouse anti-β-actin (Cwbio, Jiangsu, China, CW0096, 1:2000 dilution) monoclonal antibodies.

2.4. Transfection/Infection Assays

The BHK-21 cells seeded in 12 well plates were first transfected with the vector alone or the plasmid encoding desmin (1.0 µg) for 24 h, and subsequently infected with RABV (RC-HL strain) at an MOI of 0.1. Supernatants and cell extracts were harvested at 24, 36, and 48 h post-infection. The released RABV virions were titrated in duplicate via fluorescent focus unit (FFU) assay on the BSR cells. Cellular and viral proteins were detected by Western blotting using the appropriate antibodies. The total RNA from the infected cells was extracted via standard TRIzol (Invitrogen, MA, USA) RNA extraction protocol for quantitative RT-PCR.

2.5. siRNA Knockdown

The BHK-21 cells seeded in 12 well plates were first transfected with random siRNA or desmin specific siRNA at a final concentration of 100 nM using Lipofectamine 2000 (Invitrogen, MA, USA). At 24 h post siRNA transfection, the cells were infected with RABV (RC-HL strain) at an MOI of 0.1. The supernatants, cell extracts, and total RNA were harvested at 24, 36, and 48 h post-infection and analyzed by virus titration, Western blotting, and qRT-PCR analyses. The control siRNA sequences include sense-UUCUCCGAACGUGUCACGUTT and antisense-ACGUGACACGUUCGGAGAATT. The desmin siRNA sequences used in this study include siRNA[803–824], sense-GCAGAAUCGAAUCCCUCAATT, antisense-UUGAGGGAUUCGAUUCUGCTT; siRNA[1369–1417], sense-GCUCUCAACUUCCGAGAAA-TT, antisense-UUUCUCGGAAGUUGAGAGCTT.

2.6. Quantitative RT-PCR

qRT-PCR was performed using the SYBR Green method (Takara, Japan, RR420). Briefly, 1µg of total RNA served as the template for the first-strand cDNA synthesis in a reaction using an oligo (dT) primer and MMLV reverse transcriptase (Promega, VI, USA, M1701), following the manufacturer's instructions. The quantitative assessment of RABV N and M mRNAs, and desmin mRNA under standard cycling conditions were performed on LightCycler® 480 System (Roche Diagnostic, IN, USA). The β-actin gene mRNA expression was assessed as a control for all reactions. The primer sequences used in this study includes RABV-N: forward-GGCATTGGCAGATGATGGAACT, reverse-GGCTTGATGATTGGAACTGACTGA; RABV-M: forward-TGATTCCAGGGGCCCTC-TTG, reverse-AAGAGACATGTCAGACCA; Desmin: forward-CCTGGAGCGCAGAATCGAAT, reverse-TGAGTCAAGTCTGAAACCTTGGA.

2.7. Indirect Immunofluorescence Assay

The BHK-21 cells expressing RABV-M were washed with cold PBS and fixed with 4% formaldehyde for 15 min at room temperature, then permeabilized with 0.2% Triton X-100 for 10 min. After blocking, the cells were incubated with rabbit anti-desmin (Proteintech,

IL, USA, 16520-1-AP, 1:400 dilution) and mouse anti-myc (RABV-M) antibodies for 1 h at room temperature, and then stained with Alexa Fluor 555-anti-Rabbit IgG and Alexa Fluor 488-anti-mouse-IgG secondary antibodies (Beyotime, Shanghai, China). The nuclei were visualized via DAPI staining (Beyotime, Shanghai, China). Microscopy was performed using a Leica SP5 FLIM inverted confocal microscope. Serial optical planes of focus were taken, and the collected images were merged into one by using the Leica microsystems (LAS AF, 5.1.0, Hesse, Germany) software.

2.8. In Vivo Studies

All animal experiments were conducted according to the Guidelines for the ethical review of laboratory animal welfare People's Republic of China (National Standard GB/T 35892-2018) and approved by the Animal Experiment Committee of Guangxi University with the approval number GXU2019-021. The four-week-old female Kunming mice (obtained from Guangxi Medical University, Nanning, China) were intracranially (i.c.) inoculated with DMEM (mock) or 30 μL attenuated RABV RC-HL strain (1000 FFU). At 4 and 7 days post-infection, the mice (two groups of 6) were euthanized, and the brains were collected for total protein and RNA extraction and immunohistochemistry. The total protein and RNA were subjected to Western blotting and qRT-PCR analyses, respectively.

2.9. Immunohistochemistry

The brains from the mock and RABV (RC-HL strain) infected mice were collected at 4 and 7 days post infection and fixed in 4% paraformaldehyde, followed by paraffin embedding. The hippocampus was cut into thin sections (5 μm) followed by deparaffinization/rehydration. The slides were submersed into 3% citrate solution and heat-induced epitope retrieval is performed using a pressure cooker for 10 min. The staining steps include washing the slides in dH_2O for 5 min and incubation in 3% hydrogen peroxide for 20 min, then rewashing the slides in dH_2O for 5 min and blocking in 5% BSA solution for 30 min. The slides were then incubated with anti-desmin antibody (Proteintech, IL, USA, 16520-1-AP, 1:400 dilution) overnight at 4 °C. The antigen was detected via HRP Polymer and DAB substrate.

2.10. Statistical Analysis

Significances were calculated from ≥ 3 independent experiments in GraphPad Prism (8.0.0, MA, USA) by either the One-way ANOVA-Dunnett's T3 multiple comparison test or unpaired *t*-test with Welch's correction, as indicated in the legend.

3. Results

3.1. Identification of Desmin as a Host Interactor with RABV Matrix Protein

To identify host proteins that interact with the RABV matrix protein, we first expressed and purified the GST-tagged RABV-M and constructed a bait by conjugating the GST-M with GSH beads. The baits were subsequently incubated with whole cell lysates from BHK 21 cells to precipitate the RABV-M interacting host proteins, followed by a tandem mass spectrometry (MS) to identify these potential host interactors. We identified desmin in the GST-M samples but not the GST protein control, suggesting that desmin is potential interactor with RABV-M (please see the Data Availability Statement). To verify this finding, a GST-pulldown assay was performed to assess the M-desmin direct interaction in vitro. Briefly, the whole cell lysates from BHK 21 cells were incubated with the purified GST protein alone or the GST-tagged M constructed from the different RABV strains, including the street strain GX01 [22], attenuated strain RC-HL, and standard strain CVS. Western blotting (WB) analysis was utilized to detect desmin, GST-M, and GST proteins in the precipitates. In agreement with our MS analysis, we detected desmin in all of the GST-M precipitates, but not in the GST controls (Figure 1A, compare lanes 1–4), suggesting that desmin interacts with the M protein from various RABV strains.

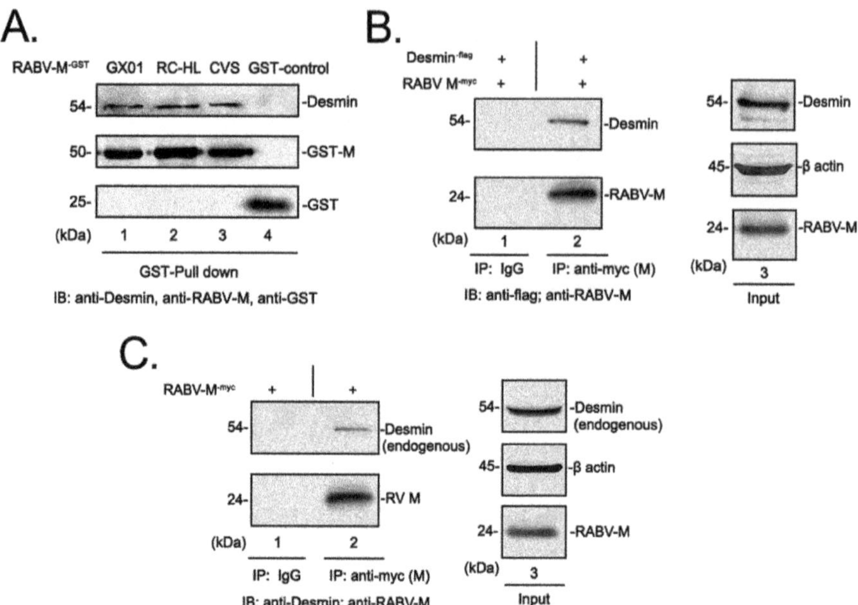

Figure 1. Identification of desmin as a host interactor with RABV matrix protein. (**A**) GST pulldown assays of desmin and M protein from different RABV strains. Purified GST and GST-RABV-M fusion proteins that were incubated with BHK-21 whole cell lysate, desmin, RABV-M, and GST control in pulldown samples were detected by Western blotting (WB) using appropriate antibodies. (**B**) Extracts from HEK293T cells transfected with the indicated plasmids were immunoprecipitated (IP) with either normal mouse IgG or anti-myc antibody, and the precipitated proteins were analyzed by WB using anti-flag (exogenous desmin) or anti-RABV-M antisera. Expression controls for exogenous desmin, RABV-M, and β-actin are shown. (**C**) BHK-21 cells were transfected with plasmid encoding RABV-M alone, cell extracts were IP with either normal mouse IgG or anti-myc antibody, endogenous desmin precipitated with RABV-M were detected by WB using anti-desmin and anti-RABV-M antisera.

We next asked whether the RABV-M interacts with desmin in vivo. Briefly, HEK-293T cells were transfected with plasmids encoding myc-tagged M and flag-tagged desmin, and whole cell lysates were incubated with either anti-myc antibody or normal mouse IgG, followed by WB analysis. We detected desmin in the M protein immunoprecipitates, but not in the corresponding IgG control precipitates (Figure 1B, compare lanes 1 and 2), suggesting that the exogenous desmin interacts with the RABV-M in HEK 293T cells. Then, we sought to further verify that RABV-M interacts with endogenously expressed desmin in a biologically relevant cell line. To this end, the BHK-21 cells were transfected with myc-tagged M alone, and the cell lysate was subjected to a co-IP followed by WB analysis. We consistently detected endogenous desmin in the RABV-M immunoprecipitates (Figure 1C). Taken together, these results demonstrated that the intermediate filament protein desmin is a specific host interactor with the RABV matrix protein.

3.2. Desmin Facilitates RABV Infection

We next sought to investigate whether the physical M-desmin interaction plays a functional role during RABV infection. Towards this end, BHK 21 cells were first transfected with the vector or the plasmid encoding flag-tagged desmin for 24 h and subsequently infected with RABV (RC-HL strain) at an MOI of 0.1. The total RNA, cell lysates, and supernatants from the infected cells were harvested at 24, 36, and 48 h post infection and subjected to qRT-PCR, WB, and virus titration analyses, respectively. We found that the exogenous expression of desmin modestly enhanced the RABV-N (Figure 2C) and RABV-M (Figure 2D) mRNA expression levels at 24 h post infection, but did not alter the viral protein expression levels (Figure 2B). Intriguingly, we observed a significant boost in the yield of progeny virions upon the addition of the exogenous desmin (Figure 2A). Indeed, the virus titration analysis showed that the overexpression of desmin increased virus budding by 10-fold at 48 h post infection, implying that the addition of exogenous desmin facilitates the RABV infection, and that this positive effect is likely due to the enhancement of later stages in the virus lifecycle, such as assembly and budding.

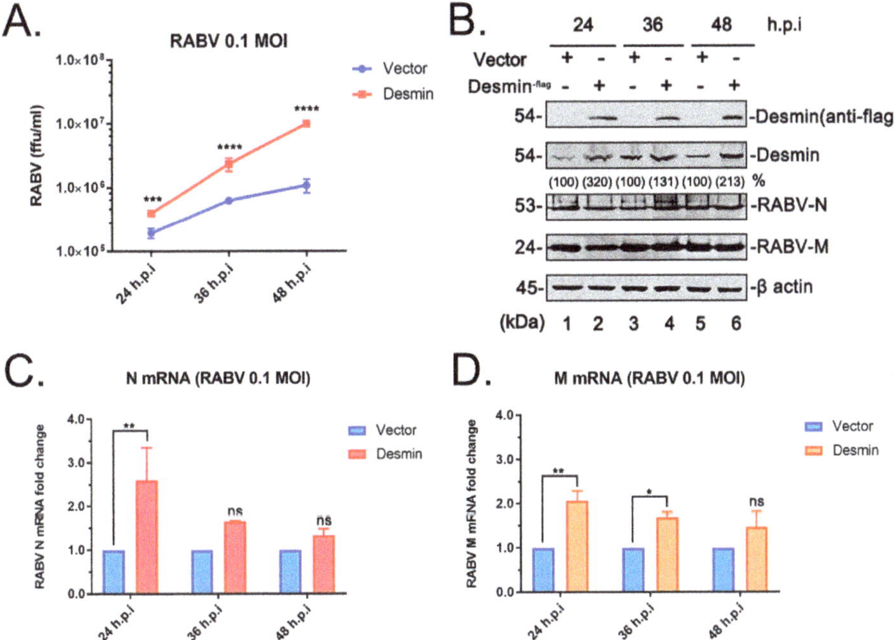

Figure 2. Exogenously expressed desmin facilitates RABV infection. BHK-21 cells were transfected with vector or plasmid encoding flag-tagged desmin, followed by infection with RABV (RC-HL strain) at an MOI of 0.1. (**A**) Virus titration of the supernatants from infected cells at 24, 36, and 48 h post infection (h.p.i). (**B**) Western blotting analysis of desmin and viral proteins RABV-N and M, with β-actin serving as a control, the indicated proteins were quantified using NIH Image-J software (1.53m 28, MD, USA). The infected cell extracts were harvested at 24, 36, and 48 h.p.i. (**C,D**) RABV-N (**C**) and RABV-M (**D**) mRNA expression levels were analyzed via qRT-PCR at 24, 36, and 48 h.p.i. Data were normalized to β-actin mRNA expression and are presented as relative fold change vs. controls. Statistical significance of bar graphs was analyzed by unpaired *t*-test with Welch's correction. ns: not significant, *: $p < 0.05$, **: $p < 0.01$, ***: $p < 0.001$, ****: $p < 0.0001$.

3.3. Suppression of Endogenous Desmin Impairs RABV Infection

We next sought to determine whether endogenous desmin is required for efficient RABV infection. Toward this end, we first designed and evaluated the siRNAs targeting desmin and selected two combinations of siRNA (Figure 3A). The BHK-21 cells were transfected with the random siRNA control or the desmin-specific siRNAs, followed by the RABV (RC-HL strain) infection at an MOI of 0.1. Again, the total RNA, cell lysates, and supernatants were harvested at 24, 36, and 48 h post infection for qRT-PCR, WB, and virus titration analyses, respectively. We did not detect a significant change in the mRNA levels of RABV-N or M genes (Figure 3C,D) among the cells that received the control or desmin-specific siRNAs, while the suppression of endogenous desmin showed an overall marginal impact on the expression levels of viral proteins (Figure 3B). Notably, we observed a significant decrease in the virus titers in the supernatants from the cells receiving desmin specific siRNAs, compared to those receiving siRNA controls. Interestingly, while the virion yield showed a step-by-step increase along with the infection time from the control cells, the knockdown of the endogenous desmin substantially impaired the virus release (Figure 3E). Taken together, these complementary results confirm that desmin interacts with the RABV matrix protein, and positively regulates the virus infection, likely by supporting the efficient virus release.

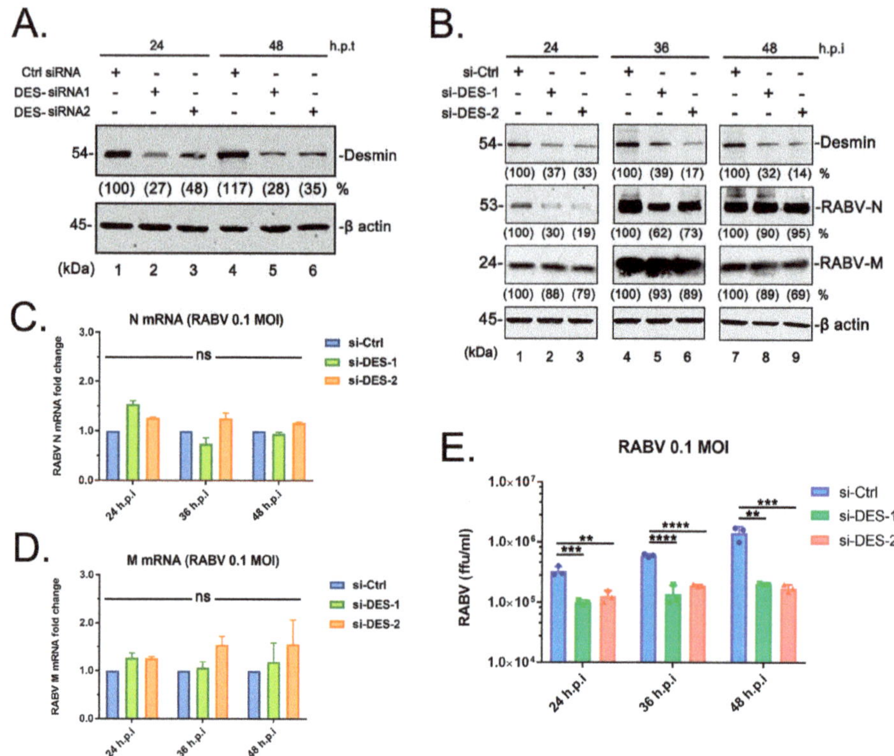

Figure 3. siRNA knockdown of endogenous desmin inhibits RABV infection. (**A**) BHK-21 cells were transfected with control or specific desmin siRNAs, the suppression of endogenous desmin levels was verified via WB and quantified using NIH Image-J software (1.53m 28, MD, USA). (**B**) BHK-21 cells received control or desmin siRNAs were infected with RABV (RC-HL strain) at an MOI of 0.1. The cell extracts were harvested at 24, 36, and 48 h.p.i, followed by WB analysis of desmin and viral proteins RABV-N and M, and β-actin served as a control. The indicated proteins were quantified using NIH Image-J software (1.53m 28, MD, USA). (**C,D**) RABV-N (**C**) and RABV-M (**D**) mRNA expression levels in infected cells were analyzed via qRT-PCR at 24, 36, and 48 h.p.i. Data were normalized to β-actin mRNA expression and are presented as relative fold change vs. controls. (**E**) Virus titration of the supernatants from infected cells at 24, 36, and 48 h.p.i. Statistical significance of bar graphs was analyzed by a one-way ANOVA. ns: not significant, **: $p < 0.01$, ***: $p < 0.001$, ****: $p < 0.0001$.

3.4. RABV Matrix Protein Localizes to the Desmin-Formed Intermediate Filament Bundles

To address the mechanism by which desmin supports RABV egress, we sought to determine whether the RABV-M colocalized with desmin in specific subcellular compartments. For this purpose, we performed an immunofluorescence assay on BHK-21 cells expressing the myc-tagged RABV-M, followed by confocal microscopy to visualize the intracellular patterns of RABV-M and endogenous desmin. It should be noted that desmin is a subunit of intermediate filaments (IFs)—one of the cytoskeletal structural components. As expected, we observed that the endogenous desmin assembles as IF bundles [23] (Figure 4, Desmin), while the RABV-M distributes throughout the cytoplasm (Figure 4, RABV-M), and localizes to the cell periphery (Figure 4, RABV-M, insert panel) [24,25]. Interestingly, we found that a portion of the M protein form bundle-like structures (Figure 4, RABV-M, triangles), and the bundle-like forms of RABV-M overlay with the desmin assembled IFs (Figure 4, Merge). This colocalization of M and desmin in the IF bundles (Figure 4, Zoom, arrows) implies that RABV-M may utilize the IFs during the intracellular transportation, and that desmin may play a role in the M protein-mediated virion assembly and budding.

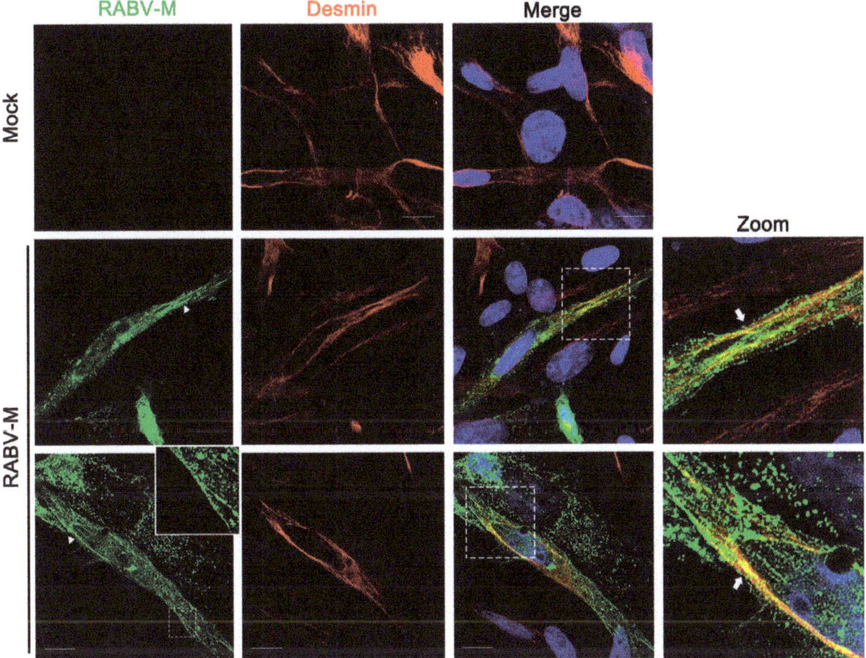

Figure 4. RABV-M co-localizes with desmin in intermediate filament bundles. The RABV-M (green), endogenous desmin (red), and cell nuclei (blue) in BHK-21 cells were visualized via confocal microscopy. The insert panel in the first lane shows the RABV-M localizes to the cell periphery (plasma membrane), while the triangles indicate the bundle-like M protein. The zoom panels highlight the colocalization of RABV-M with desmin in IF bundles (indicate as arrows). Scale bars = 10 μm.

3.5. RABV Infection Enhances Desmin Expression in Mouse Brain

The above data imply that desmin is a novel host interactor with RABV-M that positively regulates the virus infection. We therefore reasoned that RABV may modulate desmin expression to facilitate the virus replication. To address this hypothesis, two groups of six 4-week-old Kunming mice were mock-challenged or challenged with 1000 fluorescent focus-forming units (FFUs, as determined on BSR cells) of RABV (RC-HL strain). Three animals from each group were euthanized on day 4 and day 7 post challenge, and brain samples were collected and assessed by WB, qRT-PCR, and immunohistochemistry (IHC) analyses. Firstly, the tissue proteins were isolated and the levels of RABV-N and M, as well as desmin, were determined via WB (Figure 5A). Both viral proteins were detected at day 4 post challenge and their expression levels increased over time (Figure 5A, lanes 3–4).

Interestingly, we found that the levels of desmin were elevated upon the RABV challenge (Figure 5A, compare lanes 1–4). Next, the total RNA from the mouse brain was isolated and desmin mRNA was quantified via qRT-PCR analysis. We observed a significant increase in the levels of desmin mRNA expression (Figure 5B).

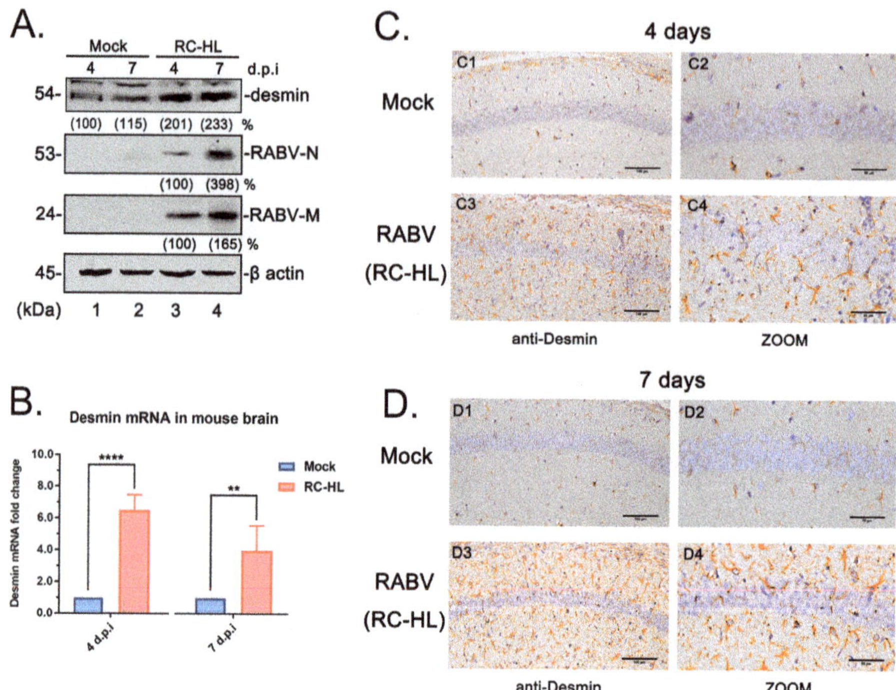

Figure 5. RABV induces desmin expression in the mouse brain. (**A**) Two groups of six 4-week-old Kunming mice were intracranially infected with mock (DMEM) or 1000 FFU RABV (RC-HL strain). The brain tissues from each group were collected at day 4 and 7 post challenge, followed by the WB analysis of desmin and viral proteins N and M, β-actin served as a control. The indicated proteins were quantified using NIH Image-J software (1.53m 28, MD, USA). (**B**) The qRT-PCR analysis of desmin mRNA expression in mouse brain. Data were normalized to β-actin mRNA expression and are presented as relative fold change vs. mock. Statistical significance of bar graphs was analyzed by unpaired *t*-test with Welch's correction. **: $p < 0.01$, ****: $p < 0.0001$. (**C,D**) Immunohistochemistry of desmin in the hippocampus of the mouse brain. The brain samples from mock and RABV-infected mice were collected at day 4 (**C**) and 7 (**D**) post infection. The hippocampus sections were subjected to IHC analysis via staining with the anti-desmin antibody. Scale bars = 100 μm in left panels and 50 μm in zoom panels.

Next, four random thin sections from the hippocampus of the mock and RABV infected mouse brains collected at day 4 and 7 post challenge were used for immunohistochemistry (IHC) analysis via immunostaining with anti-desmin antibody. In agreement with the above qRT-PCR and WB analyses, we observed that the RABV infection markedly induced the desmin expression in the mouse hippocampus at both day 4 and 7 post challenge (Figure 5C,D compares the Mock and RABV samples). Moreover, the IHC analysis showed an increase in desmin staining at day 7 compared with that at day 4 in the RABV infected mouse hippocampus (Figure 5, compare panels C3 and D3; C4 and D4), implying that desmin may be further induced during the RABV infection and transmission. In comparison, there was no appreciable change of desmin staining in the mock samples at either day

4 or 7 post challenge (Figure 5, compare panels C1 and D1; C2 and D2). In sum, these data suggest that the expression of desmin in mouse brains is significantly enhanced during RABV infection, further supporting the possible role of desmin as a host interactor that positively regulates the virus replication and transmission in vivo.

4. Discussion

As a neurotropic pathogen, RABV causes neuronal structural damage, neuronal dysfunction, and fatal encephalitis. After being bitten or scratched by an infected animal, RABV is transmitted from the wound site to the central nervous system (CNS) through the neuromuscular junction. It is thought that RABV can be transported in the infected neurons via either retrograde or anterograde trafficking. Once the virus reaches the CNS, it establishes a productive infection without interference by the peripheral immune system [1,26–28]. Over the past few decades, a series of host receptors of RABV have been identified, such as NCAM, p75NTR, and nAChR. Following attachment mediated via the glycoprotein (G), RABV is internalized into host cells via clathrin-mediated endocytosis, and the virus replication begins [1,18,28]. The cytoskeleton components, actin filaments, microtubules, and other related host proteins are targeted by RABV to facilitate the virus infection. First, the internalization and uptake of RABV requires an intact actin network, and the depolymerization of actin filaments inhibits the virus infection. Next, RABV utilizes the microtubules and the motor protein dynein/kinesin for intracellular transport [18–20,29,30]. During this stage, the M protein enhances the expression of cellular histone deacetylase 6 (HDAC6) to depolymerize the microtubules and facilitate viral RNA synthesis [31]. In addition, RABV induces the phosphorylation/inactivation of host cofilin to promote the polymerization of actin filaments, thus facilitating the M protein-mediated virus budding [32]. Here, we have identified the third member of the main cytoskeletal network, intermediate filaments (IFs), as also playing a role in RABV infection (Figure 6).

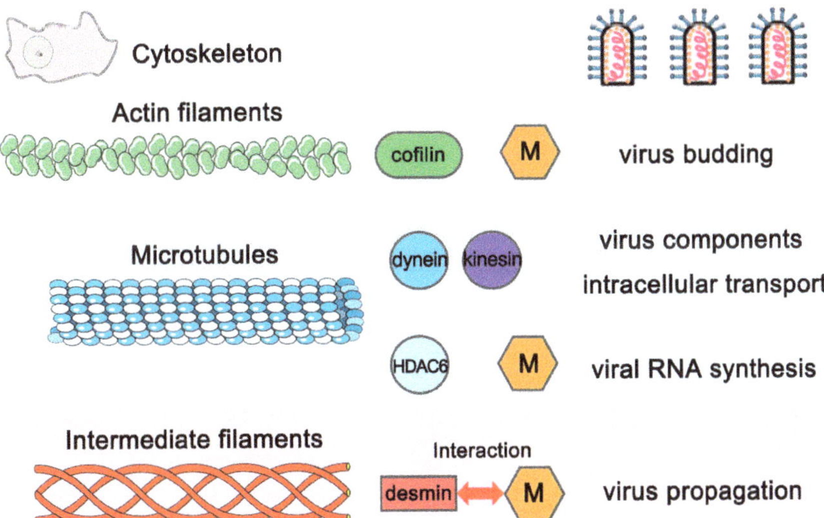

Figure 6. Graphical abstract illustrating that RABV-M exploits host proteins and cytoskeleton components to facilitate RABV infection. While the RABV utilizes cofilin and actin filament for efficient M mediated virus budding, the M protein regulates HDAC6 expression and microtubule depolymerization to enhance viral RNA synthesis. Here, we report a novel physical and functional interplay between RABV-M and intermediate filament protein desmin that promotes RABV propagation.

We found that the IFs component desmin is an interactor with the RABV-M protein that positively regulates the virus infection. It should be noted that the M protein from different RABV strains, including the attenuated strain RC-HL, standard strain CVS, and street strain GX-01 (Figure 1A), all interact with desmin, suggesting that desmin is a conserved target for the M protein of RABV strains, exhibiting different biological characteristics. Next, we showed that this physical interaction between RABV-M and desmin has an important functional role during the virus lifecycle. The overexpression of desmin in BHK-21 cells facilitated virus infection. Specifically, we observed a moderate increase in the mRNA expression levels of RABV N and M genes at 24 h post infection, however, the overexpression of desmin did not alter the intracellular proteins levels of RABV N and M. In comparison, the egress of progeny virions was significantly enhanced upon the addition of exogenous desmin at all three time points tested (Figure 2A). Correspondingly, the siRNA mediated suppression of endogenous desmin resulted in a significant reduction in the virus yields at all time points tested (Figure 3E). Taken together, these findings imply that desmin likely plays a role in the later stages of the virus lifecycle, such as virion assembly and budding. Indeed, we observed that endogenous desmin forms IF bundles throughout the cytoplasm of BHK-21 cells, and as expected, the RABV-M can distribute to the cell periphery and plasma membrane, in agreement with its role as the main driver of virus budding. Intriguingly, we found that a portion of the cytoplasmic RABV-M colocalizes on the desmin formed IF bundles (Figure 4). It is worth noting that RABV virion assembly begins with the M protein encapsulating the nucleocapsid core, followed by the M protein-mediated budding of the matured virions [1,5]. Thus, we hypothesize that the host IFs network, such as the other two cytoskeletal components, may also be usurped by the RABV M/desmin interaction, which facilitates the intracellular trafficking of RABV components.

The desmin-formed IFs are structural components of the sarcomeres in muscle cells and connect the intercellular junctions, such as the desmosomes, with the cytoskeleton [33–35]. Importantly, desmin plays an essential role in maintaining the structural and functional integrity of neuromuscular junction [36,37]. Since the neuromuscular junction is crucial for RABV transmission [1,26–28], it is tempting to speculate that IFs component desmin is beneficial to RABV transmission and pathogenesis. In support of this hypothesis, the in vivo experiments in mice indicated that the RABV challenge increases the expression level of the desmin mRNA and protein in the mouse brain. More importantly, the hippocampus is one of the cerebral regions that usually contains a high load of viral antigens [38,39], and our IHC analysis of the mouse hippocampus showed an obvious induction of desmin expression upon the RABV challenge.

IFs are involved in diverse cellular processes including vesicular transport, cell growth, cell migration, stress-related signaling pathways, and providing mechanical support to the cells. There are more than 70 genes encoding various IFs-proteins which can be classified into six types based on their structural, biochemical characteristics, and distribution [17,40–42]. Among them, type I IFs-protein keratin serve as a barrier in the epithelial tissues against microbial infection [43]; type V IFs-protein Lamin A/C are phosphorylated and reorganized by herpesvirus for the nuclear egress of viral nucleocapsid [44,45]; and type III IFs, which comprise desmin, vimentin, glial fibrillary acidic protein (GFAP), peripherin, and syncoilin, are particularly interesting due to their distinct distribution to specific cell types [17,42]. The vimentin-formed IFs (VIFs) have been linked to multiple stages of virus infection. For example, cell surface vimentin is involved in the invasion of SARS-CoV, SARS-CoV-2, PRRSV, DENV, and JEV; and the VIFs are required for the efficient replication and release of Enterovirus, FMDV, and SARS-CoV-2 [46–48]. In addition, it has been reported that desmin and vimentin are associated with the infection of TMEV, a neurotropic murine picornavirus. The TMEV virions are localized to the IFs, and the IFs network was rearranged to encompass the viral inclusion body during infection [49]. It is worth noting that type III IFs can be formed as either homopolymers or heteropolymers. While our current study reports that desmin is involved in the propagation of RABV, it is possible that other members in the type III IFs family also play roles in the RABV lifecycle, as well as desmin. Thus, the possible role and importance of the other type III IF proteins during the RABV infection remains to be addressed.

To sum up, our studies provide important and fundamental information regarding the impact of a RABV M/desmin interaction on the RABV lifecycle; however, the precise mechanism underlying the involvement of IFs during the RABV lifecycle still remains to be determined.

Author Contributions: Conceptualization, J.L. and T.R.L.; methodology, W.Z., Y.L. and J.L.; formal analysis, W.Z. and J.L.; investigation, W.Z., Y.L., M.L., J.Z. and J.L.; resources, T.R.L.; writing—original draft preparation, J.L.; writing—review and editing, J.L. and T.R.L.; supervision, X.L. and T.R.L.; project administration, X.L. and J.L.; funding acquisition, J.L. and T.R.L. All authors have read and agreed to the published version of the manuscript.

Funding: This research was funded by the National Natural Science Foundation of China (No. 31902311) and the Natural Science Foundation of Guangxi Province (No. 2018GXNSFBA281170) to J.L., and National Natural Science Foundation of China (No.32070161) to T.R.L.

Institutional Review Board Statement: The animal study protocol was approved by the Animal Experiment Committee of Guangxi University with the approval number GXU2019-021.

Informed Consent Statement: Not applicable.

Data Availability Statement: The raw data of GST pull-down/MS have been deposited in the figshare database (https://doi.org/10.6084/m9.figshare.21937601.v1, accessed on 25 January 2023.). All the other relevant data are within the manuscript.

Acknowledgments: The authors thank Ronald N. Harty, University of Pennsylvania, for his helpful comments and manuscript editing.

Conflicts of Interest: The authors declare no conflict of interest. The funders had no role in study design, data collection and analysis, decision to publish, or preparation of the manuscript.

References

1.	Fisher, C.R.; Streicker, D.G.; Schnell, M.J. The spread and evolution of rabies virus: Conquering new frontiers. *Nat. Rev. Microbiol.* **2018**, *16*, 241–255. [CrossRef] [PubMed]
2.	World Health Organization. *WHO Expert Consultation on Rabies: Third Report*; World Health Organization: Geneva, Switzerland, 2018.
3.	Smith, T.G.; Wu, X.; Franka, R.; Rupprecht, C.E. Design of future rabies biologics and antiviral drugs. *Adv. Virus Res.* **2011**, *79*, 345–363.
4.	Du Pont, V.; Plemper, R.K.; Schnell, M.J. Status of antiviral therapeutics against rabies virus and related emerging lyssaviruses. *Curr. Opin. Virol.* **2019**, *35*, 1–13. [CrossRef] [PubMed]
5.	Okumura, A.; Harty, R.N. Rabies virus assembly and budding. *Adv. Virus Res.* **2011**, *79*, 23–32.
6.	Harty, R.N.; Paragas, J.; Sudol, M.; Palese, P. A proline-rich motif within the matrix protein of vesicular stomatitis virus and rabies virus interacts with WW domains of cellular proteins: Implications for viral budding. *J. Virol.* **1999**, *73*, 2921–2929. [CrossRef]
7.	Wirblich, C.; Tan, G.S.; Papaneri, A.; Godlewski, P.J.; Orenstein, J.M.; Harty, R.N.; Schnell, M.J. PPEY motif within the rabies virus (RV) matrix protein is essential for efficient virion release and RV pathogenicity. *J. Virol.* **2008**, *82*, 9730–9738. [CrossRef]
8.	Komarova, A.V.; Real, E.; Borman, A.M.; Brocard, M.; England, P.; Tordo, N.; Hershey, J.W.; Kean, K.M.; Jacob, Y. Rabies virus matrix protein interplay with eIF3, new insights into rabies virus pathogenesis. *Nucleic Acids Res.* **2007**, *35*, 1522–1532. [CrossRef]
9.	Luo, J.; Zhang, Y.; Zhang, Q.; Wu, Y.; Zhang, B.; Mo, M.; Tian, Q.; Zhao, J.; Mei, M.; Guo, X. The Deoptimization of Rabies Virus Matrix Protein Impacts Viral Transcription and Replication. *Viruses* **2019**, *12*, 4. [CrossRef]
10.	Liu, X.; Li, F.; Zhang, J.; Wang, L.; Wang, J.; Wen, Z.; Wang, Z.; Shuai, L.; Wang, X.; Ge, J.; et al. The ATPase ATP6V1A facilitates rabies virus replication by promoting virion uncoating and interacting with the viral matrix protein. *J. Biol. Chem.* **2021**, *296*, 100096. [CrossRef]
11.	Ben Khalifa, Y.; Luco, S.; Besson, B.; Sonthonnax, F.; Archambaud, M.; Grimes, J.M.; Larrous, F.; Bourhy, H. The matrix protein of rabies virus binds to RelAp43 to modulate NF-kappaB-dependent gene expression related to innate immunity. *Sci. Rep.* **2016**, *6*, 39420. [CrossRef] [PubMed]
12.	Besson, B.; Sonthonnax, F.; Duchateau, M.; Ben Khalifa, Y.; Larrous, F.; Eun, H.; Hourdel, V.; Matondo, M.; Chamot-Rooke, J.; Grailhe, R.; et al. Regulation of NF-kappaB by the p105-ABIN2-TPL2 complex and RelAp43 during rabies virus infection. *PLoS Pathog.* **2017**, *13*, e1006697. [CrossRef] [PubMed]
13.	Sonthonnax, F.; Besson, B.; Bonnaud, E.; Jouvion, G.; Merino, D.; Larrous, F.; Bourhy, H. Lyssavirus matrix protein cooperates with phosphoprotein to modulate the Jak-Stat pathway. *Sci. Rep.* **2019**, *9*, 12171. [CrossRef] [PubMed]

14. Zan, J.; Liu, J.; Zhou, J.W.; Wang, H.L.; Mo, K.K.; Yan, Y.; Xu, Y.B.; Liao, M.; Su, S.; Hu, R.L.; et al. Rabies virus matrix protein induces apoptosis by targeting mitochondria. *Exp. Cell Res.* **2016**, *347*, 83–94. [CrossRef]
15. Kojima, I.; Onomoto, K.; Zuo, W.; Ozawa, M.; Okuya, K.; Naitou, K.; Izumi, F.; Okajima, M.; Fujiwara, T.; Ito, N.; et al. The Amino Acid at Position 95 in the Matrix Protein of Rabies Virus Is Involved in Antiviral Stress Granule Formation in Infected Cells. *J. Virol.* **2022**, *96*, e0081022. [CrossRef] [PubMed]
16. Kojima, I.; Izumi, F.; Ozawa, M.; Fujimoto, Y.; Okajima, M.; Ito, N.; Sugiyama, M.; Masatani, T. Analyses of cell death mechanisms related to amino acid substitution at position 95 in the rabies virus matrix protein. *J. Gen. Virol.* **2021**, *102*. [CrossRef]
17. Snider, N.T.; Omary, M.B. Post-translational modifications of intermediate filament proteins: Mechanisms and functions. *Nat. Rev. Mol. Cell. Biol.* **2014**, *15*, 163–177. [CrossRef] [PubMed]
18. Guo, Y.; Duan, M.; Wang, X.; Gao, J.; Guan, Z.; Zhang, M. Early events in rabies virus infection-Attachment, entry, and intracellular trafficking. *Virus Res.* **2019**, *263*, 217–225. [CrossRef]
19. Zandi, F.; Khalaj, V.; Goshadrou, F.; Meyfour, A.; Gholami, A.; Enayati, S.; Mehranfar, M.; Rahmati, S.; Kheiri, E.V.; Badie, H.G.; et al. Rabies virus matrix protein targets host actin cytoskeleton: A protein-protein interaction analysis. *Pathog. Dis.* **2021**, *79*, ftaa075. [CrossRef]
20. Liu, X.; Nawaz, Z.; Guo, C.; Ali, S.; Naeem, M.A.; Jamil, T.; Ahmad, W.; Siddiq, M.U.; Ahmed, S.; Asif Idrees, M.; et al. Rabies Virus Exploits Cytoskeleton Network to Cause Early Disease Progression and Cellular Dysfunction. *Front Vet. Sci.* **2022**, *9*, 889873. [CrossRef] [PubMed]
21. Tang, H.B.; Lu, Z.L.; Wei, X.K.; Zhong, T.Z.; Zhong, Y.Z.; Ouyang, L.X.; Luo, Y.; Xing, X.W.; Liao, F.; Peng, K.K.; et al. Viperin inhibits rabies virus replication via reduced cholesterol and sphingomyelin and is regulated upstream by TLR4. *Sci. Rep.* **2016**, *6*, 30529. [CrossRef]
22. Tang, H.B.; Pan, Y.; Wei, X.K.; Lu, Z.L.; Lu, W.; Yang, J.; He, X.X.; Xie, L.J.; Zeng, L.; Zheng, L.F.; et al. Re-emergence of rabies in the Guangxi province of Southern China. *PLoS Negl. Trop. Dis.* **2014**, *8*, e3114. [CrossRef]
23. Frank, E.D.; Tuszynski, G.P.; Warren, L. Localization of vimentin and desmin in BHK21/C13 cells and in baby hamster kidney. *Exp. Cell Res.* **1982**, *139*, 235–247. [CrossRef] [PubMed]
24. Pollin, R.; Granzow, H.; Kollner, B.; Conzelmann, K.K.; Finke, S. Membrane and inclusion body targeting of lyssavirus matrix proteins. *Cell. Microbiol.* **2013**, *15*, 200–212. [CrossRef]
25. Zhao, J.; Zeng, Z.; Chen, Y.; Liu, W.; Chen, H.; Fu, Z.F.; Zhao, L.; Zhou, M. Lipid Droplets Are Beneficial for Rabies Virus Replication by Facilitating Viral Budding. *J. Virol.* **2022**, *96*, e0147321. [CrossRef] [PubMed]
26. Lafon, M. Evasive strategies in rabies virus infection. *Adv. Virus Res.* **2011**, *79*, 33–53. [PubMed]
27. Feige, L.; Zaeck, L.M.; Sehl-Ewert, J.; Finke, S.; Bourhy, H. Innate Immune Signaling and Role of Glial Cells in Herpes Simplex Virus- and Rabies Virus-Induced Encephalitis. *Viruses* **2021**, *13*, 2364. [CrossRef]
28. Schnell, M.J.; McGettigan, J.P.; Wirblich, C.; Papaneri, A. The cell biology of rabies virus: Using stealth to reach the brain. *Nat. Rev. Microbiol.* **2010**, *8*, 51–61. [CrossRef] [PubMed]
29. Xu, J.; Gao, J.; Zhang, M.; Zhang, D.; Duan, M.; Guan, Z.; Guo, Y. Dynein- and kinesin- mediated intracellular transport on microtubules facilitates RABV infection. *Vet. Microbiol.* **2021**, *262*, 109241. [CrossRef]
30. Bauer, A.; Nolden, T.; Nemitz, S.; Perlson, E.; Finke, S. A Dynein Light Chain 1 Binding Motif in Rabies Virus Polymerase L Protein Plays a Role in Microtubule Reorganization and Viral Primary Transcription. *J. Virol.* **2015**, *89*, 9591–9600. [CrossRef]
31. Zan, J.; Liu, S.; Sun, D.N.; Mo, K.K.; Yan, Y.; Liu, J.; Hu, B.L.; Gu, J.Y.; Liao, M.; Zhou, J.Y. Rabies Virus Infection Induces Microtubule Depolymerization to Facilitate Viral RNA Synthesis by Upregulating HDAC6. *Front. Cell. Infect. Microbiol.* **2017**, *7*, 146. [CrossRef]
32. Zan, J.; An, S.T.; Mo, K.K.; Zhou, J.W.; Liu, J.; Wang, H.L.; Yan, Y.; Liao, M.; Zhou, J.Y. Rabies virus inactivates cofilin to facilitate viral budding and release. *Biochem. Biophys. Res. Commun.* **2016**, *477*, 1045–1050. [CrossRef]
33. Milner, D.J.; Weitzer, G.; Tran, D.; Bradley, A.; Capetanaki, Y. Disruption of muscle architecture and myocardial degeneration in mice lacking desmin. *J. Cell Biol.* **1996**, *134*, 1255–1270. [CrossRef]
34. Paulin, D.; Li, Z. Desmin: A major intermediate filament protein essential for the structural integrity and function of muscle. *Exp. Cell. Res.* **2004**, *301*, 1–7. [CrossRef]
35. Joanne, P.; Hovhannisyan, Y.; Bencze, M.; Daher, M.T.; Parlakian, A.; Toutirais, G.; Gao-Li, J.; Lilienbaum, A.; Li, Z.; Kordeli, E.; et al. Absence of Desmin Results in Impaired Adaptive Response to Mechanical Overloading of Skeletal Muscle. *Front. Cell Dev. Biol.* **2021**, *9*, 662133. [CrossRef]
36. Goebel, H.H. Desmin-related neuromuscular disorders. *Muscle Nerve* **1995**, *18*, 1306–1320. [CrossRef]
37. Eiber, N.; Fröb, F.; Schowalter, M.; Thiel, C.; Clemen, C.S.; Schröder, R.; Hashemolhosseini, S. Lack of Desmin in Mice Causes Structural and Functional Disorders of Neuromuscular Junctions. *Front. Mol. Neurosci.* **2020**, *13*, 567084. [CrossRef]
38. Lewis, P.; Lentz, T.L. Rabies virus entry into cultured rat hippocampal neurons. *J. Neurocytol.* **1998**, *27*, 559–573. [CrossRef]
39. Stein, L.T.; Rech, R.R.; Harrison, L.; Brown, C.C. Immunohistochemical study of rabies virus within the central nervous system of domestic and wildlife species. *Vet. Pathol.* **2010**, *47*, 630–633. [CrossRef]
40. Chung, B.-M.; Rotty, J.D.; Coulombe, P.A. Networking galore: Intermediate filaments and cell migration. *Curr. Opin. Cell Biol.* **2013**, *25*, 600–612. [CrossRef]
41. Margiotta, A.; Bucci, C. Role of Intermediate Filaments in Vesicular Traffic. *Cells* **2016**, *5*, 20. [CrossRef]
42. Etienne-Manneville, S. Cytoplasmic Intermediate Filaments in Cell Biology. *Annu. Rev. Cell Dev. Biol.* **2018**, *34*, 1–28. [CrossRef]

43. Geisler, F.; Leube, R.E. Epithelial Intermediate Filaments: Guardians against Microbial Infection? *Cells* **2016**, *5*, 29. [CrossRef]
44. Hamirally, S.; Kamil, J.P.; Ndassa-Colday, Y.M.; Lin, A.J.; Jahng, W.J.; Baek, M.C.; Noton, S.; Silva, L.A.; Simpson-Holley, M.; Knipe, D.M.; et al. Viral mimicry of Cdc2/cyclin-dependent kinase 1 mediates disruption of nuclear lamina during human cytomegalovirus nuclear egress. *PLoS Pathog.* **2009**, *5*, e1000275. [CrossRef]
45. Milbradt, J.; Webel, R.; Auerochs, S.; Sticht, H.; Marschall, M. Novel mode of phosphorylation-triggered reorganization of the nuclear lamina during nuclear egress of human cytomegalovirus. *J. Biol. Chem.* **2010**, *285*, 13979–13989. [CrossRef]
46. Cortese, M.; Lee, J.Y.; Cerikan, B.; Neufeldt, C.J.; Oorschot, V.M.J.; Köhrer, S.; Hennies, J.; Schieber, N.L.; Ronchi, P.; Mizzon, G.; et al. Integrative Imaging Reveals SARS-CoV-2-Induced Reshaping of Subcellular Morphologies. *Cell. Host Microbe* **2020**, *28*, 853–866.e5. [CrossRef]
47. Patteson, A.E.; Vahabikashi, A.; Goldman, R.D.; Janmey, P.A. Mechanical and Non-Mechanical Functions of Filamentous and Non-Filamentous Vimentin. *BioEssays News Rev. Mol. Cell. Dev. Biol.* **2020**, *42*, e2000078. [CrossRef]
48. Amraei, R.; Xia, C.; Olejnik, J.; White, M.R.; Napoleon, M.A.; Lotfollahzadeh, S.; Hauser, B.M.; Schmidt, A.G.; Chitalia, V.; Mühlberger, E.; et al. Extracellular vimentin is an attachment factor that facilitates SARS-CoV-2 entry into human endothelial cells. *Proc. Natl. Acad. Sci. USA* **2022**, *119*, e2113874119. [CrossRef]
49. Nédellec, P.; Vicart, P.; Laurent-Winter, C.; Martinat, C.; Prévost, M.C.; Brahic, M. Interaction of Theiler's virus with intermediate filaments of infected cells. *J. Virol.* **1998**, *72*, 9553–9560. [CrossRef]

Article

Putative Mitoviruses without In-Frame UGA(W) Codons: Evolutionary Implications

Andrés Gustavo Jacquat [1,2,*], Martín Gustavo Theumer [3,4] and José Sebastián Dambolena [1,2,*]

[1] Facultad de Ciencias Exactas Físicas y Naturales (FCEFyN), Universidad Nacional de Córdoba (UNC), Córdoba 5000, Argentina

[2] Instituto Multidisciplinario de Biología Vegetal (IMBIV), Consejo Nacional de Investigaciones Científicas y Técnicas (CONICET), Avenida Vélez Sarsfield 1611, Córdoba 5016, Argentina

[3] Departamento de Bioquímica Clínica, Facultad de Ciencias Químicas (FCQ), Universidad Nacional de Córdoba (UNC), Córdoba 5000, Argentina

[4] Centro de Investigaciones en Bioquímica Clínica e Inmunología (CIBICI), Consejo Nacional de Investigaciones Científicas y Técnicas (CONICET), Córdoba 5000, Argentina

* Correspondence: andresgjacquat@gmail.com (A.G.J.); jdambolena@imbiv.unc.edu.ar (J.S.D.); Tel.: +54-351-5353800 (ext. 30007) (J.S.D.)

Abstract: Mitoviruses are small vertically transmitted RNA viruses found in fungi, plants and animals. Taxonomically, a total of 105 species and 4 genera have been formally recognized by ICTV, and recently, 18 new putative species have been included in a new proposed genus. Transcriptomic and metatranscriptomic studies are a major source of countless new virus-like sequences that are continually being added to open databases and these may be good sources for identifying new putative mitoviruses. The search for mitovirus-like sequences in the NCBI databases resulted in the discovery of more than one hundred new putative mitoviruses, with important implications for taxonomy and also for the evolutionary scenario. Here, we propose the inclusion of four new putative members to the genus *Kvaramitovirus*, and the existence of a new large basally divergent lineage composed of 144 members that lack internal UGA codons (subfamily "Arkeomitovirinae"), a feature not shared by the vast majority of mitoviruses. Finally, a taxonomic categorization proposal and a detailed description of the evolutionary history of mitoviruses were carried out. This in silico study supports the hypothesis of the existence of a basally divergent lineage that could have had an impact on the early evolutionary history of mitoviruses.

Keywords: public databases; narna-levi; *Mitoviridae*; *Kvaramitovirus*; Arkeomitovirinae; Mitovirinae; UGA codon; UGG codon; origin; evolution; LECA

Citation: Jacquat, A.G.; Theumer, M.G.; Dambolena, J.S. Putative Mitoviruses without In-Frame UGA(W) Codons: Evolutionary Implications. *Viruses* **2023**, *15*, 340. https://doi.org/10.3390/v15020340

Academic Editors: Yiping Li, Yuliang Liu and Ioly Kotta-Loizou

Received: 25 October 2022
Revised: 22 January 2023
Accepted: 24 January 2023
Published: 25 January 2023

1. Introduction

Viruses belonging to the family *Mitoviridae* have a small (2151–4955 nt) non-segmented positive-sense single-stranded (+ss) RNA genome which codes for a single protein, namely, the RNA-dependent RNA polymerase (RdRp) [1,2]. These non-virion-forming viruses do not have an extracellular phase, as they are vertically transmitted RNA elements. Although it is likely that they can migrate to the cytoplasm of host cells [3,4], their replicative cycle occurs within the mitochondrial matrix of their hosts [1,5–8]. Mitovirus sequences have been found in the genome (nuclear and/or mitochondrial) of some eukaryotic organisms, especially in embryophytes [6,9,10], indicating their ability to endogenize in their plant hosts. However, the search for mitovirus sequences within the genome of other taxa has been unsuccessful [11–15]. To date, mitoviruses have been found to replicate in fungi [1,16], embryophytes [6,7] and animals [8]. Those that replicate in fungal and animal mitochondria have a similar UGA/UGG codon ratio to that of host core mitochondrial genes [5,8], while mitoviruses that replicate in plant mitochondria do not have internal UGA codons, since this codon encodes a translation stop signal in plastids [6,7]. The UGA and UGG codons are two synonymous codons for the amino acid (aa) tryptophan (W) in some genetic codes, such as the mitochondrial genetic codes of invertebrates and fungi. However, UGA is a translation termination signal for the standard genetic code, while only the UGG codon encodes for W [17].

Mitoviruses and narnaviruses (family *Narnaviridae*), which are other vertically transmitted naked RNA cytoplasmatic elements, are considered to be the simplest RNA viruses [2]. According to phylogenetic reconstructions, members of both these taxonomic families form a monophyletic lineage at the tree root of the global RNA virome. Moreover, mitoviruses and narnaviruses share a recent common ancestor with +ssRNA bacteriophages (class *Leviviricota*), forming a basal evolutionary lineage of the global RNA virome named "Narna-Levi" [18], "Branch 1" [2] or "phage-related (+)RNA viruses" [19]. Taxonomically, a total of 105 species and 4 genera (*Unuamitovirus*, *Duamitovirus*, *Triamitovirus* and *Kvaramitovirus*) of mitoviruses have been recognized by the International Committee on Taxonomy of Viruses (ICTV) [20], and recently, a new mitovirus genus named "Kvinmitovirus" has also been proposed [21]. Nevertheless, the systematic classification of the mitoviruses remains under constant review.

RNA sequencing and metatranscriptomics studies are a major source of hundreds of new "Narna-Levi"-like sequences, and consequently, these new sequences are continually being added to open databases [18,22–26]. Moreover, RNA sequencing studies in organisms for purposes other than virus identification are a very valuable source for identifying new putative mitoviruses. For example, a clade of putative mitoviruses associated with animals was recently identified from public sequence databases [21]. In addition, although empirical evidence is still needed for its confirmation, the *in silico* approach based on similarities between protein sequences and close evolutionary relationships has revealed the existence of a new mitovirus lineage (clade "Kvinmitovirus") [21]. This lineage was previously hidden from phylogenetic approaches based on hundreds or thousands of sequences obtained in metatranscriptomic studies. Following a similar approach to the one mentioned above, the aim of this article was to examine mitovirus-like sequences from different open databases to try to reveal new phylogenetic associations and to evaluate their potential contribution to the current knowledge about the evolutionary history of the family *Mitoviridae*.

2. Materials and Methods

2.1. Open Access Databases

In the present study, two public libraries of nucleotide (nt) and protein sequences were explored: the Transcriptome Shotgun Assembly (TSA) database and the non-redundant (nr) protein sequences database, both from the National Center for Biotechnology Information (NCBI, U.S. National Library of Medicine, Bethesda, MD, USA; accessed on 10 June 2022).

2.2. Bioinformatic Tools

Mitovirus-like sequences were scanned using the tBLASTn and BLASTp algorithms implemented by default at the web server of the NCBI Basic Local Alignment Search Tool (BLAST) [27] program: https://blast.ncbi.nlm.nih.gov/Blast.cgi (accessed on 15 September 2022).

Hypothetical coding regions, aa sequences and codon composition were predicted using ORFfinder (translation table numbers 1 and 4) on the web server https://www.ncbi.nlm.nih.gov/orffinder/ (accessed on 15 September 2022) and the Translate program [28]. The conserved protein sequence of Mitovirus RpRd (NCBI accession code: pfam05919: Mitovir_RNA_pol) was corroborated against CDD v. 3.20, using the NCBI CD-Search service [29], implemented by default at https://www.ncbi.nlm.nih.gov/Structure/cdd/wrpsb.cgi (accessed on 15 September 2022). A positive coincidence was considered for an e-value $\leq 1.00 \times 10^{-10}$.

To estimate the overall percent of pairwise sequence identity for all deduced proteins included in the phylogenetic studies, the Sequence Demarcation Tool (SDT) v. 1 program [30] was used (the ClustalW algorithm was chosen to compute the identity score for each pair of sequences). For multiple sequence alignments, several programs were utilized: (i) CLUSTAL-Omega v. 1.2.4 [31] at the https://www.ebi.ac.uk/Tools/msa/ (accessed on 15 September 2022) web server, (ii) Multiple Alignment using Fast Fourier Transform (MAFFT) v. 7.505 [32] at https://mafft.cbrc.jp/alignment/server/index.html (accessed on 15 September 2022) [33], and (iii) Profile Multiple Alignment with Predicted Local Structure (PROMALS) [34] at the http://prodata.swmed.edu/promals/promals.php web server. Access to these servers was between October 1, 2022 and January 18, 2023.

Phylogenetic trees by the Maximum Likelihood method were generated with IQ-TREE v. 1.6.11 [35]. To determine the optimal aminoacidic substitution matrix (model of molecular evolution) according to the Bayesian Information Criterion, the ModelFinder program [36] was used. The statistical support of branches was estimated by the Ultrafast Bootstrap 2 (UFBoot2) [37] method. These three programs were run on the Los Alamos lab web server [38]. For mitoviruses (*Mitoviridae*) phylogenetic reconstructions, five members from the genus *Unuamitovirus*, five from the genus *Duamitovirus*, five from the genus *Triamitovirus*, one from the genus *Kvaramitovirus* and five from the clade "Kvinmitovirus" were arbitrarily included. Henceforth, throughout this article, we refer to these mitoviruses as "formal mitoviruses", since they were formalized by the ICTV [20]. For simplicity, "formal mitoviruses" will include the members recently incorporated in the proposed new genus "Kvinmitovirus" [21]. To perform a deeper phylogenetic analysis, 25 members of the family *Narnaviridae*, 23 members of *Botourmiaviridae*, 20 members of "Narliviridae" and 20 members of the family "Leviviridae" (*Fiersviridae*) were included in this study. Details of all included sequences are given in Table S1. As an outgroup to rooted phylogenetic trees, the 20 RNA replicase sequences from "Leviviridae" family were chosen, taking into account the most complete and robust phylogenetic studies [2,19]. All ML-trees were displayed as generalized midpoint-rooted rectangular phylograms, using MEGA X v. 10.1.5 software [39].

3. Results and Discussion

3.1. Search for Putative Mitoviruses in the Open Databases

First of all, assembled contigs with mitovirus-like sequences were scanned on the NCBI TSA database using the tBLASTn mode of the NCBI BLAST web program. We used the tBLASTn strategy since its algorithm is more sensitive than BLASTn for matching putative mitoviral genomes with a high degree of divergence with respect to the query. In order to increase the sensibility for finding mitovirus domains in the tBLASTn search, we used an arbitrarily selected short query that only included the aa sequence of the conserved protein domain (CPD) of the mitovirus RdRps (highest evolutionary conserved region among known mitoviruses), which is composed of six conserved aa sequence motifs (I, II, III, IV, V and VI [40]). This query was the conserved region of the RNA replicase coded by Sclerotinia sclerotiorum mitovirus 3 (NCBI accession code: AGC24232.1; coordinates 265 aa–463 aa), the formal member of the viral species *Duamitovirus scsc 3*. Using this query, tBLASTn searches were performed on databases for different clades of eukaryotes (clade-limited tBLASTn search). These searches included all major taxa of eukaryotes: *Metazoan, CruMs, Malawimonadidae, Fornicata, Parabasalia, Heterolobosea, Euglenozoa, Rhodophyta, Glaucophyceae, Charophytes, Chlorophyta, Stramenopiles, Alveolata, Rhizaria, Haptista, Ancyromonadida and Cryptomonadida* (Table S2). In total, 247 TSA sequences were obtained, which were manually curated. Only those sequences with a hypothetical single open reading frame (ORF) of >1000 nt and with a hypothetical deduced protein (translation table #4 or #1) containing the Mitovirus RdRp CPD (NCBI accession conde pfam05919; e-value $\leq 1.00 \times 10^{-10}$) were included in our study. In addition, to avoid the inclusion of redundant sequences, contigs with a very high sequence similarity (pairwise identity percentage greater than 90%) were eliminated, with only one of these being preserved. These sequences, with a high degree of similarity, were considered as putative strains of the same putative viral species or as truncated contigs generated from the same original transcript. These selection criteria for mitovirus-like TSA sequences were previously applied by Jacquat et al. [21].

After sequence curing, only 54 TSA sequences were considered to be putative near-complete mitovirus genomes. These were then subjected to an alignment with formal mitoviruses (see Materials and Methods section) using the MAFFT (strategy: L-INS-i) aligner. The obtained midpoint-rooted phylogram showed several highly-supported clusters (Figure 1). However, in order to simplify the analyses in the present article, we focused only on those that were the most taxonomically and phylogenetically relevant, according to some considerations described below. Surprisingly, the obtained results revealed an interesting clustering between Ophiostoma mitovirus 7 (OnuMV7), a unique member of the species *Kvaramitovirus opno7* belonging to the monospecific genus *Kvaramitovirus* and the TSA sequence under the NCBI accession code GIFJ01468976.1. Curiously, the last two revisions reported by the Mitoviridae Study Group (SG) [20] and Jacquat and colleagues [21] found great topological instabilities of OnuMV7 for all phylogenetic reconstructions, indicating the genus *Kvaramitovirus* to be a "lonely" and "small" lineage with no clear evolutionary relationships. Therefore, the clustering reported here could be revealing the existence of a new representative member of the genus *Kvaramitovirus*. This association was analyzed throughout the present article, and provisionally, this two-member clade was referred to as "Clade A".

The above-mentioned ML-tree construction also showed another interesting clade that deserves to be explored in depth (Figure 1). This was composed of eight TSA sequences that branched at the base of the tree to form a robust basally divergent putative lineage. This clade was analyzed throughout this study, and provisionally, this was referred to as "Clade B". This branching pattern could reflect the existence of a sister lineage to existing mitoviruses, with an early evolutionary origin in the history of the viral family. This hypothesis was studied in the present work (see below) to investigate the evolutionary stage of mitoviruses, especially in the earliest evolutionary stages of the family.

In order to tentatively expand the number of members in the obtained clades "A" and "B", a search of putative mitoviruses for aa sequence similarity was performed using BLASTp software. Thus, we searched for aa sequences recorded at the NCBI nr Protein Sequence database. Regarding our attempt to expand "Clade A", the deduced protein from the sequence recorded under the NCBI accession code GIFJ01468976.1 was used as the query (query "a": Figure 1), with this query being the only putative virus to branch along with OnuMV7. Furthermore, a deduced protein that was clustered within clade "B" was arbitrarily chosen as the query to try to expand "Clade B". This was the TSA sequence with the NCBI code GFTX01082149.1 (query "b": Figure 1). The resulting first 250 hits of each BLASTp search were examined for sequences that were phylogenetically clustered within the preset clades "A" and "B".

In addition to these searches, a data set of 8469 hypothetical proteins generated from 442 RNA sequencing libraries (NCBI BioProject accession code: PRJNA716119) reported by Sadiq et al. [41] and Chen et al. [42] were also inspected to be able to tentatively expand the number of members of the obtained clades "A" and "B". These sequences were filtered to retain only hypothetical proteins with a Mitovirus RdRp CPD. This filtering enabled 574 deduced proteins to be identified as putative members of *Mitoviridae*. This set of 574 mitoviral proteins was then scanned and analyzed for members that evolutionarily fit within the clades "A" and "B". Finally, the sequences obtained from the BLASTp searches and the bioproject PRJNA716119 that clustered within clades "A" and "B" were inspected to eliminate redundant sequences (pairwise alignments with an identity greater than 90%) and sequences lacking one protein motif (I–VI) of Mitovirus RdRp. The obtained results demonstrated that a total of four new putative mitovirus were clustered into "Clade A", while 144 new putative mitovirus sequences were clustered into "Clade B". All sequences are described in the supplementary Table S3.

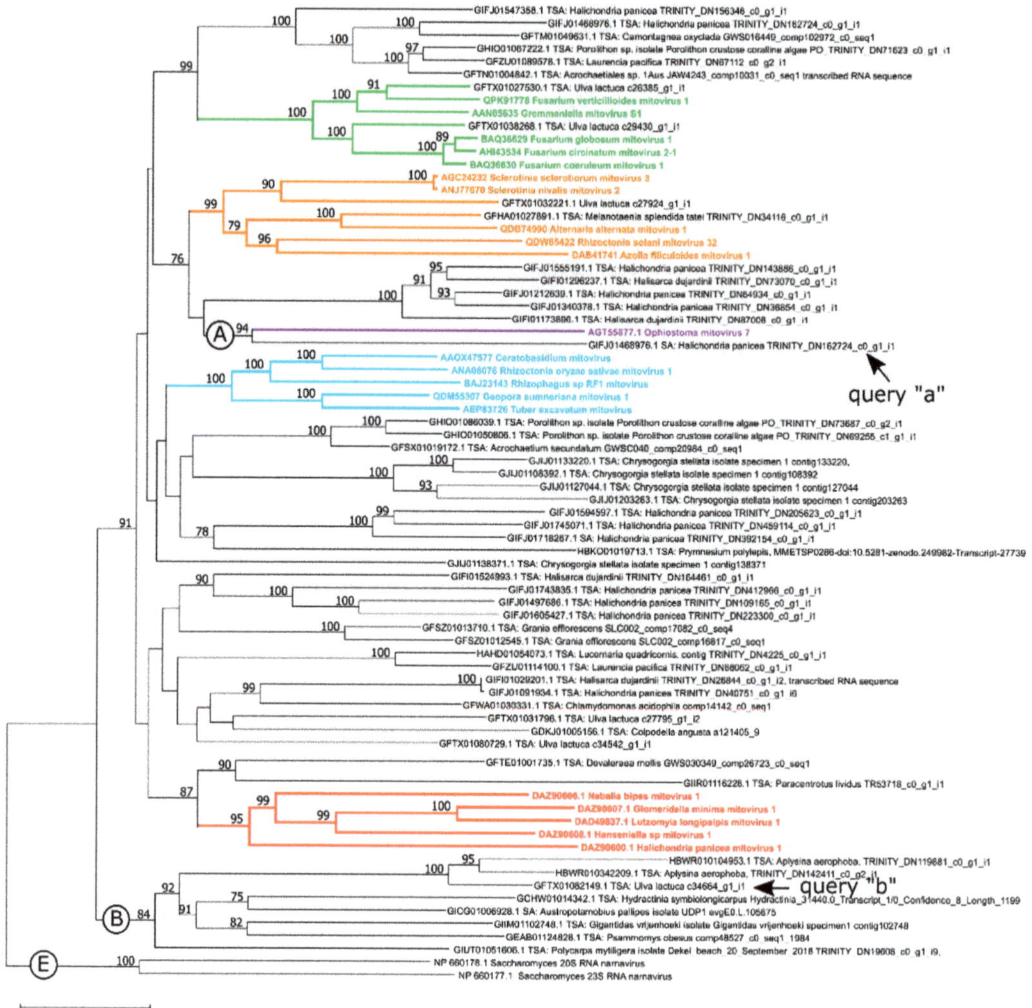

Figure 1. Generalized midpoint rooted tree of RNA replicase aa sequences. The deduced Mitovirus RdRps identified in the TSA database are labeled with the NCBI accession number, and the mitovirus formally accepted by the ICTV (colored branches) are labeled with the NCBI accession code and virus name. The color-coding of the branches: green, orange, cyan and purple indicates a member of the genus *Unuamitovirus*, *Duamitovirus*, *Triamitovirus* and *Kvaramitovirus*, respectively. Red branches indicate members of the recently proposed genus "kvinmitovirus". Formal members are indicated in colored bold type. Clades discussed in the main text are indicated with a letter ("A" and "B"). Letter E indicates the external group (two *Narnaviridae* members). An arrow indicates the sequence used as the query to search for similar sequences deposited in open databases (details in the article). Sequence alignment: MAFFT (L-INS-i). Tree construction method: Maximum Likelihood. Evolutive model: LG+F+R6. Node support values are displayed as percentages (only values ≥ 75% are shown). The bar indicates one substitution (estimated median number) per alignment site.

3.2. New Putative Mitoviruses in the Phylogenetic Context of the Phylum Lenarviricota

A deep evolutionary reconstruction including members of the "Narna-Levi" clade (formally phylum *Lenarviricota*) was performed in order to assess the evolutionary relationships of the two new clades studied in the present work. It is worth mentioning that RNA replicases encoded by "Narna-Levi" viruses have a low overall similarity among the different taxonomic families which complicates the construction of a realistic tentative phylogenetic tree [41,43]. In particular, the aa sequences included in the present study (Table S1) showed a global pairwise alignment identity of 27.91% (SD: 9.58%), indicating a technical limitation [44]. To confront this difficulty, we conducted a phylogenetic reconstruction using the PROMALS aligner that generates a probabilistic consistency-based progressive multiple structure-sequence alignment [34]. The probabilistic consistency-based scoring combines both aa similarity information and secondary structure similarity information through the profile-profile comparison using the hidden Markov model. PROMALS offers certain advantages over traditional MSA algorithms for distantly related protein homologs since it was designed to align divergent sequences [34,45].The obtained phylogram is shown in Figure 2. To support further deductions based on the tree derived from PROMAL, we also evaluated other alignment methods to construct trees: CLUSTAL-O and MAFFT / FFT-NS-i (see Figure S1).

The obtained phylogenic tree showed three main clades whose phylogenetic relationships were consistent with a recent comprehensive phylogenetic study of the phylum *Lenarviricota* [41] (Figure 2). *Narnaviridae*, *Botourmiaviridae* and "Narliviridae" form a sister clade to *Mitoviridae*, with "Leviviridae" being a basal group. This three-way topology was also observed in the phylograms generated from the alignments with CLUSTAL-O and MAFFT / FFT-NS-i (Figure S1). In our phylogenetic reconstructions, the resolution at the level of the major lineages within the Narna-Levi clade was greater than at the level of the genera. The clustering of members of the mitovirus genera was not consistent among the three trees (Figure 2 and Figure S1). For example, the genus "Kvinmitovirus" nested within the genus *Triamitovirus* (Figure 2) or a small subclade of "arkeomitoviruses" branched within the formal mitoviruses clade (Figure S1). This apparent inconsistency between members of the mitovirus genera was clarified in subsequent reconstructions, as described in the Section 3.7, in which the monophyly of "Arkeomitovirinae", formal mitoviruses, and mitoviruses genera were retained according to the ICTV report [20] and our previous study [21]. It should be remarked that the discrimination between mitoviruses, narnaviruses, narliviruses, botourmiaviruses and leviviruses shown in the present ML-trees (Figures 2 and S1) was consistent with the work of Sadiq and colleagues [41].

The 144 curated sequences that fit into "Clade B" were found to be monophyletic and constituted a sister clade to formal mitoviruses, sharing a hypothetical recent common ancestor. In addition, the branch length of "Clade B" was shorter than the branch of the formal mitoviruses clade (*Mitoviridae*), suggesting a smaller number of evolutionary change events (0.1 and 0.5, respectively, of the estimated average number of substitutions per site) with respect to the hypothetical aa sequence of the most recent common ancestor of all mitoviruses. This branch length pattern was not reflected in the phylograms obtained by CLUSTAL-O and the FAFFT/FFT-NS-I alignment (Figure S1). The tree of the Figure 2 shows that new putative mitoviruses that accommodated in the "Clade A" (labeled with bold letters in the sub-tree corresponding to the formal mitoviruses) were found to be monophyletic, including OnuMV7 (genus *Kvaramitovirus*). The monophyly of "Clade A" was kept with high support in all the alignment methods employed in the present study (Figure S3). The globality of the phylogenetic analyses allows us to suggest the existence of a basally divergent lineage of putative mitoviruses, with respect to the mitoviruses formally recognized by the ICTV, and to be able to tentatively expand the genus *Kvaramitovirus*.

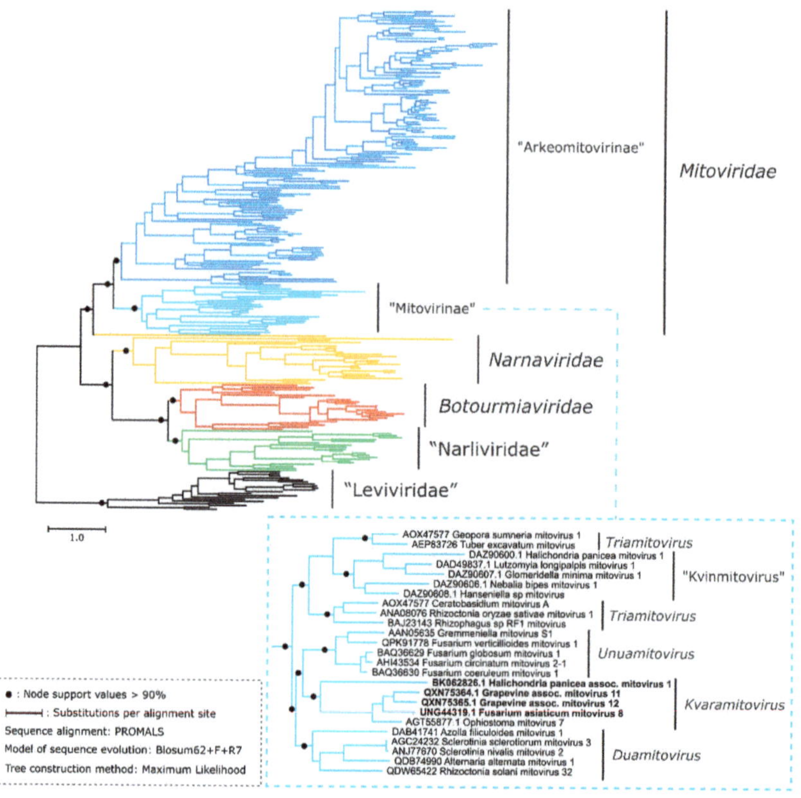

Figure 2. Generalized midpoint-rooted tree of 254 RdRp aa sequences encoded by representative members of the phylum *Lenarviricota*. Branches color-code: blue indicates the proposed subfamily "Arkeomitovirinae" ("Clade B"), cyan indicates formal mitoviruses fitted to the proposed subfamily "Mitovirinae" (formal genera *Unuamitovirus*, *Duamitovirus*, *Triamitovirus* and *Kvaramitovirus*, and also the proposed genus "Kvinmitovirus"). The family *Mitoviridae* comprises members of the blue and cyan branches. Orange, red and green indicate the families *Narnaviridae*, *Botourmiaviridae* and "Narliviridae". New members proposed for the genus *Kvaramitovirus* ("Clade A") are indicated in bold type in the sub-tree corresponding to the proposed "Mitovirinae" subfamily. The supplementary Figure S2 shows the tree displayed with the labels of the 254 members and the values of the nodes. The phylogenetic reconstructions are detailed in the Materials and Methods section.

3.3. Proposal of Supra-Generic Taxa

The members of the basally divergent clade were manually inspected in order to evaluate the genetic architecture. Surprisingly, the deduced ORFs of the putative mitoviruses included in "Clade B" did not have an internal in-frame UGA. Hence, tryptophan (W) was encoded only by the UGG codon in all 144 members. Therefore, this is a structural feature that distinguishes the representative members of this new putative mitovirus lineage from the formal mitoviruses, as the latter present several internal UGA codons. Although the lack of UGA codons has rarely been reported in fungal mitoviruses [5], it is a feature of plant mitoviruses [6]. However, phylogenetic evidence has indicated that plant mitoviruses originated by a "horizontal jump" from a fungal mitovirus to form a monophyletic clade within the genus *Duamitovirus* [6]. Moreover, the UA content in the putative near-complete nt mitovirus genomes included in "Clade B" was, in general, lower than 60% (mean: 54.6%; SD: 4.1%), with this proportion being relatively low compared to formal mitoviruses, as these have a content greater than 55%, and more commonly of 60–70% (sequences details given in Table S4). These molecular characteristics of the putative mitoviral genomes are similar to those of the narnavirus genomes. In *Narnaviridae*, the relatively low proportion of AU and the absence of UGA codons within the ORF are apparently adaptations to replication in the host cell cytoplasm. So, in order to assess whether these viruses share genomic characteristics with narnaviruses, we scanned for the presence of long uninterrupted ORFs in the antisense strand (-ssRNA), a feature detected in narnaviruses (ambigrammatic genome) [46]. Although the putative mitoviral genomes included in "Clade B" cannot be considered as complete end-to-end sequences, there is no evidence for the existence of putative large reverse ORF (1000 nt or more) encoding of functional proteins. The sequences were manually scanned (Table S3) using the NCBI-ORFfinder [28] and the NCBI-CDS [29] programs. Finally, despite the fact that these viral sequences have an architecture that is more similar to narnaviruses than authentical mitoviruses, the absence of ambigrammatic sequences, the presence of the six aa sequence motifs typical of mitoviral RdRps and the monophyly (suggesting a recent common shared ancestry with the formal mitoviruses), seem to indicate that the "Clade B" belongs to the family *Mitoviridae*. The molecular characteristics of "arkeomitoviruses" shared with narnaviruses are probably the result of an evolutionary convergence caused by selection pressure from the cytoplasmic protein biosynthesis machinery.

According to our approaches, we believe that there is sufficient evidence to propose the existence of putative mitoviruses belonging to a basally divergent lineage, with this new lineage being relatively (evolutionary speaking) distant to the already proposed mitovirus genera. These results led us to propose a taxonomic division of the family *Mitoviridae* into two supra-generic taxa: (i) subfamily "Mitovirinae", which involves the genera *Unuamitovirus*, *Duamitovirus*, *Triamitovirus* and *Kvaramitovirus*, and also the clade "Kvinmitovirus"; and (ii) subfamily "Arkeomitovirinae" (*arkeo* means ancient in the Esperanto language), which involves the putative mitoviruses initially included in "Clade B".

3.4. The genus Kvaramitovirus: A Proposal for Expansion

The exemplar OnuMV7, belonging to the monospecific genus *Kvaramitovirus*, but with an unclear phylogenetic relationship [21], was robustly clustered with four new putative mitoviruses ("Clade A") (Figure 2 and Figure S3). These putative mitoviruses were retrieved from the NCBI TSA database (GIFJ01468976.1) and the NCBI nr protein database (Grapevine-associated mitovirus 11, MW648458.1; Grapevine-associated mitovirus 12, MW648459.1 and Fusarium asiaticum mitovirus 8, MZ969058.1). The TSA sequence, identified by the NCBI accession number GIFJ01468976.1 and classified as a putative near-complete mitovirus genome in the present study, was obtained from a transcriptomic study of an exemplar of breadcrumb sponge (*Halichondria panacea, Porifera; Demospongiae*). This putative mitovirus branched at the root of the other four members and exhibited a greater number of cumulative evolutionary changes (Figure 2 and Figure S2). All these putative mitoviruses shared a high proportion of AU content: 67–73% (Table S4). The identity percentage among all combinations of pairwise alignments ranged from 24.00% to 47.00% (BLASTp alignment of protein sequences). This was lower than the species demarcation criterion (threshold of 70%) established by the ICTV Mitoviridae SG [20]. Thus, the protein sequence identity scores and the phylogenetic evidence reported here are strong reasons for proposing the following four new members, probably belonging to four new species within the genus *Kvaramitovirus*: "Halichondria panicea associated mitovirus 1" (GIFJ01468976.1), Grapevine-associated mitovirus 11 (MW648458.1), Grapevine-associated mitovirus 12 (MW648459.1) and Fusarium asiaticum mitovirus 8 (MZ969058.1) (Table S2). Halichondria panicea associated mitovirus 1 was redeposited under the GenBank Third Party Annotation (TPA) accession number BK062826.1.

3.5. On the Origin of the New Putative Mitoviruses without In-Frame UGA(W) Codons

The putative near-complete mitovirus genomes included in the new proposed subfamily called "Arkeomitovirinae" were obtained from transcriptomic or metatranscriptomic studies from the whole or a part of the presumptive host (body, tissue, an organ, portion of the body, intestinal content, excrement or body fluids/exudates) and also from environmental samples. Although an RNA-seq comes from a single organism or an anatomical part of a single organism, it is not possible to discard its origin from a mitovirus-infected symbiont or a mitovirus-infected parasite. Studies on the dinucleotide frequency of the putative mitovirus genome and the presumptive host nuclear/mitochondrial genome should be performed for proper host assignment [8]. The host assignment for these new putative mitoviruses without in-frame UGA codons requires an in-depth analysis that will be addressed in a future study. It is also important to clarify that these sequences were identified from transcriptomic studies on organisms belonging to evolutionarily unrelated taxonomic groups, such as chlorophytes, phaeophytes, cnidarians, malacostracans, bivalves, gastropods, mammals, actinopterygians, poriferans, ascidiaceans, birds (Aves), insects and tunicates. In addition to these, several sequences were obtained from diverse metatranscriptome environments (ponds, lakes, rivers, paddy sediment), animals feces and land, among other sample sites. Thus, as these new putative mitoviruses without internal UGA codon sequences were generated from independent transcriptomic studies, cross-contamination events could be discarded. Interestingly, this grouping did not involve putative mitoviruses previously identified in fungi or plants. However, due to the impossibility of accessing the samples, it was not possible to determine whether these new putative mitoviruses without in-frame UGA codons came from a cytoplasmic / mitochondrial replicating mitovirus, or whether they originated from a transcriptionally active mitovirus sequence integrated into the host genome as a non-retroviral Endogenous RNA Viral Element (NERVE).

Recently, Jacquat and colleagues [21] noted the lack of concrete evidence about mitovirus integration into the genome of non-plant organisms. In our study, the putative mitovirus genome sequences included in "Clade B" lacked in-frame UGA within the ORFs, which could be indicative of an adaptation to a translation system that does not use UGA codons—for example, the Bacterial, Archaeal and Plant Plastid Code (T.T. n° 11), Chlorophycean Mitochondrial Code (T.T. n° 16), Scenedesmus obliquus Mitochondrial Code (T.T. n° 22), Thraustochytrium Mitochondrial Code (T.T. n° 23), or the nuclear/cytoplasmatic translation system of all eukaryotic cells. The wide distribution of the geographic origin of the sequences, as well as the great diversity of environments and organisms at these sample sites, could be indicative of putative mitoviruses that infect bacteria (the UGA codon is a signal of the termination of translation with no encoding of the tryptophan aa). This hypothesis is plausible since bacteria occupy every site as they are found in all habitats on Earth [47–49].

In order to explore this last possibility, two BLAST searches were performed. Firstly, we performed a tBLASTn search on bacteria (NCBI taxid:2) entries in the TSA database, using the fragment 199 aa of the RdRp coded by Sclerotinia sclerotiorum mitovirus 3 (the same as that used in the initial search in the present work) as the query. However, this search did not generate any hits for the 23 TSA databases registered (September 2022 update). Secondly, a similar search was performed on the NCBI–Whole Genome Shotgun (WGS) database to evaluate a possible mitovirus sequence integration into the bacterial chromosome (prokaryote Bacteria: NCBI taxid:2). Here, the only hit (NCBI accession: CACLOE010000022.1) did not reveal any significant similarity (33% of the query cover and an e-value of 0.006), and therefore its mitovirus origin was discarded. Finally, a bacterial origin of the putative codonless mitovirus UGA, included in the new proposed subfamily named "Arkeomitovirinae", could be excluded. This conclusion is consistent with the extensive literature on bacterial viruses [50,51].

Fungi are considered to be the main hosts of mitoviruses, and hence, it might be expected that they are also the hosts of putative "arkeomitoviruses". This idea is plausible since the fungi are part of the microbiota of almost all multicell organisms and environments in the biosphere [48,52,53]. Only a small group of fungi, mainly the most basally divergent fungi (non-ascomycetes and basidiomycetes fungi), do not use the UGA(W) in the mitochondrial genetic codes. In agreement with this, these fungi are infected by mitoviruses, with only a few or without any UGA codons in their ORFs in an adaptation process [5]. Although this hypothesis should be considered, the fungal origin of the putative mitoviruses belonging to "arkeomitoviruses" still remains to be demonstrated by experimental procedures.

An embryophyte origin was discarded because the mitoviruses that replicated in embryophytes were shown to be a monophyletic clade which evolutionarily originated from a horizontal jump from a fungal mitovirus belonging to the genus *Duamitovirus* [6]. Moreover, the origin of the NCBI biosamples rules out this possibility (Table S3).

Finally, to summarize, we do not have information to assign a host; however, the absence of internal UGA and the relatively low proportion of AU is consistent with an adaptation to the cytoplasmic/nuclear translation system of eukaryotic cells, regardless of their host genome integration or cytoplasmic autonomous replication.

3.6. A Coherent Picture of the Evolution of Mitoviruses

All the phylogenetic reconstructions performed here indicated that "Arkeomitovirinae" is a basally divergent lineage in the mitovirus evolutionary tree. The phylogenetic evidence and the distinctive structural features (no in-frame UGA and a low proportion of AU) led us to propose the existence of a new monophyletic clade of mitoviruses that could have had an impact on the known evolutionary scenario of mitoviruses.

The first ideas on the evolutionary scenario of mitoviruses were reported by Koonin and Dolja (2014) [54] and Koonin et al. (2015) [55] (Figures 3 and 4). They proposed that ancient levivirus-like phages (inhabitants of the pre-eukaryotic world) infecting α-proteobacteria became "trapped" during eukaryogenesis. The loss (deletion) of genes associated with regressive evolution [56] may explain the transition from a multigenic (infectious levivirus-like virus) to a single gene (mito- and narna-like viruses) genome. In this way, these precursor viruses to existing mitovirus would lose their lytic capacity and become a naked replicon within the endosymbiotic organelle. In fact, this was demonstrated experimentally by Mills et al. (1967) [57], who found that an infectious virion-forming levivirus, phage Qβ, converted to a non-infectious non-capsid self-replicating ribonucleoprotein complex (mitovirus-like entity) [57]. Initially, it was thought that mitoviruses were exclusive to fungi [1,16]. Then, the discovery that all plant-replicating mitoviruses and plant-endogenous mitoviruses exhibit monophyly, and also probably originated from a horizontal jump of a glomeromycotan mitovirus, added another chapter to the evolutive scenario [6]. The evidence indicated that this was a unique evolutionary event. The ICTV Mitovirus SG classified the mitoviruses that replicate in plants within the genus *Duamitovirus*, a monophyletic clade consisting mainly of fungal mitoviruses. Another interesting finding probably expands the host range of mitoviruses beyond fungi and plants, including metazoans as the putative host [21], as evidence based on phylogenetic analysis indicated that fungal (genus *Unua-, Dua-, Tria-,* and *Kvaramitovirus*) and animal (*Kvinmitovirus*) mitoviruses probably share a recent common ancestor [21]. This hypothesis is consistent with the phylogeny of animals and fungi, which forms a monophyletic clade (Obazoa supergroup) [21]. This evolutionary scenario also proposed that current cytoplasmatic narnaviruses emerged from ancient mitovirus-like naked replicons by a "jump" to the cytoplasm (Figure 4). In light of new evidence, narnaviruses seemed to have evolved as a sister group to mitoviruses and gave rise to two other families of narna-like viruses: *Botourmiaviridae* and "Narliviridae" [41]. However, to simplify the discussion, we will only refer to the *Narnaviridae* family, which has evidence of sharing a common ancestor with mitoviruses.

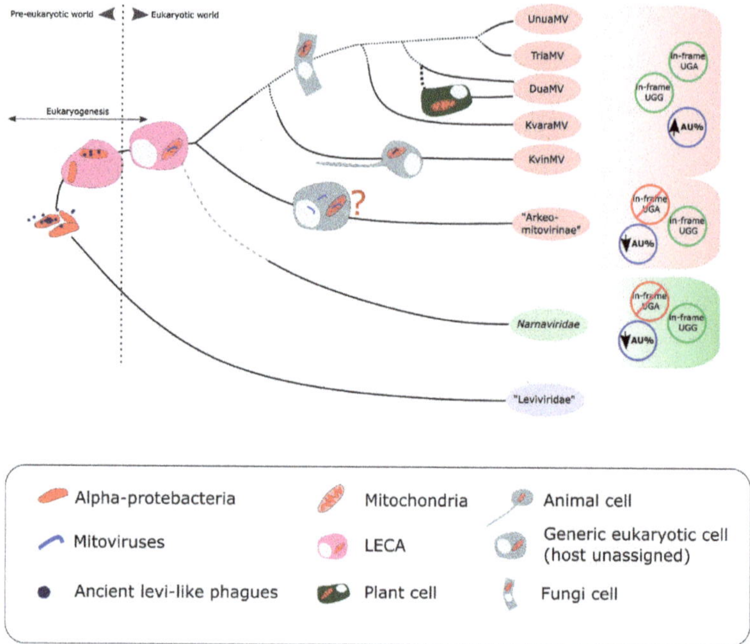

Figure 3. A comprehensive picture of the origin and evolutionary scenario of *Mitoviridae and Narnaviridae*. The scheme has temporal directionality from left (past) to right (present). Lines represent the ancestor–descendant relationship of the main mitovirus lineages based on the results reported in the present work and those reported by Mitovirus SG [20], Nibert et al. [6] and Jacquat et al. [21]. Solid lines have phylogenetic support from protein sequences. Dotted lines do not have a good topological phylogenetic resolution. The green line indicates the evolutive origin of the monophyletic lineage of plant-replicating mitoviruses (blue string within the mitochondria) and endogenized plant mitoviruses (blue string within the nucleus). This lineage together with a lineage of mitoviruses that replicates in fungi (within mitochondria) constitute the genus *Duamitovirus* (DuaMV). Moreover, the genera *Unuamitovirus* (UnuaMV), *Triamitovirus* (TriaMV) and *Kvaramitovirus* (KvaraMV) are exclusively made up of fungal mitoviruses. "Kvinmitovirus" (KvinMV) is the recently proposed mitoviruses genus. These five clades are included in the proposed subfamily "Mitovirinae", a sister clade of "Arkeomitovirinae". There is no evidence concerning the authentic host of the putative "arkeomitoviruses". The ancestor common to all mitoviruses was replicating within the mitochondria of the last eukaryotic common ancestor (LECA). This originated from levivirus-like bacteriophages that infected alpha-proteobacteria (mitochondrial precursors) in the pre-eukaryotic world. During eukaryogenesis, these levivirus-like phages became confined within the endosymbiont organelle (loss of lytic cycle) and evolved to become the ancestor of all mitoviruses by genome reduction. The main ideas of the originally proposed evolutive scenario detailed here were taken from Koonin and Dolja [54] and Koonin et al. [55].

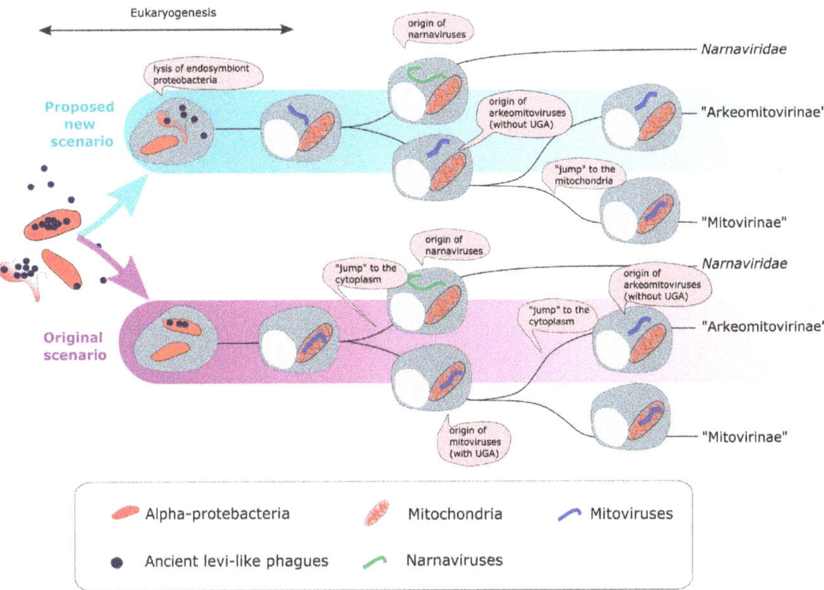

Figure 4. Currently accepted and proposed evolutionary scenarios of mitovirus origin. The scenarios are described in Sections 3.5 and 3.6 of the paper.

In our present article, a basally divergent lineage is added. Although the information we recovered is not sufficient to assign hosts to these new putative mitoviruses without in-frame UGA codons, it is sufficiently relevant to be able to predict the existence of a lineage of mitoviruses with an unusual genetic architecture that emerged early in the evolutionary history of mitoviruses. In this way, two major current mitovirus lineages would have been configured, with internal UGA ("Mitovirinae") and without internal UGA ("Arkeomitovirinae"). Although it is not possible to rule out their replication on host mitochondria or their endogenization into the host genome, the lack of in-frame UGA codons and the low proportion of AU in the genomes of the putative mitoviruses belonging to "Arkeomitovirinae" would seem to indicate their cytoplasmic location as being similar to that of narnaviruses, which have a low proportion of AU in their genomes and replicate in the cytoplasm [1]. However, this has to be demonstrated by experimental procedures.

3.7. An Alternative Parsimonious Scenario for the Origins of Mitoviruses

Although the authentic hosts and their replicative location of the "arkemitoviruses" remain to be determined, the lack of in-frame UGA codons and the low proportion of AU in the genomes would seem to indicate their cytoplasmic location as being similar to that of narnaviruses. In order to evaluate the evolutionary scenario of "arkemitoviruses" and formal mitoviruses, we analyzed the accumulated substitutions with respect to the most recent common ancestor of all mitoviruses. The phylogenic results obtained by PROMALS (designed to align divergent sequences) showed a smaller number of evolutionary change events for the "Arkeomitovirinae" clade. To corroborate this assumption, we made new phylogenetic inferences for members of the family *Mitoviridae* only. The sequences were aligned using different programs (PROMALS, CLUSTAL-O, MAFFT and EXPRESSO [58]), and prior to building the trees, each alignment was used to estimate the best-fitting substitution model. The results are shown in Figure S3. The branch length of the "Arkeomitovirinae" clade was shorter than the branch of the "Mitovirinae" clade, suggesting a lower number of substitutions per alignment site: 0.5 vs 0.6, 0.1 vs. 0.2, 0.2 vs 0.4 for PROMALS, CLUSTAL-O and EXPRESSO derived ML-trees, respectively (Figure S3). The

MAFFT-derived tree shows an inverse pattern. These differences are subtle, but it is worth noting that the evidence suggests directionality, indicating the "arkemitoviruses" to be a precursor to the mitochondrial mitoviruses. In this context, a new parsimonious scenario for the origins of *Mitoviridae* family is described below.

A levi-like phage ancestor that infected α-proteobacteria was released to the cytoplasm from a phagocytosed infected α-proteobacterium in the eukaryogenesis era. This release could have occurred by the lysis of the engulfed α-proteobacteria, with the consequent release to the host proto-eukaryotic cell cytoplasmic matrix of virions and viral components not assembled into virions (viral proteins and viral +ssRNA). This RNA viral genome of the ancestral levi-like phage could replicate in the cytoplasm since the prokaryotic genetic code (TT #11) of alpha-proteobacterium and the canonical eukaryotic genetic code (TT #1) are very similar (in both genetic codes the UGA codon is a translation stop signal). Viral RNAs would continue with replication in the cytoplasm of the first-born prokaryote, and the phages underwent genetic reduction up to becoming a non-infective cytoplasmic self-replicating entity, only maintaining the RNA replicase coding region. Under this proposed new scenario, the "arkeomitoviruses" of cytoplasmatic replication are the first forms of mitovirus to appear (see scheme in Figure 4). Then, the authentic mitochondrial replicating mitoviruses ("Mitovirinae") are a sub-lineage that migrate to the mitochondria of the common ancestor of fungi and animals (obazoans). In this way, the two major current mitovirus lineages would have been configured to the sub-families "Mitovirinae" and "Arkeomitovirinae".

The currently accepted evolutionary scenario posits that mitoviruses originated inside mitochondria (proto-mitochondria) from a levi-like phage ancestor "trapped" in the endosymbiotic organelle, and then, narnaviruses originated from a mitoviral member that jumped to the cytoplasm (Figure 4). Normally, the phages would have caused the rupture of the bacterial cell membrane (lysis). Why would it not have happened? The new parsimonious scenario for the origins of *Mitoviridae* family proposed here assumes that "arkeomitoviruses" of cytoplasmatic replication were the first forms of mitovirus to appear. The cytoplasmic origin was due to the release of levi-like phages, and the subsequent reductive evolution, into the cytoplasm caused by the "explosion" of an infected endosymbiont α-proteobacterium. That is, in this new scenario, it is proposed that the lysis would have occurred as expected. So, the hypothetical scenario (here proposed) posits that narnaviruses originated from a cytoplasmatic mitoviral ancestor. Then, a lineage of mitoviruses would have migrated into the mitochondria of the host lineage that gave rise to fungi and animals (Obazoa supergroup). In the mitochondrial environment, mitoviruses will have progressively gained internal UGA codons by selective pressure from the mitochondrial protein biosynthesis machinery ("Mitovirinae"). "Arkeomitoviruses" that remained in the cytoplasm did not gain internal UGA codons.

Phylogenetic reconstructions based on viral sequence present serious limitations for resolving deep phylogenies of small RNA viruses by their high mutation rates [59,60]. This feature explains the impossibility of establishing an unequivocal evolutionary directionality of the main lineages of the phylum *Lernaviricota* as discussed in Sadiq's paper [41], especially between narnaviruses and mitoviruses. Therefore, the evolutionary scenarios proposed in this work are hypothetical and need to be confirmed.

4. Final Comments

In an article published in 2018, it was deduced that plant mitoviruses originated from a horizontal jump from a fungal host to the evolutionary precursor of today's embryophytes. Then, another recently published work proposed the existence of a clade of putative mitoviruses that presumably replicated in animals. This in silico study was supported by previous evidence for the identification of a bona fide animal mitovirus that replicates in the mitochondria of a fly species. Here, the existence of a new monophyletic clade ("Arkeomitovirinae") of putative mitoviruses that use the canonical genetic code to replicate (presumably), which would have an impact on the evolutionary scenario of the mitoviruses, is presented. Due to its apparent basal divergence into the *Mitoviridae* phylogenetic tree, and also its genetic signatures, the "Arkeomitovirinae" group has been provisionally placed in the taxonomic rank of subfamily. Our study can be supported by the study by Neri and colleagues [61] in which they predict the existence of a large number of putative mitoviruses that are possibly adapted to the eukaryotic nuclear genetic code [61].

In this paper, the expansion of the genus *Kvaramitovirus* is proposed, which only has one formal member so far. Additionally, finally, this is the first study that has produced a detailed description of the evolutionary history of mitoviruses, a group of viruses that have directly descended from the ancestral phages of the pre-eukaryotic world.

Supplementary Materials: The following supporting information can be downloaded at https://www.mdpi.com/article/10.3390/v15020340/s1. Figure S1: Phylogenetic trees based on different alignments; Figure S2: High-resolution tree of Mitovirus RdRp amino acid sequences; Table S1: Details of sequences included in the phylogenetic reconstructions; Table S2: Result of the initial tBLASTn search of the NCBI TSA database and selection criteria; Table S3: Members of clades Kvaramitovirus and "Arkeomitovirinae"; Table S4: Molecular characteristics of mitoviruses and putative mitoviruses.

Author Contributions: A.G.J. searched and analyzed the sequences, and also performed the phylogenetic analyses. M.G.T. and J.S.D. collaborated on the data analyses and the discussion. A.G.J., M.G.T. and J.S.D. wrote the paper. All authors have read and agreed to the published version of the manuscript.

Funding: This research was supported by Grants PICT 2019-03300 and PICT-2021-CAT-I00192 from the National Agency for Scientific and Technological Promotion, Argentina (ANPCyT), PIP 11220200102478CO from the National Scientific and Technical Research Council (CONICET), and 33620180100149CB and 34020190100081CB from the Secretary of Science and Technique (SECyT-UNC).

Institutional Review Board Statement: Not applicable.

Informed Consent Statement: Not applicable.

Data Availability Statement: The virus sequences analyzed in this study and retrieved from NCBI nr protein database are available at NCBI-GenBank under accession numbers detailed in the Supplementary Material section (Tables S1 and S3). The TSA sequences reported in the present study have been redeposited into the GenBank Third Party Annotation database of the NCBI under the following accession numbers and names: NCBI accession code: BK062826 (Halichondria panicea associated mitovirus 1); NCBI accession code BK062827 (Aplysina aerophoba associated mitovirus 1); NCBI accession code BK062828 (Ulva lactuca associated mitovirus 1); NCBI accession code BK062829 (Austropotamobius pallipes associated mitovirus 1); NCBI accession code BK062830 (Gigantidas vrijenhoeki associated mitovirus 1); NCBI accession code BK062831 (Psammomys obesus associated mitovirus 1) and NCBI accession code BK062832 (Polycarpa mytiligera associated mitovirus 1).

Acknowledgments: We would like to thank native speaker, Paul Hobson, for revision of the manuscript. J.S.D. and M.G.T. are Career Members of CONICET. A.G.J. has a fellowship from CONICET.

Conflicts of Interest: The authors declare that the research was conducted in the absence of any commercial or financial relationship that could be construed as a potential conflict of interest.

References

1. Hillman, B.I.; Cai, G. The Family Narnaviridae: Simplest of RNA viruses. *Adv. Virus Res.* **2013**, *86*, 149–176. [CrossRef] [PubMed]
2. Wolf, Y.I.; Kazlauskas, D.; Iranzo, J.; Lucía-Sanz, A.; Kuhn, J.H.; Krupovic, M.; Dolja, V.V.; Koonin, E.V. Origins and Evolution of the Global RNA Virome. *Mbio* **2018**, *9*, e02329-18. [CrossRef]
3. Muñoz-Adalia, E.J.; Diez, J.J.; Fernández, M.M.; Hantula, J.; Vainio, E.J. Characterization of small RNAs originating from mitoviruses infecting the conifer pathogen Fusarium circinatum. *Arch. Virol.* **2018**, *163*, 1009–1018. [CrossRef] [PubMed]
4. Shahi, S.; Eusebio-Cope, A.; Kondo, H.; Hillman, B.I.; Suzuki, N. Investigation of Host Range of and Host Defense against a Mitochondrially Replicating Mitovirus. *J. Virol.* **2019**, *93*, e01503-18. [CrossRef] [PubMed]
5. Nibert, M.L. Mitovirus UGA(Trp) codon usage parallels that of host mitochondria. *Virology* **2017**, *507*, 96–100. [CrossRef] [PubMed]
6. Nibert, M.L.; Vong, M.; Fugate, K.K.; Debat, H.J. Evidence for contemporary plant mitoviruses. *Virology* **2018**, *518*, 14–24. [CrossRef]

7. Nerva, L.; Vigani, G.; Di Silvestre, D.; Ciuffo, M.; Forgia, M.; Chitarra, W.; Turina, M. Biological and Molecular Characterization of Chenopodium quinoa Mitovirus 1 Reveals a Distinct Small RNA Response Compared to Those of Cytoplasmic RNA Viruses. *J. Virol.* **2019**, *93*, e01998-18. [CrossRef]

8. Fonseca, P.; Ferreira, F.; Da Silva, F.; Oliveira, L.S.; Marques, J.T.; Goes-Neto, A.; Aguiar, E.; Gruber, A. Characterization of a Novel Mitovirus of the Sand Fly *Lutzomyia longipalpis* Using Genomic and Virus–Host Interaction Signatures. *Viruses* **2020**, *13*, 9. [CrossRef] [PubMed]

9. Bruenn, J.A.; Warner, B.E.; Yerramsetty, P. Widespread mitovirus sequences in plant genomes. *Peerj* **2015**, *3*, e876. [CrossRef]

10. Xu, Z.; Wu, S.; Liu, L.; Cheng, J.; Fu, Y.; Jiang, D.; Xie, J. A mitovirus related to plant mitochondrial gene confers hypovirulence on the phytopathogenic fungus Sclerotinia sclerotiorum. *Virus Res.* **2015**, *197*, 127–136. [CrossRef]

11. Katzourakis, A.; Gifford, R.J. Endogenous Viral Elements in Animal Genomes. *PLoS Genet.* **2010**, *6*, e1001191. [CrossRef]

12. Turina, M.; Ghignone, S.; Astolfi, N.; Silvestri, A.; Bonfante, P.; Lanfranco, L. The virome of the arbuscular mycorrhizal fungus *Gigaspora margarita* reveals the first report of DNA fragments corresponding to replicating non-retroviral RNA viruses in fungi. *Environ. Microbiol.* **2018**, *20*, 2012–2025. [CrossRef]

13. Blair, C.D.; Olson, K.E.; Bonizzoni, M. The Widespread Occurrence and Potential Biological Roles of Endogenous Viral Elements in Insect Genomes. *Curr. Issues Mol. Biol.* **2020**, *34*, 13–30. [CrossRef] [PubMed]

14. Picarelli, M.A.S.C.; Forgia, M.; Rivas, E.B.; Nerva, L.; Chiapello, M.; Turina, M.; Colariccio, A. Extreme Diversity of Mycoviruses Present in Isolates of Rhizoctonia solani AG2-2 LP From Zoysia japonica From Brazil. *Front. Cell. Infect. Microbiol.* **2019**, *9*, 244. [CrossRef]

15. Gilbert, C.; Belliardo, C. The diversity of endogenous viral elements in insects. *Curr. Opin. Insect Sci.* **2022**, *49*, 48–55. [CrossRef]

16. Myers, J.M.; Bonds, A.E.; Clemons, R.A.; Thapa, N.A.; Simmons, D.R.; Carter-House, D.; Ortanez, J.; Liu, P.; Miralles-Durán, A.; Desirò, A.; et al. Survey of Early-Diverging Lineages of Fungi Reveals Abundant and Diverse Mycoviruses. *Mbio* **2020**, *11*, e02027-20. [CrossRef] [PubMed]

17. Schoch, C.L.; Ciufo, S.; Domrachev, M.; Hotton, C.L.; Kannan, S.; Khovanskaya, R.; Leipe, D.; McVeigh, R.; O'Neill, K.; Robbertse, B.; et al. NCBI Taxonomy: A comprehensive update on curation, resources and tools. *Database* **2020**, *2020*, 1–21. [CrossRef]

18. Shi, M.; Lin, X.-D.; Tian, J.-H.; Chen, L.-J.; Chen, X.; Li, C.-X.; Qin, X.-C.; Li, J.; Cao, J.-P.; Eden, J.-S.; et al. Redefining the invertebrate RNA virosphere. *Nature* **2016**, *540*, 539–543. [CrossRef] [PubMed]

19. Dolja, V.V.; Koonin, E.V. Metagenomics reshapes the concepts of RNA virus evolution by revealing extensive horizontal virus transfer. *Virus Res.* **2018**, *244*, 36–52. [CrossRef] [PubMed]

20. International Committee on Taxonomy of Viruses (ICTV). Release Executive Committee 53. Code: Proposal 2021.003F. July 2021. Available online: https://talk.ictvonline.org/taxonomy/ (accessed on 1 June 2022).

21. Jacquat, A.G.; Ulla, S.B.; Debat, H.J.; Muñoz-Adalia, E.J.; Theumer, M.G.; Pedrajas, M.D.G.; Dambolena, J.S. An in silico analysis revealed a novel evolutionary lineage of putative mitoviruses. *Environ. Microbiol.* **2022**, *24*, 6463–6475. [CrossRef] [PubMed]

22. Le Lay, C.; Shi, M.; Buček, A.; Bourguignon, T.; Lo, N.; Holmes, E.C. Unmapped RNA Virus Diversity in Termites and Their Symbionts. *Viruses* **2020**, *12*, 1145. [CrossRef]

23. Wille, M.; Harvey, E.; Shi, M.; Gonzalez-Acuña, D.; Holmes, E.C.; Hurt, A.C. Sustained RNA virome diversity in Antarctic penguins and their ticks. *ISME J.* **2020**, *14*, 1768–1782. [CrossRef] [PubMed]

24. Wu, H.; Pang, R.; Cheng, T.; Xue, L.; Zeng, H.; Lei, T.; Chen, M.; Wu, S.; Ding, Y.; Zhang, J.; et al. Abundant and Diverse RNA Viruses in Insects Revealed by RNA-Seq Analysis: Ecological and Evolutionary Implications. *Msystems* **2020**, *5*, e00039-20. [CrossRef]

25. Zhang, Y.Y.; Chen, Y.; Wei, X.; Cui, J. Viromes in marine ecosystems reveal remarkable invertebrate RNA virus diversity. *Sci. China Life Sci.* **2022**, *65*, 426–437. [CrossRef]

26. Hirai, J.; Urayama, S.-I.; Takaki, Y.; Hirai, M.; Nagasaki, K.; Nunoura, T. RNA Virosphere in a Marine Zooplankton Community in the Subtropical Western North Pacific. *Microbes Environ.* **2022**, *37*, ME21066. [CrossRef]

27. Altschul, S.F.; Gish, W.; Miller, W.; Myers, E.W.; Lipman, D.J. Basic local alignment search tool. *J. Mol. Biol.* **1990**, *215*, 403–410. [CrossRef] [PubMed]

28. Artimo, P.; Jonnalagedda, M.; Arnold, K.; Baratin, D.; Csardi, G.; de Castro, E.; Duvaud, S.; Flegel, V.; Fortier, A.; Gasteiger, E.; et al. ExPASy: SIB bioinformatics resource portal. *Nucleic Acids Res.* **2012**, *40*, W597–W603. [CrossRef] [PubMed]

29. Marchler-Bauer, A.; Bryant, S.H. CD-Search: Protein domain annotations on the fly. *Nucleic Acids Res.* **2004**, *32*, W327–W331. [CrossRef]

30. Muhire, B.M.; Varsani, A.; Martin, D.P. SDT: A Virus Classification Tool Based on Pairwise Sequence Alignment and Identity Calculation. *PLoS ONE* **2014**, *9*, e108277. [CrossRef] [PubMed]

31. Sievers, F.; Wilm, A.; Dineen, D.; Gibson, T.J.; Karplus, K.; Li, W.; Lopez, R.; McWilliam, H.; Remmert, M.; Söding, J.; et al. Fast, scalable generation of high-quality protein multiple sequence alignments using Clustal Omega. *Mol. Syst. Biol.* **2011**, *7*, 539. [CrossRef]

32. Katoh, K.; Kuma, K.I.; Toh, H.; Miyata, T. MAFFT version 5: Improvement in accuracy of multiple sequence alignment. *Nucleic Acids Res.* **2005**, *33*, 511–518. [CrossRef] [PubMed]

33. Katoh, K.; Rozewicki, J.; Yamada, K.D. MAFFT online service: Multiple sequence alignment, interactive sequence choice and visualization. *Briefings Bioinform.* **2017**, *20*, 1160–1166. [CrossRef] [PubMed]

34. Pei, J.; Grishin, N.V. PROMALS: Towards accurate multiple sequence alignments of distantly related proteins. *Bioinformatics* **2007**, *23*, 802–808. [CrossRef] [PubMed]
35. Nguyen, L.-T.; Schmidt, H.A.; Von Haeseler, A.; Minh, B.Q. IQ-TREE: A Fast and Effective Stochastic Algorithm for Estimating Maximum-Likelihood Phylogenies. *Mol. Biol. Evol.* **2015**, *32*, 268–274. [CrossRef]
36. Kalyaanamoorthy, S.; Minh, B.Q.; Wong, T.K.F.; Von Haeseler, A.; Jermiin, L.S. ModelFinder: Fast model selection for accurate phylogenetic estimates. *Nat. Methods* **2017**, *14*, 587–589. [CrossRef]
37. Hoang, D.T.; Chernomor, O.; Von Haeseler, A.; Minh, B.Q.; Vinh, L.S. UFBoot2: Improving the Ultrafast Bootstrap Approximation. *Mol. Biol. Evol.* **2018**, *35*, 518–522. [CrossRef]
38. Trifinopoulos, J.; Nguyen, L.-T.; Von Haeseler, A.; Minh, B.Q. W-IQ-TREE: A fast online phylogenetic tool for maximum likelihood analysis. *Nucleic Acids Res.* **2016**, *44*, W232–W235. [CrossRef]
39. Kumar, S.; Stecher, G.; Li, M.; Knyaz, C.; Tamura, K. MEGA X: Molecular Evolutionary Genetics Analysis across Computing Platforms. *Mol. Biol. Evol.* **2018**, *35*, 1547–1549. [CrossRef]
40. Hong, Y.; Dover, S.L.; Cole, T.E.; Brasier, C.M.; Buck, K.W. Multiple Mitochondrial Viruses in an Isolate of the Dutch Elm Disease FungusOphiostoma novo-ulmi. *Virology* **1999**, *258*, 118–127. [CrossRef]
41. Sadiq, S.; Chen, Y.-M.; Zhang, Y.-Z.; Holmes, E.C. Resolving Deep Evolutionary Relationships within the RNA Virus Phylum Lenarviricota. *Virus Evol.* **2022**, *8*, veac055. [CrossRef]
42. Chen, Y.-M.; Sadiq, S.; Tian, J.-H.; Chen, X.; Lin, X.-D.; Shen, J.-J.; Chen, H.; Hao, Z.-Y.; Wille, M.; Zhou, Z.-C.; et al. RNA viromes from terrestrial sites across China expand environmental viral diversity. *Nat. Microbiol.* **2022**, *7*, 1312–1323. [CrossRef]
43. Koonin, E.V.; Dolja, V.V.; Morris, T.J. Evolution and Taxonomy of Positive-Strand RNA Viruses: Implications of Comparative Analysis of Amino Acid Sequences. *Crit. Rev. Biochem. Mol. Biol.* **1993**, *28*, 375–430. [CrossRef] [PubMed]
44. Rost, B. Twilight zone of protein sequence alignments. *Protein Eng. Des. Sel.* **1999**, *12*, 85–94. [CrossRef] [PubMed]
45. Chowdhury, B.; Garai, G. A review on multiple sequence alignment from the perspective of genetic algorithm. *Genomics* **2017**, *109*, 419–431. [CrossRef] [PubMed]
46. DeRisi, J.L.; Huber, G.; Kistler, A.; Retallack, H.; Wilkinson, M.; Yllanes, D. An exploration of ambigrammatic sequences in narnaviruses. *Sci. Rep.* **2019**, *9*, 17982. [CrossRef]
47. Flemming, H.-C.; Wuertz, S. Bacteria and archaea on Earth and their abundance in biofilms. *Nat. Rev. Microbiol.* **2019**, *17*, 247–260. [CrossRef] [PubMed]
48. He, L.; Rodrigues, J.L.M.; Soudzilovskaia, N.A.; Barceló, M.; Olsson, P.A.; Song, C.; Tedersoo, L.; Yuan, F.; Yuan, F.; Lipson, D.A.; et al. Global biogeography of fungal and bacterial biomass carbon in topsoil. *Soil Biol. Biochem.* **2020**, *151*, 108024. [CrossRef]
49. Yang, Y.; Lee, S.-H.; Jang, I.; Kang, H. Soil bacterial community structures across biomes in artificial ecosystems. *Ecol. Eng.* **2020**, *158*, 106067. [CrossRef]
50. Hatfull, G.F.; Hendrix, R.W. Bacteriophages and their genomes. *Curr. Opin. Virol.* **2011**, *1*, 298–303. [CrossRef]
51. Aiewsakun, P.; Adriaenssens, E.M.; Lavigne, R.; Kropinski, A.M.; Simmonds, P. Evaluation of the genomic diversity of viruses infecting bacteria, archaea and eukaryotes using a common bioinformatic platform: Steps towards a unified taxonomy. *J. Gen. Virol.* **2018**, *99*, 1331–1343. [CrossRef]
52. Peay, K.; Kennedy, P.G.; Talbot, J.M. Dimensions of biodiversity in the Earth mycobiome. *Nat. Rev. Genet.* **2016**, *14*, 434–447. [CrossRef] [PubMed]
53. Smith, G.R.; Steidinger, B.S.; Bruns, T.D.; Peay, K.G. Competition–colonization tradeoffs structure fungal diversity. *ISME J.* **2018**, *12*, 1758–1767. [CrossRef] [PubMed]
54. Koonin, E.V.; Dolja, V.V. Virus World as an Evolutionary Network of Viruses and Capsidless Selfish Elements. *Microbiol. Mol. Biol. Rev.* **2014**, *78*, 278–303. [CrossRef]
55. Koonin, E.V.; Dolja, V.V.; Krupovic, M. Origins and evolution of viruses of eukaryotes: The ultimate modularity. *Virology* **2015**, *479–480*, 2–25. [CrossRef] [PubMed]
56. Albalat, R.; Cañestro, C. Evolution by gene loss. *Nat. Rev. Genet.* **2016**, *17*, 379–391. [CrossRef]
57. Mills, D.R.; Peterson, R.L.; Spiegelman, S. An extracellular Darwinian experiment with a self-duplicating nucleic acid molecule. *Proc. Natl. Acad. Sci. USA* **1967**, *58*, 217–224. [CrossRef]
58. Armougom, F.; Moretti, S.; Poirot, O.; Audic, S.; Dumas, P.; Schaeli, B.; Keduas, V.; Notredame, C. Expresso: Automatic incorporation of structural information in multiple sequence alignments using 3D-Coffee. *Nucleic Acids Res.* **2006**, *34*, W604–W608. [CrossRef]
59. Holmes, E.C. What Does Virus Evolution Tell Us about Virus Origins? *J. Virol.* **2011**, *85*, 5247–5251. [CrossRef]
60. Marz, M.; Beerenwinkel, N.; Drosten, C.; Fricke, M.; Frishman, D.; Hofacker, I.L.; Hoffmann, D.; Middendorf, M.; Rattei, T.; Stadler, P.F.; et al. Challenges in RNA virus bioinformatics. *Bioinformatics* **2014**, *30*, 1793–1799. [CrossRef]
61. Neri, U.; Wolf, Y.I.; Roux, S.; Camargo, A.P.; Lee, B.; Kazlauskas, D.; Chen, I.M.; Ivanova, N.; Allen, L.Z.; Paez-Espino, D.; et al. Expansion of the global RNA virome reveals diverse clades of bacteriophages. *Cell* **2022**, *185*, 4023–4037. [CrossRef]

Article

Comparison of a Genotype 1 and a Genotype 2 Macaque Foamy Virus *env* Gene Indicates Distinct Infectivity and Cell-Cell Fusion but Similar Tropism and Restriction of Cell Entry by Interferon-Induced Transmembrane Proteins

Thomas Fricke [1], Sarah Schlagowski [1], Shanchuan Liu [1], Xiaoliang Yang [1], Uwe Fiebig [2], Artur Kaul [3], Armin Ensser [4] and Alexander S. Hahn [1,*]

[1] Junior Research Group Herpesviruses, German Primate Center—Leibniz-Institute for Primate Research, 37077 Göttingen, Germany
[2] Robert Koch Institute, Sexually Transmitted Bacterial Infections (STI) and HIV, 13353 Berlin, Germany
[3] Infection Biology, German Primate Center—Leibniz-Institute for Primate Research, 37077 Göttingen, Germany
[4] Institute for Clinical and Molecular Virology, Friedrich-Alexander-Universität Erlangen-Nürnberg, 91054 Erlangen, Germany
* Correspondence: ahahn@dpz.eu

Abstract: Foamy viruses (FVs) are naturally found in many different animals and also in primates with the notable exception of humans, but zoonotic infections are common. In several species, two different envelope (*env*) gene sequence clades or genotypes exist. We constructed a simian FV (SFV) clone containing a reporter gene cassette. In this background, we compared the *env* genes of the SFVmmu-DPZ9524 (genotype 1) and of the SFVmmu_R289hybAGM (genotype 2) isolates. SFVmmu_R289hybAGM *env*-driven infection was largely resistant to neutralization by SFVmmu-DPZ9524-neutralizing sera. While SFVmmu_R289hybAGM *env* consistently effected higher infectivity and cell-cell fusion, we found no differences in the cell tropism conferred by either *env* across a range of different cells. Infection by both viruses was weakly and non-significantly enhanced by simultaneous knockout of interferon-induced transmembrane proteins (IFITMs) 1, 2, and 3 in A549 cells, irrespective of prior interferon stimulation. Infection was modestly reduced by recombinant overexpression of IFITM3, suggesting that the SFV entry step might be weakly restricted by IFITM3 under some conditions. Overall, our results suggest that the different *env* gene clades in macaque foamy viruses induce genotype-specific neutralizing antibodies without exhibiting overt differences in cell tropism, but individual *env* genes may differ significantly with regard to fitness.

Keywords: foamy virus; interferon-induced transmembrane proteins; envelope; virus entry; fusion

Citation: Fricke, T.; Schlagowski, S.; Liu, S.; Yang, X.; Fiebig, U.; Kaul, A.; Ensser, A.; Hahn, A.S. Comparison of a Genotype 1 and a Genotype 2 Macaque Foamy Virus *env* Gene Indicates Distinct Infectivity and Cell-Cell Fusion but Similar Tropism and Restriction of Cell Entry by Interferon-Induced Transmembrane Proteins. *Viruses* **2023**, *15*, 262. https://doi.org/10.3390/v15020262

Academic Editor: Ester Ballana Guix

Received: 22 December 2022
Accepted: 11 January 2023
Published: 17 January 2023

1. Introduction

Foamy viruses are an ancient family of retroviruses that are found in many mammals and higher primates with which they co-speciate [1–3], with the notable exception of humans [4]. Endogenous foamy viruses are present in coelacanth [5], amphibia [6], fish [7], birds and serpentines [8]. SFVs have demonstrated a potential for zoonotic infection in Africa [9–11] and also in Asia [12,13]. Recently, zoonotic infections with gorilla SFV were found to be associated with hematological abnormalities and altered biochemical markers [9]. In addition, foamy viruses represent potential vector platforms for gene therapy [14–17] or oncolytic therapy [18,19], which warrants study of their cell entry mechanisms.

In several species, including in macaques, where multiple isolates or sequences have been obtained, there are at least two different sequence groups or clades that differ in the *env* gene [20–22] and most notably in a small region of the Env protein that has been proposed to act as a receptor binding domain (RBD) [23]. Three serotypes were reported to exist for feline foamy virus (FFV) [24], and different genotypes in the env locus have been characterized and were shown to correspond to these serotypes [25,26]. We recently isolated a foamy virus of rhesus macaques with a genotype 1 *env* gene, SFVmmu-DPZ9524 [20], and some time before an isolate, SFVmmu-R289hybAGM, from a rhesus monkey with a genotype 2 *env* was reported [27].

During infection with the human immunodeficiency virus, different *env* species develop that differ in coreceptor usage and cell tropism [28,29]. We were interested in whether the two envelope genotypes of rhesus macaque foamy viruses differ only in their neutralization sensitivity to antibodies elicited by viruses from their own or the respective other genotype but not in their functionality, or whether the different *env* variants confer functional differences such as altered cell tropism, which could suggest different receptor tropism or altered sensitivity to cellular restriction factors. Here, in particular, the interferon-induced transmembrane proteins (IFITM) 1-3, which restrict enveloped viruses dependent on their pathway of entry, among other determinants, at the membrane fusion step [30–33], are of interest. For FFV, human IFITMs were shown to inhibit FFV at a late stage of infection [34], hinting at additional antiviral mechanisms.

To facilitate analysis of infection, we constructed a YFP reporter virus SFV based on the SFVmmu-DPZ9524 genotype 1 isolate by replacing a part of the 3′ LTR U3 region with a YFP expression cassette. In this background, we replaced the genotype 1 *env* gene with that of the genotype 2 isolate to compare the functionality of these two *env* genes. These constructs were then analyzed with regard to susceptibility to neutralization, cell tropism, specific infectivity, and restriction by IFITMs 1-3.

2. Materials and Methods

2.1. Construction of the SFVmmu-DPZ9524 Infectious Clone

As described previously, DNA amplified from low passage virus stocks [20] was used to clone fragments of *SFVmmu-DPZ9524* into the backbone of pNCS-mNeonGreen (Allele Biotechnology, San Diego, CA, USA), replacing the mNeonGreen open reading frame. Fragments were assembled using overlapping PCR products and Gibson assembly, and checked by Sanger sequencing. Non-synonymous mutations were changed back to the reference sequence (MG051205.1), except for A to G in gag, resulting in the exchange of serine 248 to glycine as glycine was also found in a number of other foamy virus gag sequences at that position. There was an uncommonly high rate of nucleotide exchanges in the gag gene, which raises the possibility that the original isolate actually contained two viruses with slightly different gag sequences, which has been described for macaques [35]. A more trivial explanation might be that the original Illumina sequencing struggled with the quite repetitive sequence with high GC content, as this region had comparatively low read coverage, in keeping with our inability to find proof of these sequence variations by PCR. As we were only able to successfully isolate clones containing these nucleotide exchanges, we decided to proceed using them. Finally, using a clone ranging from nt1606 to the 3′ end, we amplified the LTR using primers DPZ9524 1-22 forward T7oh (TAATAC-GACTCACTATAGGGTGTGGCAGGCAGCCACTAAATG) and DPZ9524 1599-1638 LTR rev PBS oh (TTGGGCGCCAATTGTCATGGAATATTGTATATTGATTATC), amplified the rest of the genome and the vector using primers DPZ9524 1618-1642 forward (TCCATGA-CAATTGGCGCCCAACGTG) and T7 reverse (CCCTATAGTGAGTCGTATTAATTTCG) and used Gibson assembly to assemble the infectious clone AX512 pSFVmmu-DPZ9524_1. Transfection of this clone resulted in visible formation of a cytopathic effect that could be transferred, indicative of replication. The plasmid was sequenced by Next-Generation Sequencing. For library preparation, 50 ng of purified DNA and the Nextera DNA Library Preparation Kit (FC-131.1024, Illumina, San Diego, CA, USA) were used. Tagmented DNA was amplified with specific dual index primers (Integrated DNA Technologies, Coralville, IA, USA) using the NEBNext® HiFi 2 × PCR Master Mix (M0541, NEB) according to the manufacturer's protocols. The libraries were cleaned up with AMPure XP beads (A63881, Beckman Coulter, Brea, CA, USA) and quantified using the Qubit dsDNA HS Assay Kit

(Q32851, Thermo Fisher Scientific, Waltham, MA, USA). The libraries were analyzed by paired-end next-generation sequencing using a 150 cycle MiSeq reagent kit v3 (forward read 90 cycles, reverse read 60 cycles) on a MiSeq™ Instrument (Illumina, San Diego, CA, USA). The sequences were analyzed with CLC Genomics Workbench 20 (Qiagen Aarhus A/S, Aarhus, Denmark). Plasmids were fully covered with a Median Coverage between 564 and 1447 (minimum coverage 282, maximum coverage 2264). We discovered a 11nt insertion compared to the original MG051205.1 sequence; this was very likely a sequence assembly error in MG051205.1, as this sequence stretch is present in all related sequences. After next generation sequencing of the final construct, an additional mutation in the bet open reading frame, C to T, resulting in mutation of His283Tyr, was discovered. This conservative mutation was not corrected as the planned study was aimed at analyzing the function of Env and not of Bet. A summary of all SNPs in comparison to Genbank MG051205.1 is provided in Table S1.

2.2. Construction of SFV Reporter Viruses

An expression cassette driven by the immediate early promoter of the human cytomegalovirus (CMV, CMVie) was amplified from RRV-YFP [36] (MN488839.2) using primers CMVie for BetSTOP OH.

(GTGATTCTTCAGATGAAGATTAAATTAATAGTAATCAATTACGGG) and mNeon rev Delta 3′LTR oh (GTCATCAGGAGCTAATTTTACTTGTACAGCTCGTCC) and assembled together with the foamy virus backbone amplified from a construct lacking the 5′ LTR (pNCS-Foamy9524-4167-3primeEND) using primers Bet before STOP rev (ATCTTCATCTGAAGAATCACTAGAGG) and Delta 3′LTR for (TAAAATTAGCTCCTGATGACTCACG) by Gibson assembly. This construct was amplified using primers DPZ9524 9959:9983 for (TGGGAAGATCAAGAAGAATTAAGAG) and AmpQas and assembled together with the other part of the genome amplified from the full-length infectious clone AX512 pSFVmmu-DPZ9524_1 using primers DPZ9524 9958:9982 rev (TCTTAATTCTTCTTGATCTTCCCAG) and AmpQs, resulting in AX571 pSFVmmu-DPZ9524_1_CMVie-YFP.

The *env* gene of AX571 pSFVmmu-DPZ9524_1_CMVie-YFP was exchanged with that of SFVmmu_R289hybAGM by Gibson Assembly. SFVmmu_R289hybAGM *env* was amplified from cloned proviral DNA using primers 289 env for 9524oh (ACA ATGGCACCTCCAATGAC) and DPZ9524 9958:9982 rev (TCTTAATTCTTCTTGATCTTCCCAG) and joined with the AX571 pSFVmmu-DPZ9524_1_CMVie-YFP backbone, which was amplified using primers DPZ9524 9959:9983 for (TGGGAAGATCAAGAAGAATTAAGAG) and DPZ9524 7006:7027 rev (GTCATTGGAGGTGCCATTGTTC), to generate AX585 pSFVmmu-DPZ9524_1_ R289hybAGMenv_CMVie-YFP. We noticed a sequence variation in the *env* gene of our SFVmmu_R289hybAGM clone resulting in a change of amino acid 594 from Y to N in comparison to the deposited amino acid sequence (AFA44810.1). As all other sequences identified by BLAST [37] (blastp, expect value 0.05, hitlist size 100, gapcosts 11/1, matrix BLOSUM62, low complexity filter, filter string L, genetic code 1, window size 40, threshold 21, composition-based stats 2, database Jul 22, 2022 2:52 AM) from the genotype 2 clade, even from feline foamy virus, encoded N at this position, we regarded N as an at least highly plausible variant and Y as a potential sequencing artifact and did not change the sequence back from N to Y at this position.

Env expression plasmids in pcDNA6V5his (Invitrogen, Waltham, MA, USA) were constructed by GibsonAssembly. The vector backbone was amplified using primers pcDNA6 XbaI V5 for (GTCTAGAGGGCCCTTCGAAGGTAAG) and pcDNA6 KpnI ATG rev (CATGGTACCAAGCTTAAGTTTAAAC). The respective Env coding sequences were amplified from the infectious clones using genotype-specific forward primers DPZ9524 env pcDNA6 KpnI for (AACTTAAGCTTGGTACCATGGCACCTCCAATGACCTTGG) and R289hyAGM env pcDNA6 KpnI for (AACTTAAGCTTGGTACCATGGCACCTCCAATGACTTTGG) together with the reverse primer SFV env w/oSTOP rev XbaI V5 oh (CTTACCTTCGAAGGGCCCTCTAGAATTCTTCTTGATCTTCCCAGGAAG). Other plasmids that were used for this study are described in Table 1.

Table 1. Plasmids.

Plasmid	Source	Reference/Identifier
psPAX2	Addgene (kind gift from Didier Trono)	Addgene #12260
VSV-G (pMD2.G)	Addgene Addgene (kind gift from Didier Trono)	Addgene #12259
gag/pol	Addgene (kind gift from Tannishtha Reya)	Addgene #14887
pLenti CMV GFP Neo (657-2)	Lenti CMV GFP Neo (657-2) was a gift from Eric Campeau and Paul Kaufman	Addgene #17447 [38]
Gal4-TurboGFP-Luc (Gal4- driven TurboGFP-luciferase reporter plasmid)	Laboratory of Alexander Hahn	[39]
AX526 (Gal4-driven TurboGFP-luciferase reporter lentivirus, pLenti-Gal4-TurboGFP-Luc)	Laboratory of Alexander Hahn	[40]
plentiCRISPRv2 encoding sgRNAs targeting IFITM1-3 (sgIFITM1/2/3-a, sgIFITM1/2/3-b) or non-targeting sgRNAs (sgNT-a, sgNT-b)	lentiCRISRPv2 backbone from Addgene (kind gift from Feng Zhang), inserts from the laboratory of Alexander Hahn	[33,41]
Vp16-Gal4	Laboratory of Alexander Hahn	[39]
pCAGGS IAV_WSN-HA	Laboratory of Stefan Pöhlmann	[42]
pCAGGS IAV_WSN-NA	Laboratory of Stefan Pöhlmann	[42]
paMLV_env	Laboratory of Michael Farzan	[31]
pQXCIP	Laboratory of Stefan Pöhlmann	[42]
pQCXIP-IFITM1	Laboratory of Stefan Pöhlmann	[42]
pQCXIP-IFITM2	Laboratory of Stefan Pöhlmann	[42]
pQCXIP-IFITM3	Laboratory of Stefan Pöhlmann	[42]
pQCXIP-IFITM3-Y20A	Laboratory of Stefan Pöhlmann	[33]
pQCXIP-IFITM3-43AS	Laboratory of Stefan Pöhlmann	[33]

2.3. Design and Visualization

Visualization of the genome structure was performed using SnapGene Viewer (version 6.1, www.snapgene.com) and Inkscape (version 1.1, The Inkscape Project). Sequence alignment and visualization of the Env proteins was performed using MacVector (version 18.2.5, MacVector, Inc. Cary, NC, USA) and its implementation of ClustalW using the default settings.

2.4. Cell Culture

All cell lines in this study (Table 2) were incubated at 37 °C and 5% CO_2 and cultured in Dulbecco's modified Eagle medium, high glucose, GlutaMAX, 25 mM HEPES (Thermo Fisher Scientific, Waltham, MA, USA) supplemented with 10% fetal calf serum (FCS; Thermo Fisher Scientific, Waltham, MA, USA) and 50 µg/mL gentamicin (PAN-Biotech, Aidenbach, Germany) (D10) except for HUVEC, which were maintained in standard Endothelial Cell Growth Medium 2 (PromoCell, Heidelberg, Germany), and BJAB and MDA-MB-231 cells, which were maintained in RPMI (Thermo Fisher Scientific, Waltham, MA, USA) supplemented with 10% FCS and 50 µg/mL gentamicin. Interferon-alpha (IFN-α) treatment was performed by supplementing the respective culture medium with IFN-α 2b (Sigma; 5000 U/mL). For seeding and subculturing of cells, the medium was removed, the cells were washed with phosphate-buffered saline (PBS; PAN-Biotech, Aidenbach, Germany) and detached with trypsin (PAN-Biotech, Aidenbach, Germany). All transfections were performed using polyethylenimine (PEI; Polysciences) at a 1:3 ratio (µg DNA/µg PEI) mixed in Opti-MEM (Thermo Fisher Scientific, Waltham, MA, USA).

2.5. Western Blot

For detection of IFITM expression, the cells were treated with or without IFN-α (5000 U/mL) for 16 h. Thereafter, the cells were harvested and subjected to Western blot analysis. SDS polyacrylamide electrophoresis (PAGE) was performed using 8–16% gradient gels (Thermo Fisher Scientific, Dreieich, Germany). Western blotting was performed as described previously [43] using the respective antibodies (Table 3). Expression of other proteins was analyzed accordingly.

Table 2. Cell lines.

Cell Line/Cells	Cell Line Origin
HEK 293T	RRID:CVCL_0063, a kind gift from Vladan Rankovic, Göttingen, and originally purchased from the ATCC
A549	A kind gift from the laboratory of Stefan Pöhlmann, German Primate Center-Leibniz Institute for Primate Research, Göttingen, Germany
SLK	RRID:CVCL_9569, NIH AIDS Research and Reference Reagent program
Human foreskin fibroblasts (HFF)	A kind gift from the laboratory of Klaus Korn, Universitätsklinikum Erlangen, Institute for Clinical and Molecular Virology, Erlangen, Germany
Human vascular endothelial cells (HUVEC)	PromoCell
HaCaT	RRID:CVCL_0038, CLS Cell Lines Service GmbH
MDA-MB-231	ATCC HTB-26, a kind gift from the laboratory of Felix Engel
BJAB	Leibniz-Institute DSMZ (Deutsche Sammlung von Mikroorganismen und Zellkulturen GmbH)
MFB5487	*Macaca fuscata* B cell line, established from PBMC through immortalization with Herpesvirus papio. A kind gift from Ulrike Sauermann [40]
Primary rhesus monkey fibroblasts	A kind gift from the laboratory of Rüdiger Behr.

Table 3. Antibodies.

Target	Details
IFITM1	Goat, R&D (AF4827), 1:500 (secondary: Proteintech rabbit anti-goat 1:10,000)
IFITM2	Mouse, Proteintech (66137-1-lg), 1:1000 (secondary: Dianova donkey anti-mouse 1:10,000)
IFITM3	Rabbit, Cell Signal Technology (D8E8G) 1:1000 (secondary: Life Technologies goat anti-rabbit 1:10,000)
c-Myc	Mouse, Santa Cruz Biotechnology (9E10) 1:1000 (secondary: Dianova donkey anti-mouse 1:10,000)
V5	Mouse, BioRad (MCA1360), 1:1000 (secondary: Dianova donkey anti-mouse 1:10,000)
GAPDH	Mouse, GenScript, 1:5000 (secondary: Dianova donkey anti-mouse 1:10,000)
GFP	Rabbit, GenScript, 1:1000 (secondary: Life Technologies goat anti-rabbit 1:10,000)

2.6. Fusion Assay

On day one, 293T target cells were transfected in a 6-well plate overnight with 0.5 µg plasmid encoding a Gal4 response element driven TurboGFP-luciferase reporter (Gal4-TurboGFP-Luciferase). The 293T effector cells were transfected in a 6-well plate overnight either with 1 µg empty vector pNCS-mNeonGreen or AX571 SFVmmu-AX585 DPZ9524_1_CMVie-YFP (ST1) or DPZ9524_1_ R289hybAGMenv_CMVie-YFP (ST2) and 0.5 µg VP16-Gal4 transactivator plasmid. On day two, 16 h after transfection, the medium on the cells was completely removed and exchanged with fresh medium. Twenty-four hours after transfection, the effector cells were trypsinized and seeded in 96-well plates at

50,000 cells/well. On day three, the target cells were trypsinized and added to the effector cells. After 48 h, the cells were lysed in 65 µL 1× Luciferase Cell culture lysis buffer (E1531, Promega, Madison, WI, USA) for 20 min at room temperature and centrifuged for 10 min at 4 °C. Fifty microliters of each cell lysate were used to measure luciferase activity using the Beetle-Juice Luciferase Assay (PJK Biotech, Kleinblittersdorf, Germany) according to manufacturer's instructions on a Biotek Synergy 2 plate reader. Four independent experiments were performed. Each experiment was normalized to fusion signal of 293T effector cells transfected just with empty vector pNCS-mNeonGreen and VP16-Gal4 fused with 293T target cells. A paired *t*-test was performed to compare the results of the four experiments.

For testing cell-cell fusion activity of Env expressed from CMVie promoter-driven expression plasmids, A549 target cells were transduced with the lentiviral Gal4-driven TurboGFP-luciferase reporter construct as described previously [33], and were selected using 10 µg/mL of Blasticidin for three passages prior to the experiment. On 96-well, 293T cells (30,000/well) were seeded as the effector cells. After attachment, 31.25 ng of the transactivator (Gal4-VP16) plasmid were transfected together with 93.75 ng of the respective Env expression plasmid or empty vector per well. One day post transfection, A549 target cells (40,000/well) were added to 293T effector cells. The coculture was incubated for 48 h and then processed to measure luciferase activity as described above.

2.7. Retroviral Vectors, Foamy Viruses and Pseudotyped Lentiviral Particles

Retroviruses encoding the IFITM expression cassettes (pQCXIP backbone), lentiviruses (plentiCRISPRv2 backbone [41]) and lentiviral pseudotypes (pLenti CMV GFP Neo (657-2)) were produced by PEI-mediated transfection of 293T cells (see Table 1 for plasmids). For the production of SFV YFP reporter virus, 293T cells were transfected overnight in T175 flasks with either 15 µg AX571 pSFVmmu-DPZ9524_1_CMVie-YFP (ST1) or AX585 pDPZ9524_1_R289hybAGMenv_CMVie-YFP (ST2). Virus-containing cell-culture supernatant was harvested after 48 h. For retrovirus production, plasmids encoding gag/pol, pMD2.G encoding VSV-G, and the respective pQCXIP-contructs were transfected (ratio 1.6:1:1.6).

For production of lentiviruses used for transduction, psPAX2 encoding gag/pol, pMD2.G encoding VSV-G (vesicular stomatitis G protein) and the respective lentiviral construct were used (ratio 2.57:1:3.57). For lentiviral pseudotypes (LP) psPAX2, pLenti CMV GFP Neo and expression plasmids for pCAGGS IAV_WSN-HA and pCAGGS IAV_WSN-NA (encoding hemagglutinin and neuraminidase of influenza A strain WSN) for IAV-LP or paMLV_env (encoding amphotropic murine leukemia virus Env protein) for MLV-LP were used (ratio 1:1.4:2.4). Viruses were harvested twice: 24–48 h and 72–96 h after transfection.

Virus-containing supernatants were passed through a 0.45-µm CA filter and frozen at −80 °C. Transduction was performed by adding retroviruses and lentiviruses to cells for 48 h. IFITM1-, IFITM2- and IFITM3-knockout cell pools were generated by CRISPR/Cas9-mediated knockout following the protocol described by Sanjana et al. [41], except that PEI transfection was used. In short, the cells intended for knockout were transduced with lentiviruses harboring the CRISPR/Cas9 gene and sgRNAs targeting IFITM1-3 (sgIFITM1/2/3-a, sgIFITM1/2/3-b) or non-targeting sgRNAs (sgNT-a, sgNT-b). Afterwards, selection was performed using 5 µg/mL puromycin (InvivoGen; pQCXIP and pLentiCRISPRv2 constructs) for at least two passages.

2.8. Neutralization Assay

For neutralization assays, A549 cells were seeded in 96-well plates at 50,000 cells/well. At 6 h after plating, the SFV viruses were incubated for 30 min with or without heat-inactivated (56 °C for 30 min to inactivate the complement system) rhesus macaque sera before being added to the cells. The cells were harvested 48 h post-infection by brief trypsinization, followed by the addition of one volume of 5% FCS in PBS to inhibit trypsin activity and fixed by addition of one volume of PBS supplemented with 4% formaldehyde (Carl Roth). Ten thousand cells were analyzed per sample for YFP expression on an ID7000™ Spectral Cell Analyzer flow cytometer (Sony Biotechnology, San Jose, CA, USA). Data were analyzed using ID7000™ Spectral Cell Analyzer (Sony Biotechnology). Two independent experiments were performed for both SFVs. Each experiment was normalized to cells without immune sera infected with the respective virus.

2.9. Infection Experiments

Cells were seeded in 96-well plates at 90% confluency 6 h prior to infection. For IFITM ko experiments, cells were treated with IFN-α (5000 U/mL) or H2O (control) for 16 h prior to infection with either SFV ST1, SFV ST2, RRV, IAV-LP or MLV-LP. Forty-eight hours post infection, cells were trypsinized, trypsin activity was inhibited by adding 5% FCS in PBS and the cells were washed and fixed with 4% methanol-free formaldehyde (Roth) in PBS. Infection was determined by detection of GFP+/YFP+ cells (expressing the respective reporter gene of the lentiviral pseudotypes or the foamy reporter viruses) using a ID7000™ Spectral Cell Analyzer flow cytometer (Sony Biotechnology, San Jose, CA, USA).; ten thousand cells were analyzed. Curve fitting of infectivity on the different cell lines was performed using the built-in exponential equation for one phase association of GraphPad Prism version 9 for Windows (GraphPad Software, La Jolla, CA, USA) based on the Poisson distribution [43]. The span was set from 0 to 100, representing 0% or 100% infected cells, respectively, resulting in the simplified function $f(x) = 100 \times (1 - e^{-K \times x})$, with x representing input (virus dilution) and K representing the specific infectivity per input. The ratio between K_{ST1} and K_{ST2} was used to calculate the differences in infectivity between the viruses.

For spin-infection (spinfection), 293T cells were seeded in 96-well plates, 50 μL of virus-containing media was added to the cells, and the plate was spun for 2 h at 800 g. After 6 h, 50 μL of fresh culture medium were added, and the cells were incubated for 3 days before harvesting for analysis of infection by flow cytometry.

Images of infected or transduced cells were acquired on a Zeiss AxioVert.A1 cell culture microscope with LED fluorescence imaging.

2.10. Quantitative Reverse Transcription Realtime-PCR (RT-qPCR)

Viral nucleic acid was extracted from cell culture supernatants using the Viral RNA Mini kit (Qiagen, Hilden, Germany) after (or without) digestion with DNAseI (NEB) overnight, and RNA was isolated from infected cells using RNAzol and the Direct-zol RNA Miniprep kit (Zymo, Irvine, CA, USA) according to the manufacturer's instructions. qPCR with or without reverse transcription was performed using the Sensifast OneStep Probe Hi-Rox Kit (Bioline, Division of Meridian Bioscience, Cincinnati, OH, USA) on a StepOnePlus system (Thermo Fisher Scientific, Waltham, MA, USA). PCR conditions were 45 °C for reverse transcription, 2 min 95 °C initial denaturation, then 50 cycles of 95 °C for 5 s followed by 60 °C for 20 s. For the detection of the *pol* gene, primers pol for (CCCAAGGTAGT-TATGTGGTTCAT), pol rev (ATGTCCTTGTAGCAACTGATCC) and probe pol prb (/56-FAM/AATACCACT/ZEN/CCAAGCCTGGATGCA/3IABkFQ/) were used, for detection of the spliced Bet mRNA primer bet for (TGGGTACCAGACCCTTCA), bet rev (CTAG-GATTAGCGCGACGTTT), and probe bet prb (/56-FAM/ATTAGCCTC/ZEN/GAAGGA ACTCGGCTC/3IABkFQ/) were used and for the detection of the hypoxanthine phosphoribosyltransferase (HPRT) housekeeping gene, primers hprt for TGCTGAGGATTTG-GAAAGGG and hprt rev ACAGAGGGCTACAATGTGATG and probe hprt prb /56-FAM/TCGAGATGT/ZEN/GATGAAGGAGATGGGAGG-3/IABkFQ/ were used (/56-FAM indicates 5′ 6-FAM (Fluorescein) modification; ZEN indicates the internal quencher proprietary to IDT, 3IABkFQ indicates 3′ IowaBlack fluorescene quencher modification). All oligonucleotides were purchased from IDT.

Relative expression was calculated using the $2^{-\Delta Ct}$ method for each sample, then normalizing to the respective control as described by Schmittgen et al [44].

2.11. Statistical Analysis

Statistical analysis and data visualization were performed with GraphPad PRISM version 9.3.1.

3. Results

3.1. Generation of an Infectious Clone of SFVmmu-DPZ9524

We first cloned the genome of SFVmmu-DPZ9524 into a plasmid backbone to obtain an infectious clone, AX512 pSFVmmu-DPZ9524_1. Transfection of this plasmid into 293T cells, transfer of the supernatant to primary rhesus monkey fibroblasts and repeat passage (Figure 1A) resulted in expression of different viral RNAs (Figure 1B) as well as readily visible cytopathic effect (CPE) and formation of syncytia (Figure 1C), all indicative of viral replication. We compared supernatants from 293T cells transfected with AX512 pSFVmmu-DPZ9524_1, with the R289hybAGM plasmid and untransfected 293T supernatant in repeat passage experiments. Passaging of pSFVmmu-DPZ9524_1-derived virus resulted in stronger CPE and higher expression of both the pol RNA fragment and also of the spliced Bet transcript, as compared to passaging of virus derived from the R289hybAGM clone, at least under these conditions (Figure 1B,C).

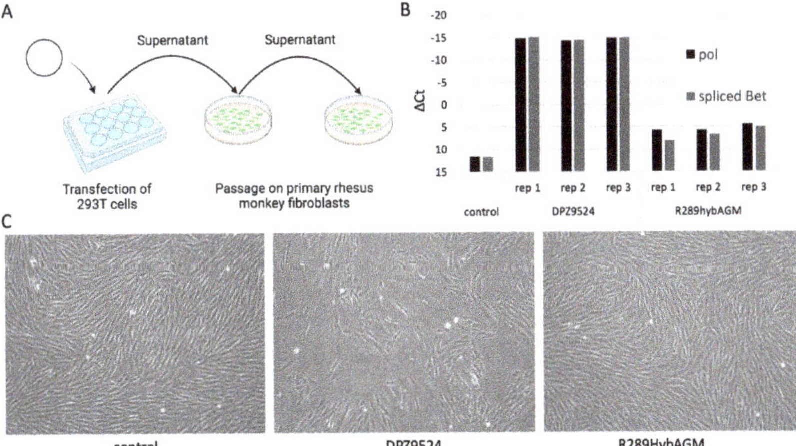

Figure 1. The pSFVmmu-DPZ9524_1 infectious clone yields infectious foamy virus. (**A**) After transfection of pSFVmmu-DPZ9524_1 or a proviral clone of R289hybAGM into 293T cells, cell-free supernatant from transfected cells or untransfected cells (control) was transferred to primary rhesus monkey fibroblasts. (Schematics created with BioRender.com.) (**B**) Expression of viral RNA (pol and spliced Bet mRNA) in primary rhesus monkey fibroblasts. $\Delta Ct = Ct_{\text{viral transcript}} - Ct_{\text{HPRT}}$. Ct was set to 40 cycles for negative samples (control). Values for three biological replicates (rep 1–3) and one control using supernatant from untransfected 293T cells are shown. (**C**) Representative phase contrast images of primary rhesus monkey fibroblasts after inoculation with the respective plasmid-derived foamy virus.

3.2. Generation of a SFV Reporter Virus with Genotype 1 and Genotype 2 env

Quantification of infection directly via immunofluorescence or qPCR is laborious, expensive and does not allow for the analysis of a high number of samples as is necessary for the comparison of two different envelope genes under different conditions. We therefore generated a reporter virus based on that clone by inserting a CMV immediate early promoter-driven YFP expression cassette into the U3 region of the 3' LTR, thereby deleting 709 bp of the LTR. We chose this strategy after fusion of bet to mNeonGreen did not result in recoverable infectivity (not shown). This construct with the CMV promoter-driven YFP expression cassette (Figure 2A) yielded appreciable infectivity after transfection of 293T producer cells as evidenced by transduction of the *YFP* reporter gene into fresh 293T cells. The construct might technically be replication-competent as the U3 region contains a CMV promoter and no additional polyA site, and such foamy virus clones have been found to replicate in the form of replication-competent recombinant revertants from self-inactivating foamy virus vectors [45]; however, these recombinant viruses exhibited strongly reduced replication. In line with this previous report, we did not observe visible spread of the virus in culture after infection (not shown), indicating that the construct can be used as a single-cycle foamy virus for practical purposes, presumably because the *YFP* open reading frame disrupts the LTR and the additional *YFP* expression cassette after duplication of the LTR during reverse transcription might impact packaging. The envelope gene of SFVmmu_R289hybAGM, which differs considerably from that of SFVmmu-DPZ9524 in its amino acid sequence (Figure 2B) was inserted into this construct in order to generate two otherwise isogenic *YFP* reporter gene foamy virus proviral clones, AX571 pSFVmmu-DPZ9524_1_CMVie-YFP (ST1) and AX585 pDPZ9524_1_R289hybAGMenv_CMVie-YFP (ST2).

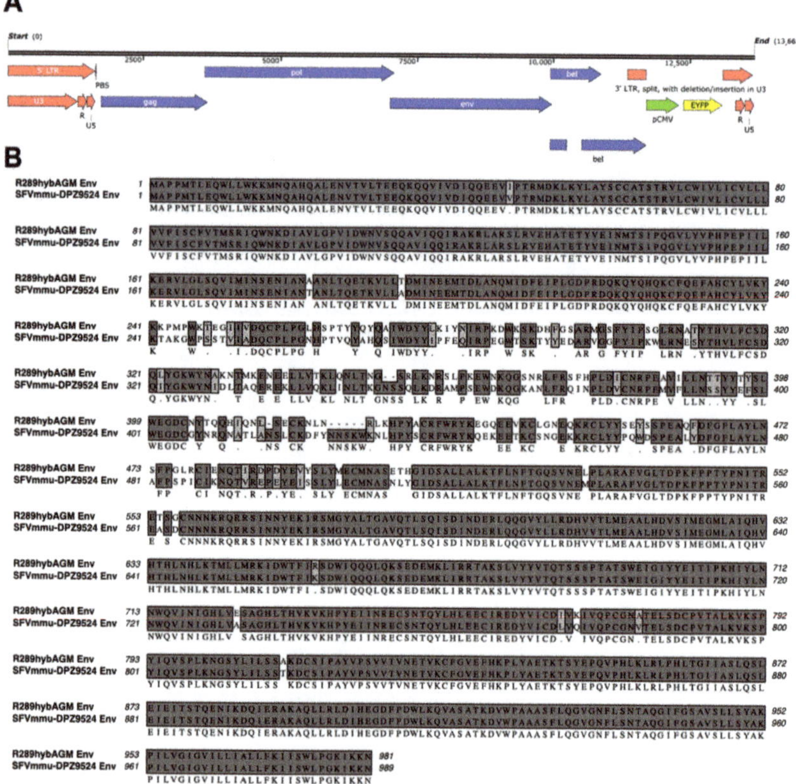

Figure 2. Generation and characterization of SFV reporter viruses with genotype 1 and genotype 2 *env*. (**A**) Genome structure of the SFV YFP reporter viruses. The CMV immediate-early promoter (pCMV)-driven YFP (enhanced YFP, EYFP) expression cassette was inserted into the U3 region of the 3′ LTR, thereby deleting 709 bp of the LTR. The envelope gene of SFVmmu_R289hybAGM was inserted into this construct in order to generate two otherwise isogenic YFP reporter gene foamy virus proviral clones, AX571 pSFVmmu-DPZ9524_1_CMVie-YFP (ST1) and AX585 pDPZ9524_1_R289hybAGMenv_CMVie-YFP (ST2). The unspliced bel transcript designates the Tas mRNA, the spliced bel transcript designates the Bet mRNA. (**B**) Sequence alignment of the Env proteins of ST1 and ST2.

3.3. The SFVmmu_R289hybAGM env Gene Is Associated with Significantly Higher Cell-Cell Fusion Than That of SFVmmu-DPZ9524

Upon transfection of the two YFP reporter virus clones, we noticed strong syncytia formation with the SFVmmu_R289hybAGM Env-expressing ST2 construct, but less so with the original SFVmmu-DPZ9524 Env-expressing ST1 (Figure 3A). Therefore, we set up a fusion assay with reporter cells that were expressing a Gal4-driven TurboGFP-luciferase construct and effector cells transfected with ST1 or ST2 together with a VP16-Gal4 transactivator. These cells were mixed and cocultured, and luciferase activity was quantified as a readout for fusion. After mixing, in line with the increased syncytia formation visible upon transfection with ST2 compared to ST1, increased luciferase activity was observed with ST2-transfected effector cells (Figure 3B). Expression of YFP, which runs at a lower molecular weight than the TurboGFP-luciferase fusion protein (below detection limit, molecular weight approx. 85 kDa) and is encoded by ST1 and ST2, as detected by Western blot was comparable (Figure 3C). A change in *env* might affect the internal promoter that drives transcription of the unspliced Tas and the spliced Bet transcripts [46]. We therefore tested expression of the Bet transcript by RT-qPCR, which allows clear differentiation from genomic RNA/DNA or plasmid DNA through its splice site. We found no difference in expression at 24h post transfection (Figure 3D).

To ascertain that the observed difference in cell-cell fusion between ST1 and ST2 stems from the envelope protein itself, we also tested ST1 and ST2 Env in an isolated cell-cell fusion assay by expressing both proteins from their cloned cDNA sequences under the control of the cytomegalovirus immediate early promoter (Figure 3E). The ST2 Env-encoding plasmid in that assay drove readily detectable cell-cell fusion, whereas the ST1 Env-encoding plasmid was not significantly active over background levels. In Western blot analysis of these fusion reactions using the V5 tag fused to the C-terminus of each Env protein, ST2 Env was observed to be expressed to considerably higher levels, most likely indicative of higher synthesis, higher stability or a combination thereof (Figure 3F). In addition, the two proteins exhibited a different pattern of proteolytic fragments, suggestive of differential proteolytic processing.

3.4. Low Frequency of Cross-Neutralization between Sera Neutralizing Genotype 1 or Genotype 2

We next tested whether the two Env proteins really belonged to different serological groups and whether they were resistant to neutralization by sera that neutralize the respective other genotype. We obtained 18 rhesus macaque sera, two of them negative for foamy virus according to prior serological testing [47]. Fifteen sera neutralized ST1 (Figure 4A,C) and three sera neutralized ST2 (Figure 4B,D). Only two sera, #158 and #160, neutralized both ST1 and ST2. At present, it is unclear whether this represents some degree of cross-neutralization or double infection of the animals. It should be noted that the two animals with sera neutralizing both ST1 and ST2 and the one neutralizing exclusively ST2, were housed together in the same group of animals (animals 158/19-162/19), compatible with infection with a genotype 2 SFV in that group of animals but not in all other animals, which form another group of animals with potential contact. Thirteen out of fifteen sera with neutralizing capacity against ST1 were without effect on ST2, arguing for at best minimal cross-neutralization between these two genotypes, at least in this direction.

3.5. No Evidence for env-Associated Differences in Cell Tropism

Tissue tropism of viruses is often determined by the interaction of their surface proteins with cellular receptors. The cellular receptors for foamy viruses so far remain unknown, even if heparan sulfate was described as an attachment factor [48]. Nevertheless, a putative receptor binding domain has been described [23], and the differences between genotypes 1 and 2 are predominantly found in that region of Env. We therefore hypothesized that the different *env* genes might confer different cellular tropism. To test this hypothesis, we infected several different cell lines and primary cells with ST1 and ST2 and compared the relative susceptibility of different cells to these two viruses (Figure 5). We consistently observed higher infectivity with ST2 than with ST1, even though both viruses were produced identically and in parallel. Nevertheless, as this was observed in all cells tested, the relative difference in infectivity between the two viruses, which was also calculated as ratio, was highly similar on all cells that were tested, indicating no major differences in tropism. In the one cell line where the two viruses seemed to perform more similarly (MFB5487), this was mostly due to 2 data points at the end of the dilution series and not consistently observed.

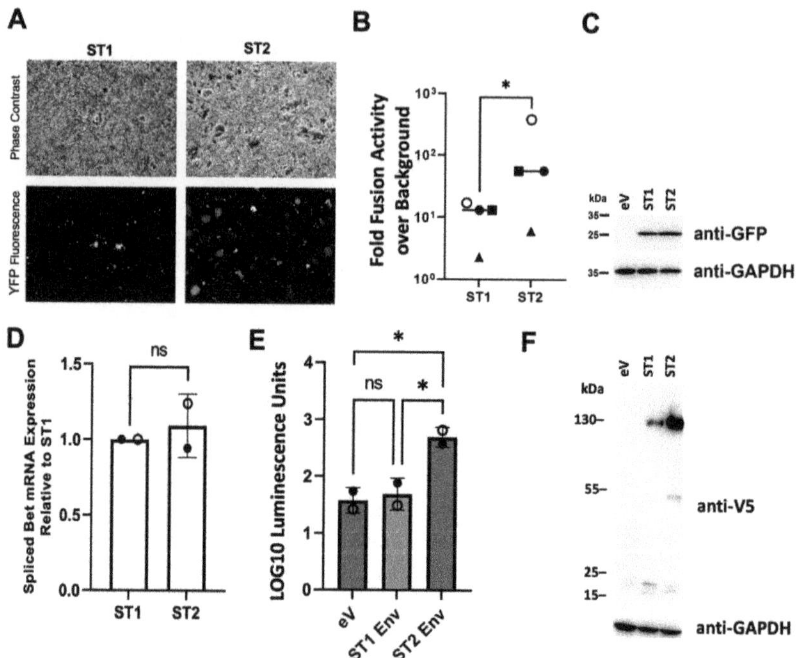

Figure 3. The R289hybAGM-derived *env* gene effects increased cell-cell fusion in the same genetic background. (**A**) Syncytia formation in 293T cells that were transfected with SFVmac ST1 and ST2 5 days post transfection. (**B**) Cell-cell fusion assay with ST1 and ST2 transfected 293T effector cells. The data show values normalized to background activity; (eV = empty vector). Four independent experiments were performed in triplicates. Luciferase activity in ST1-containing cocultures was significantly lower than in ST2-containing cocultures ($p \leq 0.05$, *; paired *t*-test comparing the log-transformed fusion activity. Identical symbols represent mean values from the same experiment.). (**C**) Representative Western blot of ST1 and ST2 transfected effector cells. Expression of the YFP reporter gene product encoded by the two viruses was determined using anti-GFP antibody. GAPDH served as loading control. (**D**) Expression of the spliced Bet mRNA 24 h post transfection of 293T cells with ST1 or ST2 as measured by RT-qPCR. The data show expression normalized to ST1, averaged from two independent experiments (open and filled circles), each in biological triplicates. Error bars show the standard deviation. ns: non-significant, ratio paired *t*-test. (**E**) Fusion activity upon expression of ST1 and ST2 Env from their respective cDNAs under the control of the CMV immediate early promoter. The data show log10-transformed, averaged arbitrary luminescence units obtained in two independent experiments, one in biological triplicates, one in six biological replicates (open and filled circles represent mean values from the same experiment). eV = empty vector. $p > 0.05$, ns: non-significant; $p \leq 0.05$, *; Ordinary one-way ANOVA. (**F**) Western blot analysis of lysates used for the determination of fusion activity in (**E**). The respective Env proteins were detected using a monoclonal anti-V5-epitope antibody recognizing the C-terminal V5-6His-tag.

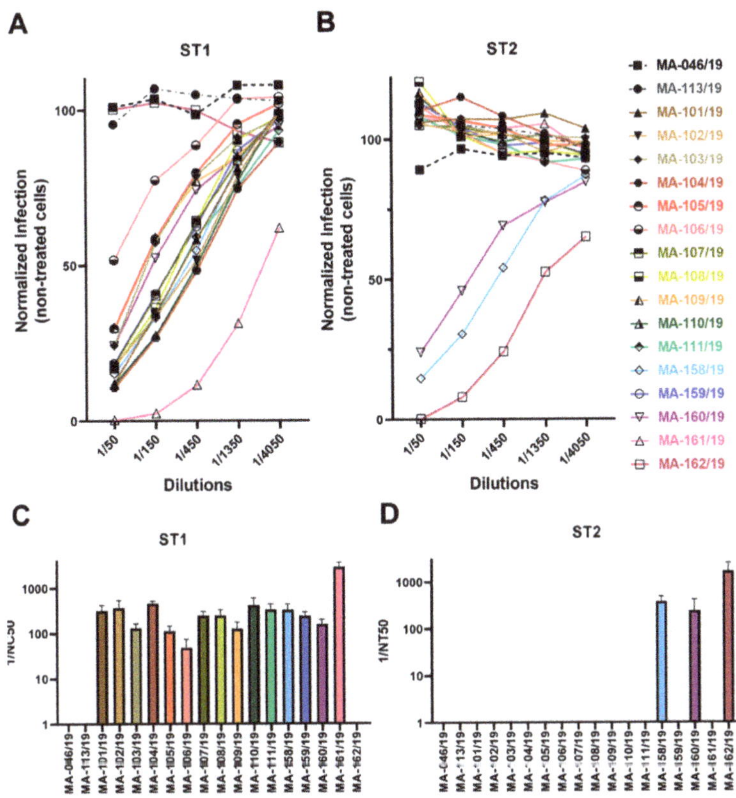

Figure 4. Low frequency of cross neutralization by sera from different animals. Neutralizing activity of rhesus macaque sera against ST1 (**A**) and ST2 (**B**). Infection was measured using flow cytometry to detect expression of the YFP reporter gene. The data show values normalized to infection without rhesus macaque serum. The experiment was performed twice with congruent results. Inverse neutralizing titers (1/NC50 values) were calculated for the individual rhesus macaque sera against ST1 (**C**) and ST2 (**D**) (nonlinear fit for different NC50, inhibitor vs. normalized response model). Error bars represent the 95% confidence interval.

3.6. SFVmmu_R289hybAGM env Effects Lower Particle Release and Higher Specific Infectivity

To follow up on the differences in infectivity between ST1 and ST2 preparations observed in Figure 5, we performed six independent transfections of ST1 and ST2 into 293T cells and first measured the release of DNAse-resistant viral nucleic acid by RT-qPCR and qPCR of the cell culture supernatants (Figure 6A). As observed previously [20,49], most of the foamy virus genome exists in the form of DNA within viral particles as evidenced by minimal differences in copy number values between PCR reactions with and without prior reverse transcription. Supernatant of ST2 producing cells contained on average less than half of the amount of viral nucleic acid as compared to supernatants from ST1 producing cells. Conversely, when these very same supernatants were tested for infectivity, ST2-containing supernatants produced approximately twice the number of YFP-positive cells (Figure 6B). Overall, when assuming the relationship of infectious particles to infected cells as linear for a multiplicity of infection below 0.01, specific infectivity of ST2 particles based on DNAse-resistant viral DNA (calculated individually for each sample from Figure 6A,B and averaged) was approximately 7.8 times higher than that of ST1 particles (Figure 6C).

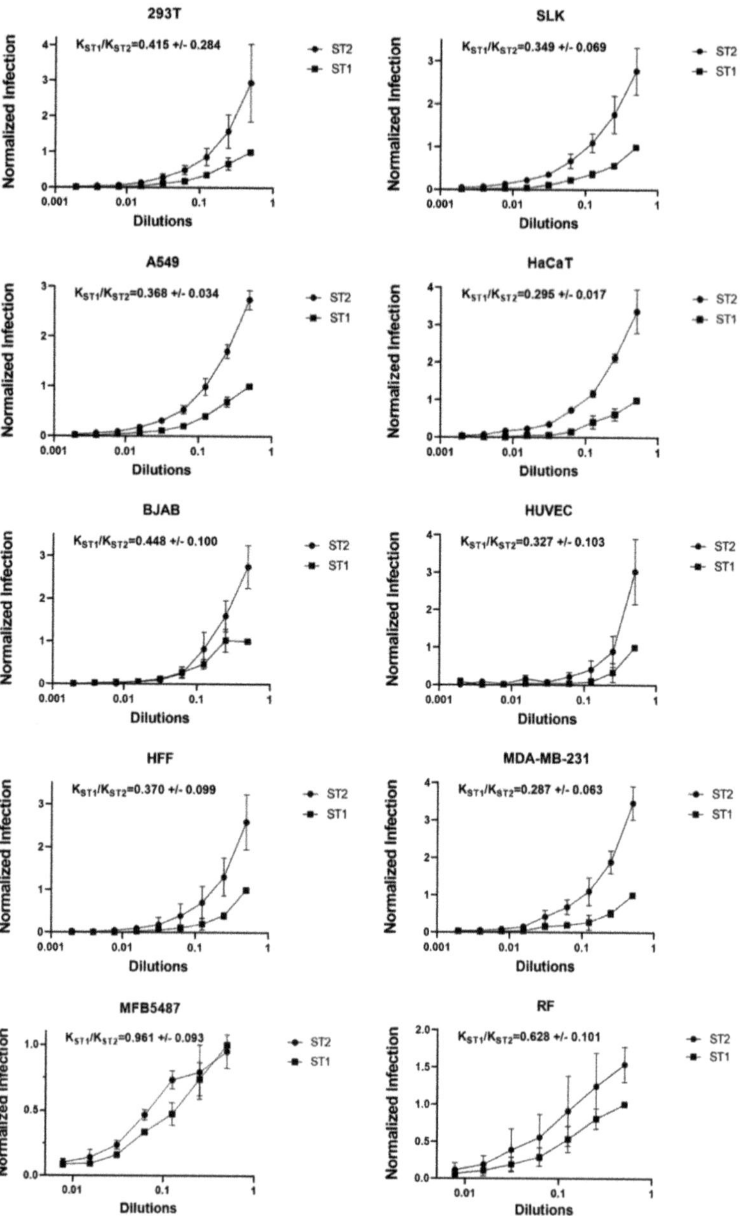

Figure 5. No evidence for *env*-associated differences in cell tropism. Infection of the indicated cells and cell lines after inoculation with increasing concentrations of ST1 or ST2. Infection was measured using flow cytometry to detect expression of YFP. The data show values normalized to infection with ST1 at the highest concentration. Curve fitting of infectivity on the different cell lines/cultures was performed using the built-in exponential equation for one phase association of GraphPad Prism version 9 to calculate K_{ST1} and K_{ST2} from the non-normalized data. The ratio between K_{ST1} and K_{ST2} represents the differences in infectivity between the viruses. Values represent the average of three experiments, error bars represent the standard deviation.

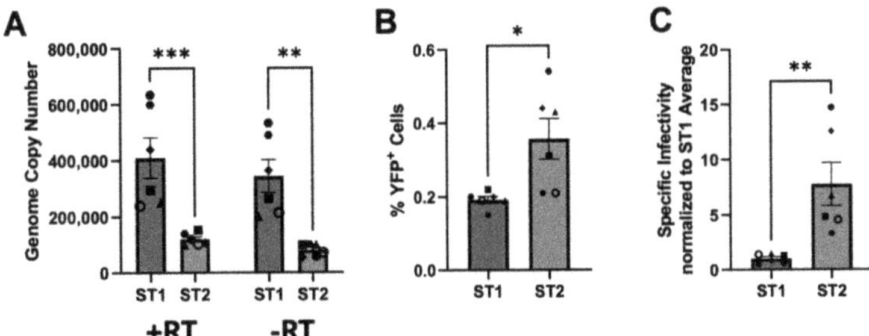

Figure 6. ST2 particles exhibit higher infectivity. (**A**) Viral genome copy number values of DNAse-treated supernatant of ST1 or ST2 producing 293T cells. Nucleic acid was extracted and analyzed by qPCR targeting the pol gene with or without prior reverse transcription (RT). Six biological replicates (rep) were performed. $p \leq 0.01$, **; $p \leq 0.001$, ***; 2-way ANOVA, Sidak's correction for multiple comparisons. (**B**) The exact same supernatants as in (**A**) were used to infect fresh 293T indicator cells by spinfection. $p \leq 0.05$, *; unpaired t-test. (**C**) Specific infectivity was calculated as the ratio of YFP-expressing cells over DNA genome copy number (-RT) for each sample and was normalized to the average ST1 infectivity. $p \leq 0.01$, **; unpaired t-test. Error bars represent the standard error of the mean.

3.7. No Evidence for env-Associated Differences in Susceptibility to IFITM-Mediated Restriction at the Cell Entry Step

We next analyzed the susceptibility of genotype 1 and genotype 2 *env*-driven cell entry to restriction by IFITM expression. Very similar to our results regarding cellular tropism, there was no difference between the two *env* genes in that entry driven by either *env* gene product was susceptible to recombinant overexpression of IFITM proteins, which resulted in reduction in infection for overexpression of all IFITM proteins. The reduction in infection by both viruses was significant for IFITM3 (Figure 7A,B). For ST1, inhibition of infection by IFITM1, IFITM2 and the IFITM3 43AS mutant, which shows broader subcellular localization than IFITM3 [50], was significant (Figure 7A). Expression of the IFITM proteins was verified by Western blot (Figure 7E). The IFITM3 Y20A sorting motif mutant, also described to exhibit wider sub-cellular distribution [51], was slightly less potent than the 43AS mutant, hinting at subtle differences between these mutants and their effects. These results suggest that sufficiently high IFITM expression levels could at least in theory affect SFV infection. Comparison with lentiviral particles pseudotyped with either the IFITM-resistant murine leukemia virus (MLV) glycoprotein or the IFITM-sensitive influenza A virus glycoproteins (IAV), demonstrated that the effect of IFITM overexpression on foamy virus infection lies somewhere between the two, but the observed degree of inhibition was much less pronounced than that of IAV glycoprotein-driven infection.

We next tested the effect of IFITM knockout under conditions with and without prior induction of IFITM expression through treatment with IFN-α (Figure 7F). Simultaneous knockout of all three IFITM proteins in A549 cells resulted in a slight enhancement of infection for both ST1 and ST2, but this did not reach significance. It should also be noted that the enhancement was approximately 1.31 fold for ST1 under conditions without prior interferon stimulation, and this barely changed to 1.35 fold when the cells had been previously stimulated with interferon, with a similarly negligible change from 1.25 fold to 1.3 fold for ST2, despite strong induction of IFITM expression (Figure 7G). This is similar to observations with the MLV envelope-pseudotyped lentivirus, which was enhanced by 1.22 fold and 1.11 fold, respectively, upon IFITM knockout (Figure 7F). For comparison, the enhancement by IFITM knockout changed from 1.37 fold to 2.04 fold enhancement for influenza HA/NA-pseudotyped lentivirus upon interferon pretreatment (Figure 7F). These results strongly argue against appreciable effects of IFITMs at expression levels reached after interferon stimulation, at least in A549 cells. Further, neither of the two different *env* genes was associated with increased susceptibility to IFITM-mediated restriction in this assay.

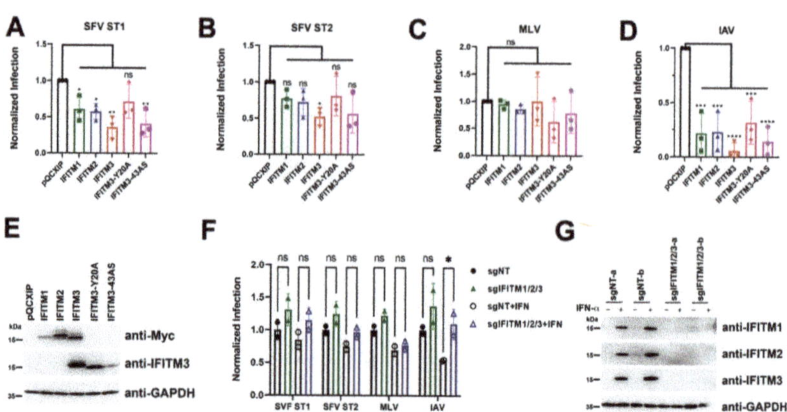

Figure 7. No evidence for *env*-associated differences in susceptibility to IFITM-mediated restriction. (**A–D**) A549 cells were transduced with pQCXIP-constructs to express IFITM1, IFITM2, IFITM3, IFITM3-Y20A, IFITM3-43AS or pQCXIP (empty vector). IFITM overexpressing cells were infected with SFV ST1, SFV ST2, MLV lentiviral pseudotype (MLV-LP) or IAV lentiviral pseudotype (IAV-LP). Infection was measured using flow cytometry to detect expression of the fluorescent reporter gene. The data show values normalized to pQCXIP (empty vector), and error bars represent standard error of the mean of five experiments, each performed in triplicates. Statistical significance was determined by ordinary one-way ANOVA ($p > 0.05$, ns; $p \leq 0.05$, *; $p \leq 0.01$, **; $p \leq 0.001$, ***; $p \leq 0.0001$, ****). (**E**) Representative Western blot of IFITM overexpressing cells. Expression of myc-tagged IFITMs was determined using anti-myc antibody. Expression of IFITM 3 was determined using anti-IFITM3 antibody. GAPDH served as loading control. (**F**) IFITM1/2/3 triple-knockout. A549 cells were transduced with lentiviral vectors encoding Cas9 and the sgRNAs IFITM-knockout (sgIFITM1/2/3-a, sgIFITM1/2/3-b) or control cells (sgNT-a, sgNT-b) treated with IFN-α (5000 U/mL) or H₂O (control) and infected with SFV ST1, SFV ST2, MLV lentiviral pseudotype (MLV-LP) or IAV lentiviral pseudotype (IAV-LP). Infection was measured using flow cytometry to detect expression of the fluorescent reporter gene. The data show values normalized to the average of sgNT-a, sgNT-b and error bars represent standard error of the mean of five experiments, each performed in triplicates. Statistical significance was determined by 2-way ANOVA ($p > 0.05$, ns; $p \leq 0.05$, *). (**G**) Representative Western blot of control (sgNT-a or sgNT-b) cells or IFITM knockout (sgIFITM1/2/3-a or sgIFITM1/2/3-b) treated with or without IFN-α (5000 U/mL). IFITM expression was detected with anti-IFITM1, anti-IFITM2 or anti-IFITM3; GAPDH served as loading control.

4. Discussion

The SFVmmu-DPZ9524-based infectious clone and YFP reporter virus generated in this study (Figures 1 and 2A) will facilitate further research on SFVs by adding to the available repertoire of viruses [19,52]. Using this system, we found to our surprise the genotype 2 *env* gene of SFVmmu_R289hybAGM endows SFVmmu-DPZ9524, originally a type 1 virus, with significantly higher fusion activity (Figure 3), and with higher infectivity across a range of target cells (Figures 5 and 6), which was coupled with lower particle release (Figure 6A). Whether the higher fusion activity of the SFVmmu_R289hybAGM *env* gene product is due to higher expression levels, as suggested by our Western blot results, higher intrinsic fusogenicity, altered surface expression, different proteolytic processing or a combination of these factors is currently an open question. Nevertheless, our findings may be helpful for the construction of foamy virus-based gene vectors, as transduction efficiency is always an issue for viral vectors, and the SFVmmu_R289hybAGM envelope would clearly be a superior choice for such an application. Generally, our observations suggest that the different foamy virus envelope genes from genotypes circulating within a species may differ considerably with regard to their ability to mediate entry and cell-cell fusion, something so far only described for foamy virus *env* genes from different species [53]. As we only tested a single *env* from each genotype, these findings should not be generalized, and each one of the two tested *env* genes could represent an outlier or have acquired a particularly advantageous or deleterious mutation that may not be found in similar isolates from other animals. Nevertheless, our results suggest that foamy viruses may be able to quickly acquire enhanced fitness through acquisition of a different *env* gene through recombination, which would, in the case studied here, result in increased cell-cell fusion and at the same time release of higher absolute infectivity and higher per-particle infectivity. Given the known very high propensity to template switching of foamy viruses [54], the possibility of such recombination events should be kept in mind in light of the well-known zoonotic potential [9–13,52]. Frequent exchange of the variant parts of *env* has been proposed as one possible explanation for the high evolutionary stability of the foamy virus genome and its diversity in the *env* gene [22]. However, it is currently unclear how often such recombination events occur in co-infected animals.

Interestingly, our plasmid-derived infectious clone of SFVmmu-DPZ9524 replicated measurably better in primary rhesus monkey fibroblasts than the plasmid-derived SFVmmu _R289hybAGM clone (Figure 1), despite its comparatively suboptimal *env* gene. This suggests that other factors in these two viral clones may balance or, in this case, rather overcompensate the effects of *env*.

Our findings regarding neutralization clearly also confirm for these two rhesus macaque SFV *env* genes that the *env* genes cluster into genotypes that almost certainly have been selected for by the humoral immune response. All but two out of 15 genotype 1-neutralizing sera did not affect genotype 2 Env-driven infection (Figure 4B,D). This is similar to observations in an earlier report [55]. Only three sera neutralized ST2, and we found only one that exclusively neutralized ST2 and not ST1 (Figure 4A,C). Even so, we suspect double infection and not cross-neutralization for the two dual-reactive sera, as there was absolutely no cross-neutralization visible for the serum that exclusively neutralized ST2 and double infection was reported to be frequent in macaques [55]. The degree of resistance to cross-neutralization observed with most sera is remarkable in that the differences in the Env amino acid sequences lie mostly within the short, proposed receptor binding domain (Figure 2B). This demonstrates that large parts of the foamy envelope protein are not targeted by neutralizing antibodies during natural infection.

Our findings regarding IFITM-mediated restriction suggest that entry of macaque SFVs may be somewhat sensitive to IFITM3 if expressed at sufficiently high levels (Figure 7A,B), in line with an endosomal route of entry as reported for foamy viruses [53,56]. The observed sensitivity to inhibition by IFITMs was highly similar between ST1 and ST2. In light of the notion that IFITMs primarily restrict at the membrane fusion step [32,33], one might have expected that the more fusogenic ST2 would be more resistant to IFITM-mediated restriction. Nevertheless, this was not readily observed, even if ST1 may have exhibited slightly increased sensitivity to overexpression of individual IFITMs, but not in a meaningfully different manner. While convenient as an experimental system and important to establish whether individual IFITMs in principle possess the ability to restrict a pathogen, overexpression of individual IFITM genes by means of retroviral transduction may lead to aberrant localization and expression levels that exceed those normally found in interferon-stimulated cells [30]. In line with this notion, ST1 and ST2 infection was only minimally altered by simultaneous knockout of IFITMs 1-3 in the A549 cell line (Figure 7F), which is often used to study the effect of IFITM expression [30,33,51], and the effect did not reach significance, neither with nor without prior interferon induction of IFITM expression. In addition, the relative increase in infection was not profoundly different between these conditions, different from what was observed for IAV HA/NA-driven infection and different from what would be expected after strong, interferon-mediated induction of IFITM expression. While this manuscript underwent revision, IFITM3 was reported to restrict the entry of prototype foamy virus [57], in principle compatible with our findings in overexpression experiments testing the two macacine Env proteins. We focused our studies here on the entry process and did not analyze whether IFITMs act at a later stage of infection as shown for FFV [34] and now also for prototype foamy virus [57], and if at such a later stage there are differences between the two env genes. This might be a topic for future studies. At least for cells that are infected through entry pathways similar to those used for infection of A549, our results suggest that IFITM-mediated restriction may play a minor role in the control of macaque foamy virus infection at the cell entry step as compared to the IFITM-mediated restriction observed for IAV glycoprotein-driven entry.

Supplementary Materials: The following supporting information can be downloaded at: https://www.mdpi.com/article/10.3390/v15020262/s1, Table S1: Nucleotide variants.

Author Contributions: Conceptualization, T.F. and A.S.H.; methodology, T.F. and A.S.H.; validation, T.F., A.E. and A.S.H.; formal analysis, T.F. and A.S.H.; investigation, T.F., S.S., S.L. and X.Y.; resources, A.K. and U.F.; data curation, T.F, A.E. and A.S.H.; writing—original draft preparation, A.S.H.; writing—review and editing, A.S.H. and T.F.; visualization, T.F.; supervision, A.S.H.; project administration, A.S.H.; funding acquisition, A.S.H. All authors have read and agreed to the published version of the manuscript.

Funding: This research was funded by an Exploration Grant from the Boehringer-Ingelheim Foundation to A.S.H. X.Y. was funded by China Petroleum Central Hospital and has no conflict of interest with the article. S.L. was funded by a scholarship of the China Scholarship Council (CSC), file number 202106300006.

Institutional Review Board Statement: The German Primate Center is permitted to house and breed non-human primates according to §11 of the German Animal Welfare Act. Permit was issued under license number 392001/7 by local veterinary authorities. Animals were housed under conditions in accordance with the German Animal Welfare Act guidelines on the use of nonhuman primates for biomedical research by the European Union and the Weatherall report.

Data Availability Statement: Not applicable.

Acknowledgments: We thank Felix Engel, Michael Farzan, and Stefan Pöhlmann for sharing reagents. We thank Rüdiger Behr for sharing primary rhesus monkey fibroblasts.

Conflicts of Interest: The authors declare no conflict of interest.

References

1. Switzer, W.M.; Salemi, M.; Shanmugam, V.; Gao, F.; Cong, M.-E.; Kuiken, C.; Bhullar, V.; Beer, B.E.; Vallet, D.; Gautier-Hion, A.; et al. Ancient Co-Speciation of Simian Foamy Viruses and Primates. *Nature* **2005**, *434*, 376–380. [CrossRef]
2. Katzourakis, A.; Gifford, R.J.; Tristem, M.; Gilbert, M.T.P.; Pybus, O.G. Macroevolution of Complex Retroviruses. *Science* **2009**, *325*, 1512. [CrossRef]
3. Han, G.-Z.; Worobey, M. Endogenous Viral Sequences from the Cape Golden Mole (*Chrysochloris asiatica*) Reveal the Presence of Foamy Viruses in All Major Placental Mammal Clades. *PLoS ONE* **2014**, *9*, e97931. [CrossRef] [PubMed]
4. Meiering, C.D.; Linial, M.L. Historical Perspective of Foamy Virus Epidemiology and Infection. *Clin. Microbiol. Rev.* **2001**, *14*, 165–176. [CrossRef]
5. Han, G.-Z.; Worobey, M. An Endogenous Foamy-Like Viral Element in the Coelacanth Genome. *PLoS Pathog.* **2012**, *8*, e1002790. [CrossRef]
6. Chen, Y.; Zhang, Y.-Y.; Wei, X.; Cui, J. Multiple Infiltration and Cross-Species Transmission of Foamy Viruses across the Paleozoic to the Cenozoic Era. *J Virol.* **2021**, *95*, e0048421. [CrossRef] [PubMed]
7. Naville, M.; Volff, J.-N. Endogenous Retroviruses in Fish Genomes: From Relics of Past Infections to Evolutionary Innovations? *Front. Microbiol.* **2016**, *7*, 1197. [CrossRef]
8. Aiewsakun, P. Avian and Serpentine Endogenous Foamy Viruses, and New Insights into the Macroevolutionary History of Foamy Viruses. *Virus Evol.* **2020**, *6*, vez057. [CrossRef]
9. Buseyne, F.; Betsem, E.; Montange, T.; Njouom, R.; Bilounga Ndongo, C.; Hermine, O.; Gessain, A. Clinical Signs and Blood Test Results among Humans Infected with Zoonotic Simian Foamy Virus: A Case-Control Study. *J. Infect. Dis.* **2018**, *218*, 144–151. [CrossRef] [PubMed]
10. Gessain, A.; Montange, T.; Betsem, E.; Bilounga Ndongo, C.; Njouom, R.; Buseyne, F. Case-Control Study of the Immune Status of Humans Infected with Zoonotic Gorilla Simian Foamy Viruses. *J. Infect. Dis.* **2020**, *221*, 1724–1733. [CrossRef]

11. Lambert, C.; Couteaudier, M.; Gouzil, J.; Richard, L.; Montange, T.; Betsem, E.; Rua, R.; Tobaly-Tapiero, J.; Lindemann, D.; Njouom, R.; et al. Potent Neutralizing Antibodies in Humans Infected with Zoonotic Simian Foamy Viruses Target Conserved Epitopes Located in the Dimorphic Domain of the Surface Envelope Protein. *PLoS Pathog.* **2018**, *14*, e1007293. [CrossRef] [PubMed]

12. Engel, G.A.; Small, C.T.; Soliven, K.; Feeroz, M.M.; Wang, X.; Kamrul Hasan, M.; Oh, G.; Rabiul Alam, S.M.; Craig, K.L.; Jackson, D.L.; et al. Zoonotic Simian Foamy Virus in Bangladesh Reflects Diverse Patterns of Transmission and Co-Infection. *Emerg. Microbes Infect.* **2013**, *2*, e58. [CrossRef] [PubMed]

13. Huang, F.; Wang, H.; Jing, S.; Zeng, W. Simian Foamy Virus Prevalence in Macaca Mulatta and Zookeepers. *AIDS Res. Hum. Retroviruses* **2012**, *28*, 591–593. [CrossRef]

14. Olszko, M.E.; Trobridge, G.D. Foamy Virus Vectors for HIV Gene Therapy. *Viruses* **2013**, *5*, 2585–2600. [CrossRef]

15. Rajawat, Y.S.; Humbert, O.; Kiem, H.-P. In-Vivo Gene Therapy with Foamy Virus Vectors. *Viruses* **2019**, *11*, 1091. [CrossRef]

16. Trobridge, G.D. Foamy Virus Vectors for Gene Transfer. *Expert Opin. Biol. Ther.* **2009**, *9*, 1427–1436. [CrossRef] [PubMed]

17. Williams, D.A. Foamy Virus Vectors Come of Age. *Mol. Ther.* **2008**, *16*, 635. [CrossRef]

18. Budzik, K.M.; Nace, R.A.; Ikeda, Y.; Russell, S.J. Evaluation of the Stability and Intratumoral Delivery of Foreign Transgenes Encoded by an Oncolytic Foamy Virus Vector. *Cancer Gene Ther.* **2022**, *29*, 1240–1251. [CrossRef] [PubMed]

19. Budzik, K.M.; Nace, R.A.; Ikeda, Y.; Russell, S.J. Oncolytic Foamy Virus—Generation and Properties of a Nonpathogenic Replicating Retroviral Vector System That Targets Chronically Proliferating Cancer Cells. *J. Virol.* **2021**, *95*, 10. [CrossRef]

20. Ensser, A.; Großkopf, A.K.; Mätz-Rensing, K.; Roos, C.; Hahn, A.S. Isolation and Sequence Analysis of a Novel Rhesus Macaque Foamy Virus Isolate with a Serotype-1-like Env. *Arch. Virol.* **2018**, *163*, 2507–2512. [CrossRef] [PubMed]

21. Nandakumar, S.; Bae, E.H.; Khan, A.S. Complete Genome Sequence of a Naturally Occurring Simian Foamy Virus Isolate from Rhesus Macaque (SFVmmu_K3T). *Genome Announc.* **2017**, *5*, 33. [CrossRef]

22. Aiewsakun, P.; Richard, L.; Gessain, A.; Mouinga-Ondémé, A.; Vicente Afonso, P.; Katzourakis, A. Modular Nature of Simian Foamy Virus Genomes and Their Evolutionary History. *Virus Evol.* **2019**, *5*, vez032. [CrossRef] [PubMed]

23. Duda, A.; Lüftenegger, D.; Pietschmann, T.; Lindemann, D. Characterization of the Prototype Foamy Virus Envelope Glycoprotein Receptor-Binding Domain. *J. Virol.* **2006**, *80*, 8158–8167. [CrossRef]

24. Flower, R.L.; Wilcox, G.E.; Cook, R.D.; Ellis, T.M. Detection and Prevalence of Serotypes of Feline Syncytial Spumaviruses. *Arch. Virol.* **1985**, *83*, 53–63. [CrossRef] [PubMed]

25. Winkler, I.G.; Flügel, R.M.; Löchelt, M.; Flower, R.L. Detection and Molecular Characterisation of Feline Foamy Virus Serotypes in Naturally Infected Cats. *Virology* **1998**, *247*, 144–151. [CrossRef] [PubMed]

26. Zemba, M.; Alke, A.; Bodem, J.; Winkler, I.G.; Flower, R.L.; Pfrepper, K.; Delius, H.; Flügel, R.M.; Löchelt, M. Construction of Infectious Feline Foamy Virus Genomes: Cat Antisera Do Not Cross-Neutralize Feline Foamy Virus Chimera with Serotype-Specific Env Sequences. *Virology* **2000**, *266*, 150–156. [CrossRef]

27. Blochmann, R.; Curths, C.; Coulibaly, C.; Cichutek, K.; Kurth, R.; Norley, S.; Bannert, N.; Fiebig, U. A Novel Small Animal Model to Study the Replication of Simian Foamy Virus in Vivo. *Virology* **2014**, *448*, 65–73. [CrossRef]

28. Alkhatib, G.; Combadiere, C.; Broder, C.C.; Feng, Y.; Kennedy, P.E.; Murphy, P.M.; Berger, E.A. CC CKR5: A RANTES, MIP-1alpha, MIP-1beta Receptor as a Fusion Cofactor for Macrophage-Tropic HIV-1. *Science* **1996**, *272*, 1955–1958. [CrossRef]

29. Feng, Y.; Broder, C.C.; Kennedy, P.E.; Berger, E.A. HIV-1 Entry Cofactor: Functional CDNA Cloning of a Seven-Transmembrane, G Protein-Coupled Receptor. *Science* **1996**, *272*, 872–877. [CrossRef]

30. Bailey, C.C.; Zhong, G.; Huang, I.-C.; Farzan, M. IFITM-Family Proteins: The Cell's First Line of Antiviral Defense. *Annu. Rev. Virol.* **2014**, *1*, 261–283. [CrossRef]

31. Huang, I.-C.; Bailey, C.C.; Weyer, J.L.; Radoshitzky, S.R.; Becker, M.M.; Chiang, J.J.; Brass, A.L.; Ahmed, A.A.; Chi, X.; Dong, L.; et al. Distinct Patterns of IFITM-Mediated Restriction of Filoviruses, SARS Coronavirus, and Influenza a Virus. *PLoS Pathog.* **2011**, *7*, e1001258. [CrossRef]

32. Li, K.; Markosyan, R.M.; Zheng, Y.-M.; Golfetto, O.; Bungart, B.; Li, M.; Ding, S.; He, Y.; Liang, C.; Lee, J.C.; et al. IFITM Proteins Restrict Viral Membrane Hemifusion. *PLoS Pathog.* **2013**, *9*, e1003124. [CrossRef]

33. Hörnich, B.F.; Großkopf, A.K.; Dcosta, C.J.; Schlagowski, S.; Hahn, A.S. Interferon-Induced Transmembrane Proteins Inhibit Infection by the Kaposi's Sarcoma-Associated Herpesvirus and the Related Rhesus Monkey Rhadinovirus in a Cell-Specific Manner. *mBio* **2021**, *12*, e0211321. [CrossRef]

34. Kim, J.; Shin, C.-G. IFITM Proteins Inhibit the Late Step of Feline Foamy Virus Replication. *Anim. Cells Syst.* **2020**, *24*, 282–288. [CrossRef]

35. Feeroz, M.M.; Soliven, K.; Small, C.T.; Engel, G.A.; Andreina Pacheco, M.; Yee, J.L.; Wang, X.; Kamrul Hasan, M.; Oh, G.; Levine, K.L.; et al. Population Dynamics of Rhesus Macaques and Associated Foamy Virus in Bangladesh. *Emerg. Microbes Infect.* **2013**, *2*, e29. [CrossRef]

36. Hahn, A.S.; Desrosiers, R.C. Rhesus Monkey Rhadinovirus Uses Eph Family Receptors for Entry into B Cells and Endothelial Cells but not Fibroblasts. *PLoS Pathog.* **2013**, *9*, e1003360. [CrossRef] [PubMed]

37. Altschul, S.F.; Gish, W.; Miller, W.; Myers, E.W.; Lipman, D.J. Basic Local Alignment Search Tool. *J. Mol. Biol.* **1990**, *215*, 403–410. [CrossRef]

38. Campeau, E.; Ruhl, V.E.; Rodier, F.; Smith, C.L.; Rahmberg, B.L.; Fuss, J.O.; Campisi, J.; Yaswen, P.; Cooper, P.K.; Kaufman, P.D. A Versatile Viral System for Expression and Depletion of Proteins in Mammalian Cells. *PLoS ONE* **2009**, *4*, e6529. [CrossRef] [PubMed]

39. Hörnich, B.F.; Großkopf, A.K.; Schlagowski, S.; Tenbusch, M.; Kleine-Weber, H.; Neipel, F.; Stahl-Hennig, C.; Hahn, A.S. SARS-CoV-2 and SARS-CoV Spike-Mediated Cell-Cell Fusion Differ in Their Requirements for Receptor Expression and Proteolytic Activation. *J. Virol.* **2021**, *95*, 9. [CrossRef] [PubMed]

40. Großkopf, A.K.; Schlagowski, S.; Fricke, T.; Ensser, A.; Desrosiers, R.C.; Hahn, A.S. Plxdc Family Members Are Novel Receptors for the Rhesus Monkey Rhadinovirus (RRV). *PLoS Pathog.* **2021**, *17*, e1008979. [CrossRef]

41. Sanjana, N.E.; Shalem, O.; Zhang, F. Improved Vectors and Genome-Wide Libraries for CRISPR Screening. *Nat. Meth.* **2014**, *11*, 783–784. [CrossRef]

42. Wrensch, F.; Karsten, C.B.; Gnirß, K.; Hoffmann, M.; Lu, K.; Takada, A.; Winkler, M.; Simmons, G.; Pöhlmann, S. Interferon-Induced Transmembrane Protein-Mediated Inhibition of Host Cell Entry of Ebolaviruses. *J. Infect. Dis.* **2015**, *212*, S210–S218. [CrossRef] [PubMed]

43. Großkopf, A.K.; Ensser, A.; Neipel, F.; Jungnickl, D.; Schlagowski, S.; Desrosiers, R.C.; Hahn, A.S. A Conserved Eph Family Receptor-Binding Motif on the GH/GL Complex of Kaposi's Sarcoma-Associated Herpesvirus and Rhesus Monkey Rhadinovirus. *PLoS Pathog.* **2018**, *14*, e1006912. [CrossRef]

44. Schmittgen, T.D.; Livak, K.J. Analyzing Real-Time PCR Data by the Comparative C(T) Method. *Nat. Protoc.* **2008**, *3*, 1101–1108. [CrossRef]

45. Bastone, P.; Löchelt, M. Kinetics and Characteristics of Replication-Competent Revertants Derived from Self-Inactivating Foamy Virus Vectors. *Gene Ther.* **2004**, *11*, 465–473. [CrossRef] [PubMed]

46. Meiering, C.D.; Linial, M.L. Reactivation of a Complex Retrovirus Is Controlled by a Molecular Switch and Is Inhibited by a Viral Protein. *Proc. Natl. Acad. Sci. USA* **2002**, *99*, 15130–15135. [CrossRef] [PubMed]

47. Kaul, A.; Schönmann, U.; Pöhlmann, S. Seroprevalence of Viral Infections in Captive Rhesus and Cynomolgus Macaques. *Primate Biol.* **2019**, *6*, 1–6. [CrossRef] [PubMed]

48. Plochmann, K.; Horn, A.; Gschmack, E.; Armbruster, N.; Krieg, J.; Wiktorowicz, T.; Weber, C.; Stirnnagel, K.; Lindemann, D.; Rethwilm, A.; et al. Heparan Sulfate Is an Attachment Factor for Foamy Virus Entry. *J. Virol.* **2012**, *86*, 10028–10035. [CrossRef] [PubMed]

49. Yu, S.F.; Sullivan, M.D.; Linial, M.L. Evidence That the Human Foamy Virus Genome Is DNA. *J. Virol.* **1999**, *73*, 1565–1572. [CrossRef]

50. John, S.P.; Chin, C.R.; Perreira, J.M.; Feeley, E.M.; Aker, A.M.; Savidis, G.; Smith, S.E.; Elia, A.E.H.; Everitt, A.R.; Vora, M.; et al. The CD225 Domain of IFITM3 Is Required for Both IFITM Protein Association and Inhibition of Influenza a Virus and Dengue Virus Replication. *J. Virol.* **2013**, *87*, 7837–7852. [CrossRef] [PubMed]

51. Williams, D.E.J.; Wu, W.-L.; Grotefend, C.R.; Radic, V.; Chung, C.; Chung, Y.-H.; Farzan, M.; Huang, I.-C. IFITM3 Polymorphism Rs12252-C Restricts Influenza a Viruses. *PLoS ONE* **2014**, *9*, e110096. [CrossRef] [PubMed]

52. Couteaudier, M.; Montange, T.; Njouom, R.; Bilounga-Ndongo, C.; Gessain, A.; Buseyne, F. Plasma Antibodies from Humans Infected with Zoonotic Simian Foamy Virus Do Not Inhibit Cell-to-Cell Transmission of the Virus despite Binding to the Surface of Infected Cells. *PLoS Pathog.* **2022**, *18*, e1010470. [CrossRef]

53. Stirnnagel, K.; Schupp, D.; Dupont, A.; Kudryavtsev, V.; Reh, J.; Müllers, E.; Lamb, D.C.; Lindemann, D. Differential PH-Dependent Cellular Uptake Pathways among Foamy Viruses Elucidated Using Dual-Colored Fluorescent Particles. *Retrovirology* **2012**, *9*, 71. [CrossRef]

54. Gärtner, K.; Wiktorowicz, T.; Park, J.; Mergia, A.; Rethwilm, A.; Scheller, C. Accuracy Estimation of Foamy Virus Genome Copying. *Retrovirology* **2009**, *6*, 32. [CrossRef] [PubMed]

55. O'Brien, T.C.; Albrecht, P.; Hannah, J.E.; Tauraso, N.M.; Robbins, B.; Trimmer, R.W. Foamy Virus Serotypes 1 and 2 in Rhesus Monkey Tissues. *Arch. Gesamte Virusforsch.* **1972**, *38*, 216–224. [CrossRef] [PubMed]

56. Berka, U.; Hamann, M.V.; Lindemann, D. Early Events in Foamy Virus-Host Interaction and Intracellular Trafficking. *Viruses* **2013**, *5*, 1055–1074. [CrossRef] [PubMed]

57. Wang, Z.; Tuo, X.; Zhang, J.; Chai, K.; Tan, J.; Qiao, W. Antiviral Role of IFITM3 in Prototype Foamy Virus Infection. *Virol. J.* **2022**, *19*, 195. [CrossRef]

Article

ZDHHC11 Suppresses Zika Virus Infections by Palmitoylating the Envelope Protein

Dingwen Hu [1,†], Haimei Zou [2,†], Weijie Chen [3,†], Yuting Li [4], Ziqing Luo [4], Xianyang Wang [4], Dekuan Guo [4], Yu Meng [4], Feng Liao [4], Wenbiao Wang [5], Ying Zhu [1,*], Jianguo Wu [1,3,*] and Geng Li [3,4,*]

1   State Key Laboratory of Virology, College of Life Sciences, Wuhan University, Wuhan 430072, China
2   The Second Affiliated Hospital, Guangzhou University of Chinese Medicine, Guangzhou 510120, China
3   Guangdong Provincial Key Laboratory of Virology, Institute of Medical Microbiology, Jinan University, Guangzhou 510632, China
4   Laboratory Animal Center, Guangzhou University of Chinese Medicine, Guangzhou 510006, China
5   Medical Research Center, Guangdong Provincial People's Hospital, Guangdong Academy of Medical Sciences, Guangzhou 510080, China
*   Correspondence: yingzhu@whu.edu.cn (Y.Z.); jwu@whu.edu.cn (J.W.); lg@gzucm.edu.cn (G.L.); Tel.: +86-020-39358550 (G.L.)
†   These authors contributed equally to this work.

Abstract: Zika virus (ZIKV) is an RNA-enveloped virus that belongs to the *Flavivirus genus*, and ZIKV infections potentially induce severe neurodegenerative diseases and impair male fertility. Palmitoylation is an important post-translational modification of proteins that is mediated by a series of DHHC-palmitoyl transferases, which are implicated in various biological processes and viral infections. However, it remains to be investigated whether palmitoylation regulates ZIKV infections. In this study, we initially observed that the inhibition of palmitoylation by 2-bromopalmitate (2-BP) enhanced ZIKV infections, and determined that the envelope protein of ZIKV is palmitoylated at Cys308. ZDHHC11 was identified as the predominant enzyme that interacts with the ZIKV envelope protein and catalyzes its palmitoylation. Notably, ZDHHC11 suppressed ZIKV infections in an enzymatic activity-dependent manner and ZDHHC11 knockdown promoted ZIKV infection. In conclusion, we proposed that the envelope protein of ZIKV undergoes a novel post-translational modification and identified a distinct mechanism in which ZDHHC11 suppresses ZIKV infections via palmitoylation of the ZIKV envelope protein.

Keywords: Zika virus; palmitoylation; 2-BP; envelope protein; ZDHHC11

Citation: Hu, D.; Zou, H.; Chen, W.; Li, Y.; Luo, Z.; Wang, X.; Guo, D.; Meng, Y.; Liao, F.; Wang, W.; et al. ZDHHC11 Suppresses Zika Virus Infections by Palmitoylating the Envelope Protein. *Viruses* **2023**, *15*, 144. https://doi.org/10.3390/v15010144

Academic Editors: Yuliang Liu and Yiping Li

Received: 22 November 2022
Revised: 27 December 2022
Accepted: 29 December 2022
Published: 2 January 2023

1. Introduction

The Zika virus (ZIKV) was initially detected in the Rhesus macaque in Uganda in 1947, and was endemic in the Americas from 2014 to 2015 [1]. ZIKV belongs to the *Flavivirus genus*, and, like the dengue virus, the clinical symptoms of ZIKV infections are self-limited and primarily include fever, headache, and muscle aches, among other symptoms [2]. However, ZIKV infections are also linked to microcephaly in neonates [3], Guillain–Barré syndrome in adults [4], and even male infertility [5], which has raised global concerns. Similar to other flaviviruses, ZIKV is an enveloped virus with a single-stranded RNA genome of approximately 11 kb [6]. Following infection, the viral genome is injected into susceptible host cells and is translated into a large polyprotein, which is cleaved into three structural proteins, namely, capsid, envelope, and pr-Membrane proteins, and seven non-structural (NS) proteins, namely, NS1, NS2A, NS2B, NS3, NS4A, NS4B, and NS5 [6]. The NS proteins are primarily associated with genome replication, immune evasion, and pathological damage [7–9], while the structural proteins are mainly involved in receptor binding, genome packaging, and virion assembly [10].

The envelope protein of ZIKV is a crucial multifunctional protein that forms the viral surface along with the Membrane protein, and facilitates viral entry by binding to cell surface receptors and promotes membrane fusion [11]. The envelope protein is the primary viral protein that induces host neutralizing antibodies [11]. The functions of the envelope protein are therefore strictly regulated by various post-translational modification systems of host cells. The TRIM7-mediated ubiquitination of the envelope protein facilitates viral entry and pathogenesis, and thus promotes viral infections; however, the host deubiquitination system catalyzes deubiquitination of the envelope protein via a series of ubiquitin-specific proteases, and thus exerts an antiviral effect [12]. Furthermore, a previous study reported that glycosylation of the envelope protein of ZIKV at Asn154 mediates receptor binding [13].

Palmitoylation is a reversible protein post-translational modification in which a palmitate moiety is attached to the cysteine residues of target proteins, and contributes to protein trafficking, stability, and functionality [14]. The palmitoylation of proteins is mediated by a series of enzymes that contain a zinc finger Asp–His–His–Cys (ZDHHC) motif, which are named ZDHHC1 to ZDHHC9 and ZDHHC11 to ZDHHC24 [14]. It has been demonstrated that the ZDHHC11 protein anchors to the endoplasmic reticulum (ER) and regulates the oligomerization of TRFA6 to induce the activation of NF-kB [15]; ZDHHC11 also promotes MITA-IRF3 interactions for facilitating the immune response following infections with DNA viruses [16].

It has been identified that the palmitoylation of proteins plays an essential role in the regulation of tumorigenesis, immune response, and inflammatory reactions [17–19]. Viral proteins also undergo palmitoylation for regulating viral infections. For instance, palmitoylation of the spike protein of severe acute respiratory syndrome coronavirus 2 (SARS-CoV-2) is essential for viral infectivity [20,21], while palmitoylation of the hemagglutinin (HA) protein of influenza virus is crucial for membrane fusion [22,23]. However, it remains to be investigated whether the proteins of ZIKV are modified by palmitoylation.

The findings of this study demonstrated that the presence of a palmitoylation inhibitor enhanced ZIKV infection and that the envelope protein of ZIKV is palmitoylated at Cys308. ZDHHC11 was identified as the predominant enzyme that interacts with the envelope protein and mediates its palmitoylation. Notably, the study revealed that the overexpression of ZDHHC11 suppressed ZIKV infection, while ZDHHC11 knockdown enhanced viral infection.

2. Materials and Methods

2.1. Cell Lines and Cultures

HEK293T cells, Vero cells, and C6/36 cells were purchased from the American Type Culture Collection (ATCC), and U251 cells were purchased from China Center for Type Culture Collection (CCTCC). BHK-21 cells were kindly gifted by Dr. Cheng-Feng Qin of the Beijing Institute of Microbiology and Epidemiology, China. The HEK293T cells, Vero cells, U251 cells, and BHK-21 cells were maintained in Dulbecco's modified Eagle's medium (DMEM) supplemented with 10% fetal bovine serum (FBS) and 1% Penicillin-Streptomycin at 37 °C in a humidified incubator containing 5% CO_2. The C6/36 cells were maintained in RPMI 1640 medium supplemented with 10% FBS and 1% Penicillin-Streptomycin at 30 °C in a humidified incubator containing 5% CO_2.

2.2. Antibodies and Reagents

Antibodies against Flag (F3165, 1:2000) and HA (H6908, 1:5000) were purchased from Sigma (Burlington, MA, USA). Antibodies against the envelope protein of ZIKV (GTX133314, 1:1000), NS1 protein of ZIKV (GTX133307, 1:1000), glyceraldehyde-3-phosphate dehydrogenase (GAPDH; GTX100118, 1:5000), and ZDHHC11 (GTX106800, 1:1000) were purchased from GeneTex (Hsinchu City, Taiwan, P.R.C). For the immunofluorescence studies, antibodies against the envelope protein of flaviviruses (ab214333, 1:100) were purchased from Abcam (Shanghai, China). N-ethylmaleimide (E3876), hydroxylamine hydrochloride (255580), and 2-bromopalmitate (2-BP; 21604) were obtained from Sigma (Burlington, MA, USA). BMCC-biotin (21900) was purchased from Thermo Fisher Scientific (Waltham, MA, USA). Streptavidin-horseradish peroxidase (M00091) was purchased from

Genscript (Piscataway, NJ, USA). Fluorescein (FITC)-conjugated AffiniPure goat anti-mouse immunoglobulin (Ig)G (H+L) (SA00003-1) and Cy3-conjugated AffiniPure goat anti-rabbit IgG (H+L) (SA00009-2) were purchased from ProteinTech (Rosemont, IL, USA).

2.3. Viruses

Information pertaining to ZIKV and protocols used for the culture of ZIKV in this study have been described in our previous studies [24].

2.4. Plasmid Construction

The Flag-tagged ZDHHC1–ZDHHC24 expression plasmids were purchased from WZ Biosciences (Columbia, MD, USA). The expression plasmids for the ZIKV capsid, prMembrane/membrane, and envelope protein were constructed as described in our previous study [24].

2.5. RNA Interference Studies

Short interfering RNAs (siRNAs) targeting the human ZDHHC11 mRNA were designed and synthesized by HippoBio and transfected to U251 cells using lipofectamine 2000 (Invitrogen, CA, USA, 11668019), according to the manufacturer's instructions. The following sequences were used for the RNA interference studies:

si-ZDHHC11-1: 5′-CGCAAAGAAGAGAGUUCAA-3′
si-ZDHHC11-2: 5′-GGUAUGAAGAUGUCAAGAA-3′
si-ZDHHC11-3: 5′-GCCAAGAAGAUGACCACCUUU-3′

2.6. Immunoblotting and Immunoprecipitation Studies

The cells were lysed in lysis buffer (150 mM NaCl, 50 mM Tris–HCl, 5 mM EDTA, 10% glycerol, and 1% Triton X-100) with a complete protease inhibitor cocktail. Western blot analysis was performed using the indicated antibodies. For immunoprecipitation, each lysate was incubated overnight with the indicated antibody with rotation at 4 °C, following which protein A/G agarose (Pierce) was added for capturing the antibody complex. The beads were finally washed thrice with lysis buffer and eluted through incubation at 100 °C for 10 min in 70 μL of 2× loading buffer. The eluted samples were assayed with immunoblotting.

2.7. Quantitative Real-Time Polymerase Chain Reaction (PCR)

The total RNA was extracted with Trizol reagent and quantified using real-time PCR as previously described [24]. The following primers were used for PCR:

ZIKV-forward: 5′-GGTCAGCGTCCTCTCTAATAAACG-3′
ZIKV-reverse: 3′-GCACCCTAGTGTCCACTTTTTCC-5′
GAPDH-forward: 5′-ATGACATCAAGAAG GTGGTG-3′
GAPDH-reverse: 3′-CATACCAGGAAATGAGCTTG-5′
ZDHHC11-forward: 5′-GATCTTCTCGTTCCACCTCGT-3′
ZDHHC11-reverse: 3′-TGTGCATGTTTTGATCTGTCGAA-5′

2.8. Immunofluorescence Analysis

The cells were washed with phosphate-buffered saline (PBS), supplemented with 0.1% bovine serum albumin (BSA), and subsequently fixed with 4% paraformaldehyde for 15 min. The cells were washed thrice with PBS, supplemented with 0.1% BSA, permeabilized with 0.1% Triton X-100 for 5 min at room temperature, and incubated with blocking solution (PBS, 5% BSA) for an additional 30 min. The cells were subsequently incubated overnight with the primary antibody at 4 °C, washed thrice with PBS, and further incubated with the secondary FITC-conjugated antibody or Cy3 for 30 min. The slides were washed thrice with PBS and mounted onto glass coverslips using Prolong Gold Antifade Reagent with DAPI. The processed slides were imaged using a Zeiss LSM710 confocal microscope.

2.9. Acyl-Biotin Exchange (ABE) Assay

The ABE assay was performed for determining palmitoylation, as previously described [17]. Briefly, the cells were lysed using lysis buffer, containing 50 μM N-ethylmaleimide, and the target proteins were immunoprecipitated by adding the corresponding antibody. The immunoprecipitated complexes were resuspended with 1 M hydroxylamine hydrochloride buffer for 1 h at room temperature, and subsequently washed five times with lysis buffer (pH = 6.2). The beads were then resuspended with BMCC-biotin (5 μM) buffer for 2 h at room temperature and subsequently washed five times. The complexes were analyzed with sodium dodecyl-sulfate polyacrylamide gel electrophoresis (SDS-PAGE) and detected using the indicated antibody.

2.10. Statistics Analysis

All the data were analyzed using GraphPad Prism v8.4.0. Statistical significance was determined using Student's *t* test (*, $p < 0.05$; **, $p < 0.01$; ***, $p < 0.001$).

3. Results

3.1. 2-BP Enhances ZIKV Infection in Vero Cells

We have previously reported several host factors that target Zika virus envelope protein ubiquitination [12], a kind of dynamic protein post-translational modification, to inhibit viral infection [24,25]. Palmitoylation is an important post-translational modification of proteins, and viral proteins undergo palmitoylation during viral infections. In this study, we initially investigated the role of palmitoylation in ZIKV infections. The palmitate analogue, 2-BP, is widely used in studies on protein palmitoylation; it irreversibly inhibits the enzymatic activity of all palmitoyltransferases and serves as a non-selective inhibitor of protein palmitoylation [26]. To this end, we first evaluated the cytotoxicity of 2-BP at various concentrations in Vero cells using CCK-8 assays, and the results demonstrated that 2-BP was not cytotoxic at concentrations below 40 μM (Figure 1A). The ZIKV-infected Vero cells were subsequently treated with 2-BP. Interestingly, the results of real-time quantitative PCR and immunoblotting studies demonstrated that the levels of viral RNA and expression of viral proteins increased significantly following treatment with 2-BP (Figure 1B,C). The findings were consistent with the results of plaque assays, which revealed that treatment with 2-BP enhanced ZIKV infection in Vero cells (Figure 1D,E). These findings suggested that the inhibition of palmitoylation promotes ZIKV infections, and that the palmitoylation of viral proteins suppresses ZIKV infection.

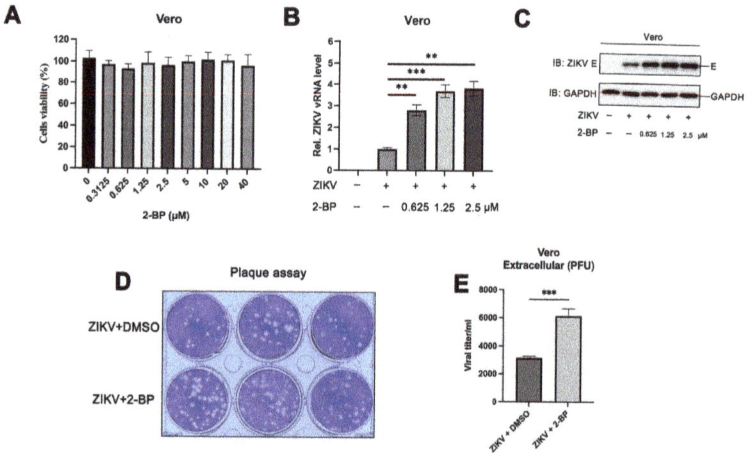

Figure 1. The palmitoylation inhibitor, 2-BP, enhances ZIKV infection in Vero cells. (**A**) Vero cells were treated with DMSO or indicated concentrations of 2-BP for 24 h, following which the cell viability was

detected using CCK-8 assays. (**B–E**) Vero cells were treated with DMSO or indicated concentrations of 2-BP for 6 h and subsequently infected with ZIKV (multiplicity of infection (MOI) = 1) using incubation for 24 h. (**B**) The levels of viral RNA were measured with qPCR. (**C**) The content of viral proteins was detected using immunoblotting studies, and (**D,E**) the ZIKV titer in the supernatants was calculated with plaque assays. (**, $p < 0.01$; ***, $p < 0.001$).

3.2. The ZIKV Envelope Protein Is Palmitoylated at Cys308

In order to determine the mechanism by which protein palmitoylation regulates ZIKV infections, we attempted to determine whether ZIKV proteins undergo palmitoylation in host cells. ABE assays were performed using HEK293T cells expressing the structural proteins of ZIKV, including the capsid, pre-Membrane, and envelope proteins. The results demonstrated that the envelope protein underwent palmitoylation, while the capsid and pre-Membrane proteins were not palmitoylated (Figure 2A). Palmitoylation typically occurs at the cysteine residues of target proteins, and the envelope protein of ZIKV contains 13 cysteine residues (Figure 2B). In order to determine the specific residue of the ZIKV envelope protein that undergoes palmitoylation, we initially used the GPS-Palm palmitoylation site prediction system [27], which revealed that Cys3, Cys116, and Cys308 were the most likely sites for palmitoylation (Figure 2C). Of these, Cys3 is located in the central-barrel like Domain I (yellow), Cys116 is located in the elongated finger-like structure Domain II (red), and Cys308 is located in the immunoglobulin-like module Domain III (blue) (Figure 2B,D). These cysteine residues were subsequently substituted with a serine, and ABE assays were conducted in HEK293T cells overexpressing the wild-type (WT) or mutated envelope proteins. The results demonstrated that the palmitoylation level of only the Cys308Ser mutant was significantly reduced compared to that of the WT envelope protein (Figure 2E). The finding suggested that the envelope protein of ZIKV is palmitoylated at Cys308.

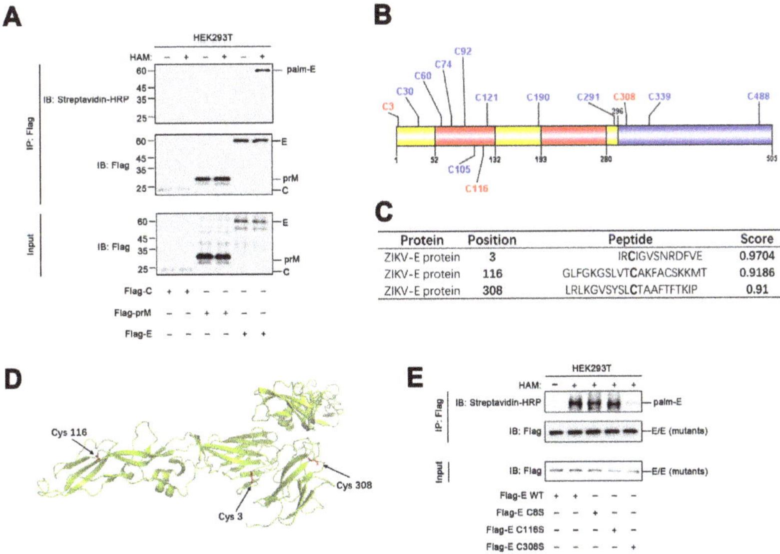

Figure 2. The envelope protein of ZIKV is palmitoylated at Cys308. (**A**) HEK293T cells were transfected with a plasmid encoding the ZIKV capsid (C), prMembrane (prM), and envelope (E) proteins for 24 h, lysed for ABE assays, and detected using immunoblotting with the indicated antibodies. (**B**) Schematic depicting all the 13 cysteine residues in the structure of the envelope protein.

(**C**) Possible palmitoylation sites in the envelope protein of ZIKV were predicted using the GPS-Palm system. (**D**) The predicted cysteine residues are highlighted in the structure of the envelope protein. (**E**) HEK293T cells were transfected with plasmids encoding the WT or mutated envelope proteins for 24 h, lysed for ABE assays, and the palmitoylation states were detected using immunoblotting with the indicated antibodies.

3.3. ZDHHC11 Mediates the Palmitoylation of the ZIKV Envelope Protein

Protein palmitoylation is generally catalyzed by a group of enzymes containing the conserved zinc finger Asp–His–His–Cys (ZDHHC) motif [14]. In order to determine the specific enzyme that catalyzes the palmitoylation of the envelope protein, we constructed expression plasmids encoding 23 palmitoyltransferases, including ZDHHC1 to ZDHHC9 and ZDHHC11 to ZDHHC24. HEK293T cells were subsequently transfected with the plasmids expressing the envelope protein and the ZDHHCs. The results of co-immunoprecipitation experiments revealed that ZDHHC9, ZDHHC11, and ZDHHC22 were associated with the envelope protein of ZIKV (Figure 3A), which suggested that ZDHHC9, ZDHHC11, and ZDHHC22 could be involved in the palmitoylation of the ZIKV envelope protein. In order to elucidate the specific enzyme that catalyzes the palmitoylation of the ZIKV envelope protein, HEK293T cells were transfected with plasmids expressing the envelope protein, ZDHHC9, ZDHHC11, and ZDHHC22, and subjected to ABE assays. The findings revealed that the level of palmitoylation of the ZIKV envelope protein was markedly enhanced by ZDHHC11 but not by ZDHHC9 and ZDHHC22 (Figure 3B). Additional ABE assays confirmed that the palmitoylation of the envelope protein was remarkably induced when the expression of ZDHHC11 was increased (Figure 3C), indicating that ZDHHC11 mediated the palmitoylation of the ZIKV envelope. As aforementioned, the level of palmitoylation of the envelope protein reduced significantly when Cys308 was substituted to serine (Figure 2E), and ZDHHC11 failed to induce palmitoylation of the mutant Cys308Ser envelope protein (Figure 3D). Taken together, these findings demonstrated that ZDHHC11 mediated the palmitoylation of the ZIKV envelope protein at Cys308.

3.4. ZDHHC11 Interacts with the ZIKV Envelope Protein

In order to determine the palmitoylation of the envelope protein by ZDHHC11, co-immunoprecipitation experiments were performed using HEK293T cells transfected with plasmids expressing the ZIKV envelope protein and ZDHHC11. The results demonstrated that ZDHHC11 interacted with the envelope protein (Figure 4A), and that the ZIKV envelope protein co-immunoprecipitated with ZDHHC11 (Figure 4B). Confocal laser scanning microscopy (CLSM) indicated that the envelope protein and ZDHHC11 co-localized and formed large spots in the cytoplasm of HEK293T cells (Figure 4C). Based on structural analyses of the ZIKV envelope protein, we constructed seven plasmids encoding different truncated envelope proteins (Figure 4G) for identifying the specific region of the ZIKV envelope that binds to ZDHHC11. The results of co-immunoprecipitation assays revealed that, similar to the full-length ZIKV envelope protein (1–505 aa), the truncated 52–505 and 132–505 aa envelope proteins could interact with ZDHHC11; however, the truncated 193–505 and 280–505 aa envelope proteins did not interact with ZDHHC11 (Figure 4D). Additionally, three truncated envelope proteins (1–193, 1–296, and 1–406 aa) interacted with ZDHHC11 (Figure 4E). These results demonstrated that the partial Domain I region (132–193 aa) of the ZIKV envelope protein mediates the envelope-ZDHHC11 interactions (Figure 4G). The results of co-immunoprecipitation studies also demonstrated that the N-terminal region (1–197 aa) of ZDHHC11 is required for binding to the ZIKV envelope (Figure 4F,H). Taken together, these results suggested that ZDHHC11 binds to the Domain I region (132–193 aa) of the ZIKV envelope via its N-terminal region (1–197 aa) to subsequently catalyze the palmitoylation of the envelope protein.

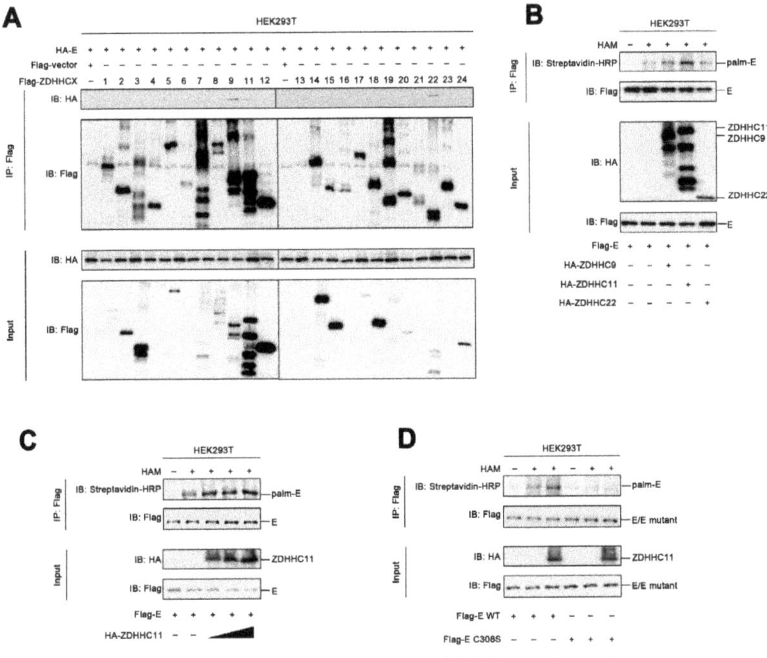

Figure 3. ZDHHC11 mediates the palmitoylation of the ZIKV envelope protein. (**A**) HEK293T cells were transfected with plasmids encoding the envelope protein and ZDHHCXs for 24 h, lysed for co-immunoprecipitation assays using the Flag antibody, and detected using immunoblotting studies with the indicated antibodies. (**B**) HEK293T cells were transfected with plasmids encoding the envelope protein, ZDHHC9, ZDHHC11, and ZDHHC22 for 24 h, lysed for ABE assays, and detected using immunoblotting with the indicated antibodies. (**C**) HEK293T cells were transfected with plasmids encoding the envelope protein and ZDHHC11 for 24 h, lysed for ABE assays, and detected using immunoblotting with the indicated antibodies. (**D**) HEK293T cells were transfected with plasmids encoding the WT or mutant envelope proteins and ZDHHC11 for 24 h, lysed for ABE assays, and detected using immunoblotting with the indicated antibodies.

3.5. Overexpression of ZDHHC11 Inhibits ZIKV Infection

The aforementioned results demonstrated that the ZIKV envelope protein is palmitoylated by ZDHHC11, and treatment with the palmitoylation inhibitor, 2-BP, promotes ZIKV infection. In this study, we also investigated the role of ZDHHC11 in ZIKV infections. The results demonstrated that the expression of the envelope protein and NS1 in ZIKV-infected U251 cells transfected with expression plasmids encoding ZDHHC11 was attenuated as the expression of ZDHHC11 increased (Figure 5A), while the levels of viral RNA reduced markedly as the ZDHHC11 mRNA levels increased (Figure 5B,C). These findings were consistent with the results of immunofluorescence staining, which revealed that the content of viral proteins decreased significantly in the presence of ZDHHC11 (Figure 5D). The viral titer in the supernatant of ZIKV-infected U251 cells was also determined and quantified using plaque assays, and the results demonstrated that ZIKV infection was dramatically attenuated following the overexpression of ZDHHC11 (Figure 5E,F). Consistent with this, the overexpression of ZDHHC11 also inhibited ZIKV infection in A549 cells (Figure 5G). We also generated an inactivated form of ZDHHC11 (ZDHHS11) by inducing a mutation in the DHHC motif, and evaluated its effect on ZIKV infection. The results of real-time quantitative PCR and immunoblotting studies demonstrated that the levels of viral RNA and the expression of viral proteins decreased significantly in the presence of ZDHHC11 but not ZDHHS11 in ZIKV-infected U251 cells (Figure 5H,I). The findings indicated that the enzymatic activity of ZDHHC11 is essential for the anti-ZIKV effect. Overall, these results suggested that the overexpression of ZDHHC11 inhibits ZIKV infection in an enzymatic activity-dependent manner.

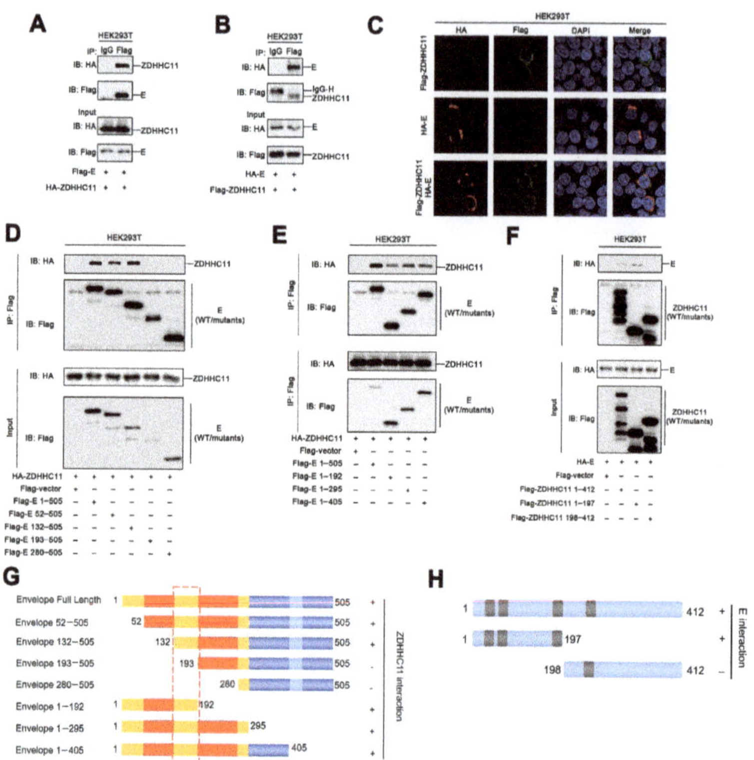

Figure 4. ZDHH11 interacts with the envelope protein of ZIKV. (**A,B**) HEK293T cells were transfected with plasmids encoding the envelope protein and ZDHHC11 for 24 h, lysed for co-immunoprecipitation assays with the indicated antibodies, and detected using immunoblotting with the indicated antibodies. (**C**) HEK293T cells were transfected with HA-E or Flag-ZDHHC11, or co-transfected with HA-E and Flag-ZDHHC11. The sub-cellular localizations of HA-E (red), Flag-ZDHHC11 (green), and the nuclear marker, DAPI (blue) were analyzed with CLSM. (**D,E**) HEK293T cells were transfected with plasmids encoding ZDHHC11 and the WT or truncated envelope proteins for 24 h, lysed for co-immunoprecipitation assays with the indicated antibodies, and detected using immunoblotting with the indicated antibodies. (**F**) HEK293T cells were transfected with plasmids encoding the envelope protein and ZDHHC11 or truncated ZDHHC11 for 24 h, lysed for co-immunoprecipitation assays with the indicated antibodies, and detected using immunoblotting with the indicated antibodies. (**G**) Diagrammatic representation of the full-length and truncated envelope protein. Domain I, II, and III are marked by yellow, red and blue. (**H**) Diagrammatic representation of the full-length and truncated ZDHHC11 enzyme.

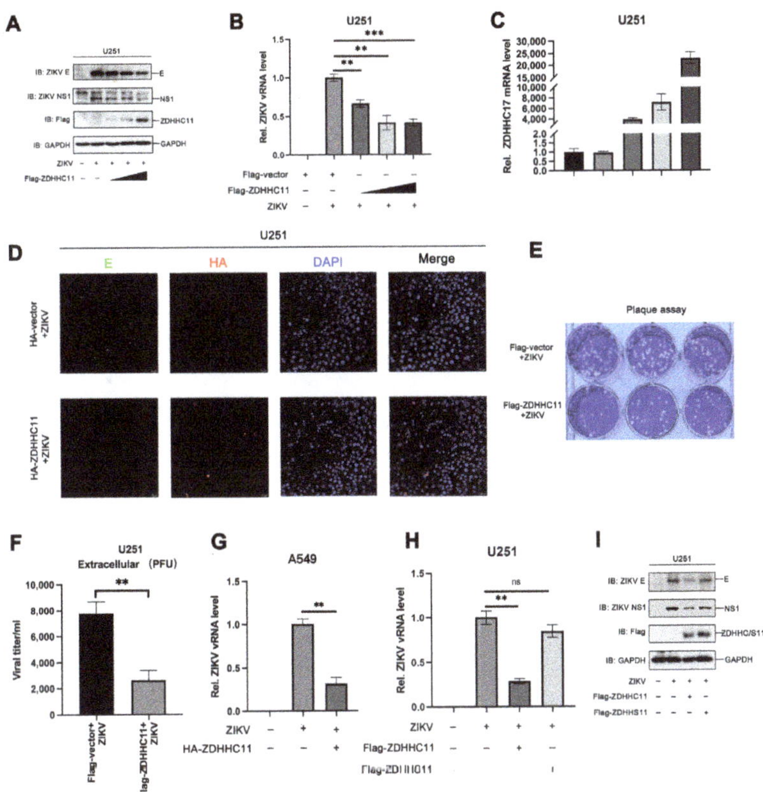

Figure 5. The overexpression of ZDHHC11 inhibits ZIKV infection. (**A–F**) U251 cells were transfected with plasmids encoding Flag-ZDHHC11 or control vector for 12 h and subsequently infected with ZIKV (MOI = 1) using incubation for 48 h. The content of viral proteins was detected by (**A**) immunoblotting and (**D**) confocal microscopy. The levels of (**B**) viral RNA and (**C**) ZDHHC11 mRNA were measured by quantitative PCR. (**E,F**) The titer of ZIKV in the supernatants was calculated using plaque assays. (**G**) A549 cells were transfected with plasmids encoding HA-ZDHHC11 or control vector for 12 h and subsequently infected with ZIKV (MOI = 1) using incubation for 48 h. The content of viral RNA was detected by qPCR. (**H,I**) U251 cells were transfected with plasmids encoding Flag-ZDHHC11, Flag-ZDHHS11, or the control vector for 12 h and subsequently infected with ZIKV (MOI = 1) using incubation for 48 h. (**H**) The levels of viral RNA were measured by quantitative PCR and (**I**) the content of viral proteins was detected using immunoblotting. (ns, no significance; **, $p < 0.01$; ***, $p < 0.001$).

3.6. ZDHHC11 Knockdown Enhances ZIKV Infection

To further confirm the role of ZDHHC11 in ZIKV infections, the endogenous ZD-HHC11 in U251 cells was knocked down using an RNA interference strategy (Figure 6A). The levels of viral RNA and expression of viral proteins, including the envelope protein and NSP1, increased significantly in ZIKV-infected U251 cells, following transfection with anti-ZDHHC11 siRNA (Figure 6B,C). The results of CLSM also demonstrated that the ZIKV envelope protein content increased substantially following ZDHHC11 knockdown (Figure 6D). The viral titer in the supernatant of ZIKV-infected U251 cells subjected to siRNA transfection was detected and quantified using plaque assays. The results demonstrated that ZIKV infection increased significantly following ZDHHC11 knockdown

(Figure 6E,F). These findings revealed that ZDHHC11 knockdown enhances ZIKV infection, and that ZDHHC11 has an anti-ZIKV effect.

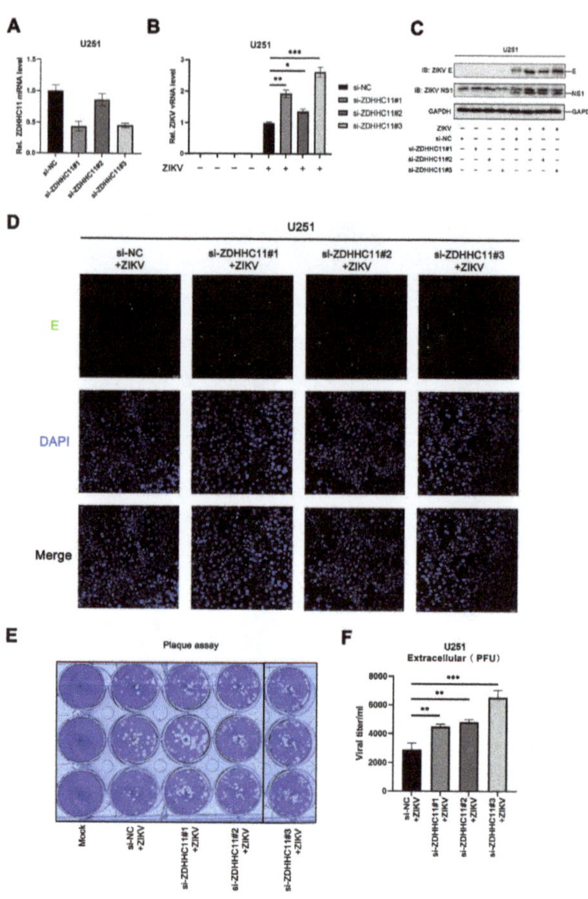

Figure 6. ZDHHC11 knockdown enhances ZIKV infection. U251 cells were transfected with anti-ZDHHC11 siRNA or the negative control, and subsequently infected with ZIKV (MOI = 1) using incubation for 24 h. The mRNA levels of (**A**) ZDHHC11 and (**B**) viral RNA were measured with quantitative PCR. The abundance of viral proteins was detected with (**C**) immunoblotting and (**D**) CLSM. (**E,F**) The ZIKV titer in the supernatants was calculated using plaque assays. (*, $p < 0.05$; **, $p < 0.01$; ***, $p < 0.001$).

4. Discussion

A comprehensive understanding of virus–host interactions is important for identifying the pathogenic mechanisms of viruses and developing novel therapeutic strategies. In this study, we observed that treatment with the palmitoylation inhibitor, 2-BP, enhanced ZIKV infection in Vero cells, which suggested that protein palmitoylation suppresses ZIKV infections. However, there are no reports on the palmitoylation of ZIKV proteins to date. In this study, we identified that the envelope protein of ZIKV undergoes palmitoylation, and subsequent computational and mutation analyses confirmed that the envelope protein is palmitoylated at Cys308. ZDHHC11 was identified as the predominant enzyme that catalyzes the palmitoylation of the envelope protein. We also found that ZDHHC11 binds to the Domain II region (132–193 aa) of the envelope protein via its N-terminal region (1–197 aa). The role of ZDHHC11 in ZIKV infection was investigated, and the results demonstrated that the overexpression of ZDHHC11 suppressed ZIKV infection in an enzymatic activity-dependent manner, while ZDHHC11 knockdown enhanced viral infections.

The envelope protein of flaviviruses plays crucial roles in viral structure, invasion, and assembly, and also serves as the primary antigen in inducing the production of host antibodies [11]. The functions of the envelope protein are tightly regulated by several post-translational modifications. The envelope proteins of several flaviviruses undergo N-linked glycosylation, in which a glycan is attached to the amide nitrogen of Asp [28]. It has been reported that the glycosylation of the envelope protein of ZIKV at Asn154 facilitates the expression and secretion of E ectodomain, production of virus like particle (VLP), and infectivity [29]. A previous study demonstrated that deficient glycosylation enhances viral entry, virion assembly, and the production of progeny viruses in C6/36 cells [30], but reduces pathogenesis in ifnar1$^{-/-}$ mice [13,31]. A recent study reported that the ZIKV envelope protein undergoes K63-linked polyubiquitination at Lys38 and Lys281, which is catalyzed by the E3-ubiquitin ligase, TRIM7 [12]. Ubiquitination of the envelope protein promotes viral entry and pathogenesis, while deubiquitylation of the envelope protein by ubiquitin-specific proteases exerts antiviral effects [24,25]. The results of this study demonstrated that the ZIKV envelope protein undergoes palmitoylation, which is a novel post-translational modification of proteins. The palmitoylation of the envelope protein possibly serves as an antiviral strategy in host cells, as ZIKV infections were enhanced following treatment with the palmitoylation inhibitor, 2-BP. Subsequently, computational and mutation analyses confirmed that palmitoylation of the envelope protein occurs at Cys308. As flaviviral envelope proteins are highly homologous and have high amino acid sequence identity, it is likely that the envelope proteins of other flaviviruses are modified by palmitoylation. Understanding the relationship between palmitoylation and viral invasion, release, and occurrence of other post-translational modifications of viral proteins may provide better insights into the effects of palmitoylation on ZIKV.

Palmitoylation is a reversible post-translational modification of proteins that regulates the localization, stability, and functions of proteins by attaching palmitic acid to the cysteine residues of target proteins. Several studies have indicated that palmitoylation regulates the functions of coronavirus proteins. Previous studies have demonstrated that palmitoylation of the spike (S) protein of SARS-CoV-2 is essential for the spike protein-mediated formation of syncytia [21], organization of membrane lipids [32], subcellular localization [33], and viral entry [20]. Therefore, targeting the palmitoylation of the spike protein of SARS-CoV-2 offers a promising therapeutic strategy against SARS-CoV-2 infection [34]. Additionally, palmitoylation of the NSP1 protein of alphaviruses is necessary for membrane binding and replication [35–37]. However, the effect of palmitoylation on ZIKV or other flaviviruses has not been investigated to date. The initial findings of this study suggested that the inhibition of palmitoylation enhances ZIKV infection, and that the envelope protein is palmitoylated at Cys308. Whether the function of nonstructural proteins NS1 to NS5 is regulated by palmitoylation modification deserves further investigation.

The findings revealed that ZDHHC11 is the primary enzyme that catalyzes the palmitoylation of the ZIKV envelope protein. A previous study demonstrated that the ER-anchored protein, ZDHHC11, positively regulates NF-kB signaling by enhancing the oligomerization of TRFA6 [15]. ZDHHC11 also promotes MITA-IRF3 interactions by binding to MITA via its C-terminal domain to promote type I interferon signaling during infections with DNA viruses, but not RNA viruses [16]. Here, we proposed that ZDHHC11 also exerted anti-viral effects upon ZIKV through binding to the envelope protein and palmitoylating the envelope protein. Considering that ZDHHC11 is a multifunctional protein, it may be possible that it limits ZIKV infection by other cellular processes besides palmitoylating envelope proteins.

5. Conclusions

We performed a preliminary investigation of the effect of palmitoylation on ZIKV infection, and the findings revealed that the ZIKV envelope protein is palmitoylated at

Cys308. ZDHHC11 was identified as the predominant enzyme that interacts with and mediates the palmitoylation of the ZIKV envelope protein. Functional analysis revealed that the overexpression of ZDHHC11 inhibited ZIKV infection, while ZDHHC11 knockdown enhanced viral infection. Overall, our study describes a novel mechanism of host regulation of ZIKV infection and identified potential novel targets of ZIKV infection.

Author Contributions: Experimental design, D.H., Y.Z., W.C., W.W., J.W. and G.L.; molecular experiments, D.H., H.Z., Y.L., Z.L., D.G., Y.M., X.W. and F.L.; writing—review and editing, D.H., W.W., Y.Z. and G.L. All authors have read and agreed to the published version of the manuscript.

Funding: This research was funded by: the National Nature Science Foundation of China, grant number 81973549 and U22A20335; the Key-Area Research and Development Program of Guangdong Province, grant number 2020B1111100002; and Guangzhou University of Chinese Medicine First-Class Universities and Top Disciplines Scientific Research Team Projects, grant number 2021XK16.

Institutional Review Board Statement: Not applicable.

Informed Consent Statement: Not applicable.

Data Availability Statement: The data supporting the findings of this study are available within the manuscript.

Acknowledgments: The authors gratefully acknowledge Cheng-Feng Qin of the Beijing Institute of Microbiology and Epidemiology, for the gift of BHK-21 cells. We also thank Lingnan Medical Research Center of Guangzhou University of Chinese Medicine for the help of providing the confocal laser scanning microscope.

Conflicts of Interest: The authors declare no conflict of interest.

References

1. Faria, N.R.; Azevedo, R.; Kraemer, M.; Souza, R.; Cunha, M.S.; Hill, S.C.; Theze, J.; Bonsall, M.B.; Bowden, T.A.; Rissanen, I.; et al. Zika virus in the Americas: Early epidemiological and genetic findings. *Science* **2016**, *352*, 345–349. [CrossRef] [PubMed]
2. Pierson, T.C.; Diamond, M.S. The emergence of Zika virus and its new clinical syndromes. *Nature* **2018**, *560*, 573–581. [CrossRef] [PubMed]
3. Johansson, M.A.; Mier-y-Teran-Romero, L.; Reefhuis, J.; Gilboa, S.M.; Hills, S.L. Zika and the Risk of Microcephaly. *N. Engl. J. Med.* **2016**, *375*, 1–4. [CrossRef] [PubMed]
4. Fontes, C.A.; Dos, S.A.; Marchiori, E. Magnetic resonance imaging findings in Guillain-Barre syndrome caused by Zika virus infection. *Neuroradiology* **2016**, *58*, 837–838. [CrossRef]
5. Govero, J.; Esakky, P.; Scheaffer, S.M.; Fernandez, E.; Drury, A.; Platt, D.J.; Gorman, M.J.; Richner, J.M.; Caine, E.A.; Salazar, V.; et al. Zika virus infection damages the testes in mice. *Nature* **2016**, *540*, 438–442. [CrossRef]
6. Musso, D.; Gubler, D.J. Zika Virus. *Clin. Microbiol. Rev.* **2016**, *29*, 487–524. [CrossRef]
7. Li, A.; Wang, W.; Wang, Y.; Chen, K.; Xiao, F.; Hu, D.; Hui, L.; Liu, W.; Feng, Y.; Li, G.; et al. NS5 Conservative Site Is Required for Zika Virus to Restrict the RIG-I Signaling. *Front. Immunol* **2020**, *11*, 51. [CrossRef]
8. Riedl, W.; Acharya, D.; Lee, J.H.; Liu, G.; Serman, T.; Chiang, C.; Chan, Y.K.; Diamond, M.S.; Gack, M.U. Zika Virus NS3 Mimics a Cellular 14-3-3-Binding Motif to Antagonize RIG-I- and MDA5-Mediated Innate Immunity. *Cell. Host Microbe* **2019**, *26*, 493–503. [CrossRef]
9. Ngueyen, T.; Kim, S.J.; Lee, J.Y.; Myoung, J. Zika Virus Proteins NS2A and NS4A Are Major Antagonists that Reduce IFN-beta Promoter Activity Induced by the MDA5/RIG-I Signaling Pathway. *J. Microbiol. Biotechnol* **2019**, *29*, 1665–1674. [CrossRef]
10. Sirohi, D.; Kuhn, R.J. Zika Virus Structure, Maturation, and Receptors. *J. Infect. Dis* **2017**, *216*, S935–S944. [CrossRef]
11. Slon, C.J.; Mongkolsapaya, J.; Screaton, G.R. The immune response against flaviviruses. *Nat. Immunol.* **2018**, *19*, 1189–1198. [CrossRef]
12. Giraldo, M.I.; Xia, H.; Aguilera-Aguirre, L.; Hage, A.; van Tol, S.; Shan, C.; Xie, X.; Sturdevant, G.L.; Robertson, S.J.; McNally, K.L.; et al. Envelope protein ubiquitination drives entry and pathogenesis of Zika virus. *Nature* **2020**, *585*, 414–419. [CrossRef] [PubMed]
13. Carbaugh, D.L.; Baric, R.S.; Lazear, H.M. Envelope Protein Glycosylation Mediates Zika Virus Pathogenesis. *J. Virol.* **2019**, *93*, e00113-19. [CrossRef] [PubMed]
14. Linder, M.E.; Deschenes, R.J. Palmitoylation: Policing protein stability and traffic. *Nat. Rev. Mol. Cell Biol.* **2007**, *8*, 74–84. [CrossRef] [PubMed]
15. Liu, E.; Sun, J.; Yang, J.; Li, L.; Yang, Q.; Zeng, J.; Zhang, J.; Chen, D.; Sun, Q. ZDHHC11 Positively Regulates NF-kappaB Activation by Enhancing TRAF6 Oligomerization. *Front. Cell Dev. Biol.* **2021**, *9*, 710967. [CrossRef] [PubMed]

16. Liu, Y.; Zhou, Q.; Zhong, L.; Lin, H.; Hu, M.M.; Zhou, Y.; Shu, H.B.; Li, S. ZDHHC11 modulates innate immune response to DNA virus by mediating MITA-IRF3 association. *Cell Mol. Immunol.* **2018**, *15*, 907–916. [CrossRef]
17. Lu, Y.; Zheng, Y.; Coyaud, E.; Zhang, C.; Selvabaskaran, A.; Yu, Y.; Xu, Z.; Weng, X.; Chen, J.S.; Meng, Y.; et al. Palmitoylation of NOD1 and NOD2 is required for bacterial sensing. *Science* **2019**, *366*, 460–467. [CrossRef]
18. Mukai, K.; Konno, H.; Akiba, T.; Uemura, T.; Waguri, S.; Kobayashi, T.; Barber, G.N.; Arai, H.; Taguchi, T. Activation of STING requires palmitoylation at the Golgi. *Nat. Commun.* **2016**, *7*, 11932. [CrossRef]
19. Ko, P.J.; Dixon, S.J. Protein palmitoylation and cancer. *Embo. Rep.* **2018**, *19*, e46666. [CrossRef]
20. Wu, Z.; Zhang, Z.; Wang, X.; Zhang, J.; Ren, C.; Li, Y.; Gao, L.; Liang, X.; Wang, P.; Ma, C. Palmitoylation of SARS-CoV-2 S protein is essential for viral infectivity. *Signal. Transduct. Target. Ther.* **2021**, *6*, 231. [CrossRef]
21. Li, D.; Liu, Y.; Lu, Y.; Gao, S.; Zhang, L. Palmitoylation of SARS-CoV-2 S protein is critical for S-mediated syncytia formation and virus entry. *J. Med. Virol.* **2022**, *94*, 342–348. [CrossRef]
22. Wagner, R.; Herwig, A.; Azzouz, N.; Klenk, H.D. Acylation-mediated membrane anchoring of avian influenza virus hemagglutinin is essential for fusion pore formation and virus infectivity. *J. Virol.* **2005**, *79*, 6449–6458. [CrossRef] [PubMed]
23. Chen, B.J.; Takeda, M.; Lamb, R.A. Influenza virus hemagglutinin (H3 subtype) requires palmitoylation of its cytoplasmic tail for assembly: M1 proteins of two subtypes differ in their ability to support assembly. *J. Virol.* **2005**, *79*, 13673–13684. [CrossRef]
24. Hu, D.; Wang, Y.; Li, A.; Li, Q.; Wu, C.; Shereen, M.A.; Huang, S.; Wu, K.; Zhu, Y.; Wang, W.; et al. LAMR1 restricts Zika virus infection by attenuating the envelope protein ubiquitination. *Virulence* **2021**, *12*, 1795–1807. [CrossRef] [PubMed]
25. Wang, Y.; Li, Q.; Hu, D.; Gao, D.; Wang, W.; Wu, K.; Wu, J. USP38 Inhibits Zika Virus Infection by Removing Envelope Protein Ubiquitination. *Viruses* **2021**, *13*, 29. [CrossRef] [PubMed]
26. Yeste-Velasco, M.; Linder, M.E.; Lu, Y. Protein S-palmitoylation and cancer. *Biochim. Biophys. Acta (BBA)-Rev. Cancer* **2015**, *1856*, 107–120. [CrossRef] [PubMed]
27. Ning, W.; Jiang, P.; Guo, Y.; Wang, C.; Tan, X.; Zhang, W.; Peng, D.; Xue, Y. GPS-Palm: A deep learning-based graphic presentation system for the prediction of S-palmitoylation sites in proteins. *Brief. Bioinform.* **2021**, *22*, 1836–1847. [CrossRef] [PubMed]
28. Carbaugh, D.L.; Lazear, H.M. Flavivirus Envelope Protein Glycosylation: Impacts on Viral Infection and Pathogenesis. *J. Virol* **2020**, *94*, 94. [CrossRef]
29. Mossenta, M.; Marchese, S.; Poggianella, M.; Slon, C.J.; Burrone, O.R. Role of N-glycosylation on Zika virus E protein secretion, viral assembly and infectivity. *Biochem. Biophys. Res. Commun.* **2017**, *492*, 579–586. [CrossRef]
30. Fontes-Garfias, C.R.; Shan, C.; Luo, H.; Muruato, A.E.; Medeiros, D.; Mays, E.; Xie, X.; Zou, J.; Roundy, C.M.; Wakamiya, M.; et al. Functional Analysis of Glycosylation of Zika Virus Envelope Protein. *Cell Rep.* **2017**, *21*, 1180–1190. [CrossRef]
31. Annamalai, A.S.; Pattnaik, A.; Sahoo, B.R.; Muthukrishnan, E.; Natarajan, S.K.; Steffen, D.; Vu, H.; Delhon, G.; Osorio, F.A.; Petro, T.M.; et al. Zika Virus Encoding Nonglycosylated Envelope Protein Is Attenuated and Defective in Neuroinvasion. *J. Virol.* **2017**, *91*, e01348-17. [CrossRef] [PubMed]
32. Mesquita, F.S.; Abrami, L.; Sergeeva, O.; Turelli, P.; Qing, E.; Kunz, B.; Raclot, C.; Paz, M.J.; Abriata, L.A.; Gallagher, T.; et al. S-acylation controls SARS-CoV-2 membrane lipid organization and enhances infectivity. *Dev. Cell* **2021**, *56*, 2790–2807. [CrossRef] [PubMed]
33. Zeng, X.T.; Yu, X.X.; Cheng, W. The interactions of ZDHHC5/GOLGA7 with SARS-CoV-2 spike (S) protein and their effects on S protein's subcellular localization, palmitoylation and pseudovirus entry. *Virol. J.* **2021**, *18*, 257. [CrossRef]
34. Ramadan, A.A.; Mayilsamy, K.; McGill, A.R.; Ghosh, A.; Giulianotti, M.A.; Donow, H.M.; Mohapatra, S.S.; Mohapatra, S.; Chandran, B.; Deschenes, R.J.; et al. Identification of SARS-CoV-2 Spike Palmitoylation Inhibitors That Results in Release of Attenuated Virus with Reduced Infectivity. *Viruses* **2022**, *14*, 531. [CrossRef] [PubMed]
35. Ahola, T.; Kujala, P.; Tuittila, M.; Blom, T.; Laakkonen, P.; Hinkkanen, A.; Auvinen, P. Effects of palmitoylation of replicase protein nsP1 on alphavirus infection. *J. Virol.* **2000**, *74*, 6725–6733. [CrossRef] [PubMed]
36. Zhang, N.; Zhao, H.; Zhang, L. Fatty Acid Synthase Promotes the Palmitoylation of Chikungunya Virus nsP1. *J. Virol.* **2019**, *93*, e01747-18. [CrossRef] [PubMed]
37. Bakhache, W.; Neyret, A.; Bernard, E.; Merits, A.; Briant, L. Palmitoylated Cysteines in Chikungunya Virus nsP1 Are Critical for Targeting to Cholesterol-Rich Plasma Membrane Microdomains with Functional Consequences for Viral Genome Replication. *J. Virol.* **2020**, *94*, e02183-19. [CrossRef] [PubMed]

Review

Host Responses to Respiratory Syncytial Virus Infection

Ayse Agac, Sophie M. Kolbe, Martin Ludlow, Albert D. M. E. Osterhaus, Robert Meineke
and Guus F. Rimmelzwaan *

Research Center for Emerging Infections and Zoonoses, University of Veterinary Medicine Hannover,
30559 Hannover, Germany; ayse.agac@tiho-hannover.de (A.A.);
sophie.madleine.kolbe@tiho-hannover.de (S.M.K.); martin.ludlow@tiho-hannover.de (M.L.);
albert.osterhaus@tiho-hannover.de (A.D.M.E.O.); robert.meineke@tiho-hannover.de (R.M.)
* Correspondence: guus.rimmelzwaan@tiho-hannover.de

Abstract: Respiratory syncytial virus (RSV) infections are a constant public health problem, especially in infants and older adults. Virtually all children will have been infected with RSV by the age of two, and reinfections are common throughout life. Since antigenic variation, which is frequently observed among other respiratory viruses such as SARS-CoV-2 or influenza viruses, can only be observed for RSV to a limited extent, reinfections may result from short-term or incomplete immunity. After decades of research, two RSV vaccines were approved to prevent lower respiratory tract infections in older adults. Recently, the FDA approved a vaccine for active vaccination of pregnant women to prevent severe RSV disease in infants during their first RSV season. This review focuses on the host response to RSV infections mediated by epithelial cells as the first physical barrier, followed by responses of the innate and adaptive immune systems. We address possible RSV-mediated immunomodulatory and pathogenic mechanisms during infections and discuss the current vaccine candidates and alternative treatment options.

Keywords: respiratory syncytial virus; innate immunity; adaptive immunity; immunopathology; immune evasion; vaccines

Citation: Agac, A.; Kolbe, S.M.;
Ludlow, M.; Osterhaus, A.D.M.E.;
Meineke, R.; Rimmelzwaan, G.F.
Host Responses to Respiratory
Syncytial Virus Infection. *Viruses*
2023, *15*, 1999. https://doi.org/
10.3390/v15101999

Academic Editors: Yiping Li and
Yuliang Liu

Received: 18 August 2023
Revised: 22 September 2023
Accepted: 23 September 2023
Published: 26 September 2023

1. Introduction

Respiratory syncytial virus (RSV) is a widely circulating pathogen in the human population and a major cause of acute lower respiratory tract infections (ALRIs) in infants and older adults. RSV is an enveloped, negative-sense, single-stranded RNA virus belonging to the *Orthopneumovirus* genus of the family *Pneumoviridae* [1]. The viral genome consists of ten genes encoding eleven proteins: nonstructural proteins 1 and 2 (NS1/2), nucleocapsid protein (N), matrix protein (M), phosphoprotein (P), small hydrophobic protein (SH), glycoprotein (G), fusion protein (F), large protein (L), and M2 with two overlapping open-reading frames leading to two proteins, M2.1 and M2.2 [1]. Two antigenic subtypes, RSV-A and -B, are distinguished, each with several genotypes [2,3]. Currently, genotypes ON1 (RSV-A) and BA-CC (RSV-B) are circulating and differ from previous genotypes by the presence of sequence duplications of 72 bp and 60 bp, respectively, in the second hypervariable region of the G gene [4]. Both subtypes may co-circulate during an outbreak, with RSV A being the predominant strain in most years, although regional and seasonal differences are common [4].

RSV infections are a major cause of severe ALRI in infants and older adults [5–8]. In 2019, approx. 33 million RSV cases in infants were reported, of which 3.2 million required hospitalization [9]. Children under six months of age are at higher risk of hospitalization and fatal outcomes, accounting for ~50% of total hospitalized ALRI cases in high-income countries [9]. In older adults, 5.2 million RSV cases of ALRI were reported in the same year, resulting in 470,000 reported hospitalizations and 33,000 in-hospital deaths [10]. During the SARS-CoV-2 pandemic, reduced activity of various respiratory viruses, including RSV, was reported [11,12], probably due to non-pharmaceutical preventive measures to limit the transmission of SARS-CoV-2. Subsequently, increased sizes and numbers of outbreaks, with different times of onset in comparison to previous years, were reported for RSV disease, mainly affecting infants and children [12]. The decrease in RSV infections during the pandemic may have reduced immunity in children, referred to as the 'immunity dept', making them more susceptible to severe infections [12,13].

Nearly all infants are infected with RSV by the age of 2 years [14]. Reinfections occur throughout life, while disease severity tends to diminish with subsequent exposures. Older adults, however, represent a population that is also at risk of more severe disease during RSV infection [15–17]. Most infections in infants are mild, leading to an upper respiratory tract illness (URTI) or 'flu-like' symptoms often accompanied by otitis media [18,19]. Severe ALRI, mainly observed in infants younger than six months of age, can lead to bronchiolitis, pneumonia, and croup [18,19]. ALRIs are characterized by rhinorrhea, dry/wheezy cough, tachypnea, and dyspnea [19]. Infants, the immunocompromised, and older adults have a higher risk of developing severe infections [5–7,20,21]. Several studies further indicated a correlation between RSV-associated hospitalization during infancy and the development of asthma and recurrent wheezing later in life [22–26]. However, other factors, like comorbidities, may also account for this [6,27,28].

Virus-neutralizing human monoclonal antibodies directed to the F protein, i.e., Synagis® (Palivizumab) and Beyfortus® (Nirsevimab), have proven effective in lowering the risk of ALRI in high-risk infants when administered prophylactically. However, vaccine development for RSV has been challenging. The first clinical trials in infants in the 1960s with a formalin-inactivated RSV vaccine (FI-RSV) led to enhanced respiratory disease (ERD) in infants following their first natural infection [29]. Since then, vaccine development has been approached more cautiously while research has focused on understanding the immune response and immunopathogenesis during infection. The immaturity of an infant's immune system as well as immunosenescence and preexisting immunity in older adults represent the most significant challenges for the development of a protective vaccine [30]. Several promising RSV vaccine candidates based on various production platforms have been developed and are currently in clinical trials. Recently, two protein-based vaccines, Arexvy (GlaxoSmithKline Biologicals (GSK), Brentford, UK) and Abrysvo (Pfizer, New York, NY, USA), were approved for use in older adults. Furthermore, a prefusion F-protein-based RSV vaccine candidate was developed for use in pregnant women to protect their infants early in life and was recently approved by the FDA.

Previous work illustrates the importance of understanding the correlation between protective immunity and immunopathogenesis in severe cases of RSV infections for the development of efficient and protective vaccines and therapeutics for those most at risk. This review focuses on the current knowledge of host immune responses to RSV infection, leading to either viral control or more severe disease due to immunopathogenesis, and the current use of this knowledge in vaccine development.

2. Innate Immune Responses to RSV Infection

Although the innate immune response toward infection with different *Pneumoviridae* members is still far from clear and may differ considerably [31–34], for RSV infection of humans many different mechanisms have been described. Epithelial cells are the primary target of infection (Figure 1) and secrete proinflammatory mediators upon RSV infection, leading to the recruitment and activation of innate immune cells. Several cell types are involved in the innate immune response, such as polymorphonuclear cells, cytotoxic lymphocytes, i.e., natural killer (NK) cells, or mononuclear phagocytes [35,36]. Tissue-resident macrophages and dendritic cells are among the first innate immune cells activated and mediate the further recruitment of inflammatory leukocytes and lymphocytes, like eosinophils, neutrophils, monocytes, and NK cells, to the site of infection by the secretion of type I IFNs and chemokines [37]. These cells subsequently begin eliminating the virus and simultaneously mediate the activation and recruitment of adaptive immune cells [36]. The initial immune response during epithelial cell infection, the protective role of innate immune cells, and their potential contribution to immunopathogenesis during RSV infections will be discussed in the following sections.

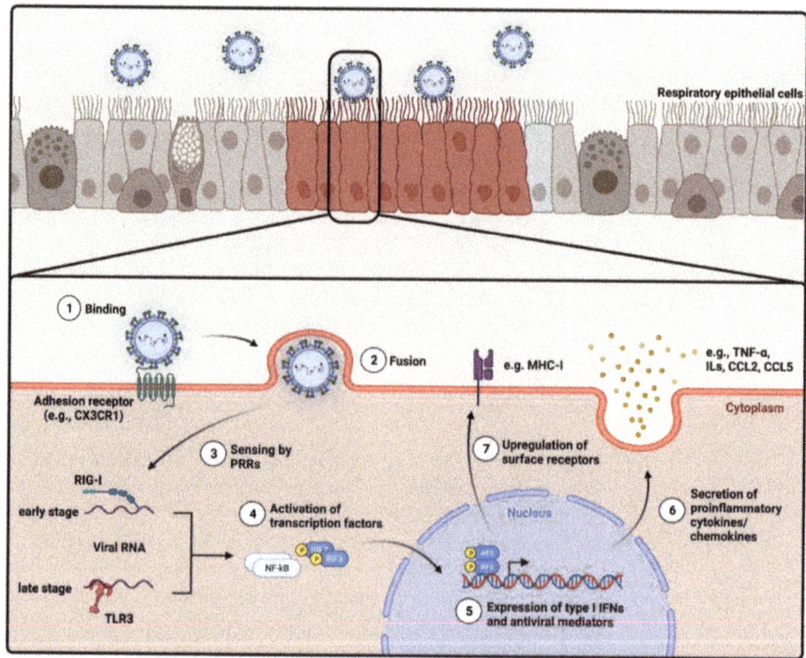

Figure 1. Overview of the intracellular mechanisms in airway epithelial cells (AECs) after respiratory syncytial virus (RSV) entry. RSV enters ciliated AECs by binding to receptors on the surface of the cells (e.g., CX3CR1), followed by internalization of the virus particle into the cytoplasm. Following infection, intracellular pattern recognition receptors, like RIG-I or TLR3, sense viral ssRNA or dsRNA intermediates resulting in the activation of transcription factors like NK-κB or IRF3. Nuclear translocation of transcription factors leads to the expression of several antiviral mediators, like cytokines, chemokines, and receptors for antigen presentation. RIG-I: retinoic-acid-inducible protein I; TLR3: toll-like receptor 3; IRF3: interferon regulatory factor 3; IFN: interferon; TNF-α: tumor necrosis factor α; IL: interleukin; MHC-I: major histocompatibility complex I. Created with BioRender.com.

2.1. Infection of Epithelial Cells

The airway epithelium and associated mucus represent the first physiochemical barrier during respiratory infections [38]. The pseudostratified epithelium consists of three major cell types, namely ciliated cells, basal cells, and non-ciliated, secretory goblet cells which are connected by tight junctions [39,40]. The epithelial cells are mainly covered by mucus, contributing to protection from inhaled pathogens. The mucus is comprised of two distinct layers: a periciliary liquid layer, allowing the movement of cilia, and a more viscous upper layer, immobilizing the pathogen. Both layers consist of mucins, defensins, lysozymes, and immunoglobulins [41]. Mucociliary clearance, resulting from the continuous beating of motile cilia and consequently the movement of the outer mucus layer toward the pharynx, forms the first innate defense mechanism against respiratory pathogens. The protective properties of the mucus layer have already been shown for various respiratory viruses, like influenza viruses or coronaviruses [42–45]. An increased susceptibility of older adults to respiratory infections may be partly explained by a decrease in mucus production, cilia movement, and mucociliary clearance resulting in reduced pathogen removal [42,46].

Once RSV has traversed the mucus layer, it primarily infects airway epithelial cells (AECs), usually via the apical surface [47]. The infection of AECs depends on the cell's differentiation status as infection increases with ciliogenesis. RSV therefore preferentially infects polarized ciliated AECs resulting in a focal distribution of the infection [48]. Hallmarks of AEC infection are cell sloughing, cilia loss, mucus hypersecretion, and syncytia formation. The loss of cilia and increased mucus production may contribute to airway obstruction during infection [38,49].

Several viral receptors have been proposed, as reviewed in [50]. Among others, nucleolin [51,52], CX3CR1 [53–56], heparan sulfate proteoglycans (HSPGs) [57–59], and intracellular adhesion molecule-1 (ICAM-1) [60] were associated with RSV attachment [50]. The interaction of RSV F and G with the HSPGs on the surface of immortalized cell lines leads to viral attachment and facilitates infection [59]. However, since HSPGs are barely expressed on the apical surface of epithelial cells *in vivo*, other receptors may be required for RSV infection [59].

CX3CR1 is expressed on various cell types, including innate and adaptive immune cells, and on motile cilia of AECs, which matches the RSV cell tropism during an infection [56,61,62]. Only two ligands are known for CX3CR1: the RSV G protein and CX3CL1 (fractalkine) [61]. CX3CL1 is a transmembrane protein on the surface of epithelial cells and contributes to the cell adhesion of leukocytes during infection [61,63]. Soluble CX3CL1 serves as a chemoattractant for T cells and monocytes via interaction with CX3CR1 on the surface of these cells [64]. RSV G contains a CX3C motif in its conserved central region, implicating a direct interaction between RSV G and CX3CR1 [65]. A study by Ha et al. showed a possible interaction between the receptor and G since a mutation in the CX3C motif (CX3C to CX4C) reduced the infection of epithelial cells in RSV-infected cotton rats [66]. However, a direct biochemical interaction between the RSV G protein and CX3CR1 has not been demonstrated yet. These data show that although G may facilitate the infection of epithelial cells, it is not strictly necessary for it [66]. However, the direct interaction of G and CX3CR1 may not only affect the infection of AECs but also interfere with the CX3CR1-CX3CL1 interaction. Soluble G may thereby act as a chemoattractant resulting in the increased recruitment of $CX3CR1^+$ immune cells to the site of infection.

With the infection of respiratory epithelial cells, antiviral and innate immune responses are initiated. The response of AECs to RSV infection is similar to that against other viral infections like influenza viruses, SARS-CoV-2, or herpesviruses [67–71]. Upon viral attachment and entrance, intra- and extracellular pattern recognition receptors (PRRs) recognize the pathogen and initiate a signaling cascade resulting in the expression of antiviral genes [72]. Among the PRRs, retinoic-acid-inducible gene I (RIG-I)-like receptors (RLRs) and toll-like receptors (TLRs) were shown to play a role in virus recognition [73]. During the early stage of infection, RIG-I senses the viral RNA and activates the transcription factor interferon regulatory factor (IRF)-3, ultimately leading to the upregulation of MHC-I, NLRC5, and IFN-β [73]. During the later stages of an active infection, intracellular TLR3 binds to the viral RNA [73] and activates the transcription factor NF-κB, which induces the expression of type I interferons (IFN) [74,75]. Type I IFN signaling modulates metabolic pathways, thereby setting the cells in an 'antiviral state' that restricts viral replication and spreading to uninfected cells. Infection of AECs with RSV consequently induces an altered expression and secretion of chemoattractants and adhesion molecules, e.g., tumor necrosis factor (TNF)-α, CXCL6, CXCL10, RANTES/CCL5, interleukin (IL) 1β, IL-6, IL-8, CCL2, macrophage inflammatory protein 1α (MIP-1α), granulocyte colony-stimulating factor (G-CSF), granulocyte-macrophage colony-stimulating factor (GM-CSF), ICAM-1, vascular cell adhesion protein 1 (VCAM-1), and major histocompatibility complex I/II (MHC-I/II) [76–83]. Subsequently, their secretion triggers the recruitment of innate and adaptive immune cells and their interaction with infected AECs.

The immune modulatory mechanisms of different RSV proteins are well-known and have been reviewed previously [84]. NS1 and NS2 play a significant role in immune modulation by interacting with components of different immune signaling pathways. For example,

NS1/2 interferes with the Janus kinase (JAK)/signal transducer and activator of transcription (STAT) signaling by suppressor of cytokine signaling proteins (SOCS)-dependent inhibition of JAK kinases and induction of proteasome-mediated STAT degradation [85,86]. *In vitro* studies have also shown the capacity of NS1 to translocate into the nucleus, where it interacts with the mediator complex and chromatin, indicating a regulatory effect of NS1 on the host's gene expression [87]. Further, the respective interactions of NS1 and NS2 with mitochondrial antiviral-signaling protein (MAVS) and RIG-I interfere with the activation of IRF3 and NF-κB [87–90]. IFN-β synthesis is impaired by RSV infection, and the expression of anti-apoptotic genes is increased [90–92]. Following this, previous *in vivo* and *in vitro* studies have shown that the modulation of immune responses by NS1/2 inhibited apoptosis but induced necroptosis in RSV-infected AECs, which is considered an inflammatory cell death variant [92–94]. The release of HMGB1 into the extracellular space during necroptosis causes the recruitment of proinflammatory and Th2-type cytokines, ultimately resulting in increased airway inflammation and disease severity [94]. Besides NS1/2, an immunomodulatory role was also proposed for RSV nucleoprotein (N). *In vitro* studies indicate that RSV N induces the formation of inclusion bodies that serve as 'traps' for MAVS and melanoma-differentiation-associated gene 5 (MDA-5) during an early stage of infection [95,96]. RSV N interacts with PKR leading to reduced phosphorylation of eIFα and reduction in PKR-mediated signaling [97]. The activity of RSV proteins in infected AECs not only interferes with intrinsic immune pathways but restricts the activation of several genes that are involved in the production of cytokines and chemokines, which are crucial for the recruitment of innate and adaptive immune cells [84].

AECs, thereby, initiate the first antiviral, innate immune response upon RSV infection by the increased expression of adhesion molecules on the surface and the secretion of a variety of cytokines and chemokines that are essential for the recruitment of innate and adaptive immune cells. However, RSV has developed ways to alter and suppress this initial immune response, possibly affecting the activation and recruitment of immune cells during the early phase of infection. As mentioned above, cilia movement and mucus production are important during the first encounter with RSV and other respiratory viruses. However, most *in vitro* models are not able to represent the airway epithelium in its physiological three-dimensional structure, cellular composition, and motility [98]. Air–liquid interfaces and organoids represent promising systems to investigate virus–host interactions and immune responses during an infection under more physiological conditions, when compared to immortalized cell lines, and are therefore of increasing importance for investigating respiratory viruses [99–101].

2.2. Eosinophils

Eosinophils are a subset of granulocytic, polymorphonuclear leukocytes originating in the bone marrow and can be detected in low numbers in the peripheral blood after maturation [102,103]. They can migrate to different tissues, where they contribute to the physiology of the tissue under homeostatic conditions [102,103]. Increased numbers of eosinophils and their recruitment to the infected tissue can be observed during viral infection, mediating the primary antiviral host defense through their phagocytic and antigen-presenting activity. The protective role of eosinophils during infection has been shown for several viruses, like Influenza A virus (IAV), SARS-CoV-2, human parainfluenza virus, and human rhinovirus [104–107].

Chemoattractants like CCL5 or MIP-1α, secreted by RSV-infected AECs, lead to the recruitment of eosinophils to the site of infection [108–110]. Following recruitment, eosinophils sense viral ssRNA in a TLR7-dependent manner [111]. This induces a signaling cascade resulting in the increased expression of the transcription factor IRF7 and, consequently, the increased expression of type I IFNs like IFN-β [111]. Upon interaction with infected AECs and viral RNA, eosinophils are activated, characterized by the increased surface expression of phagocytic surface marker CD11b (Figure 2) [111–113]. Activated eosinophils contribute to viral clearance by degranulation, defined by the secretion of RNA-degrading enzymes, such as the eosinophilic cationic protein (ECP) [109,114].

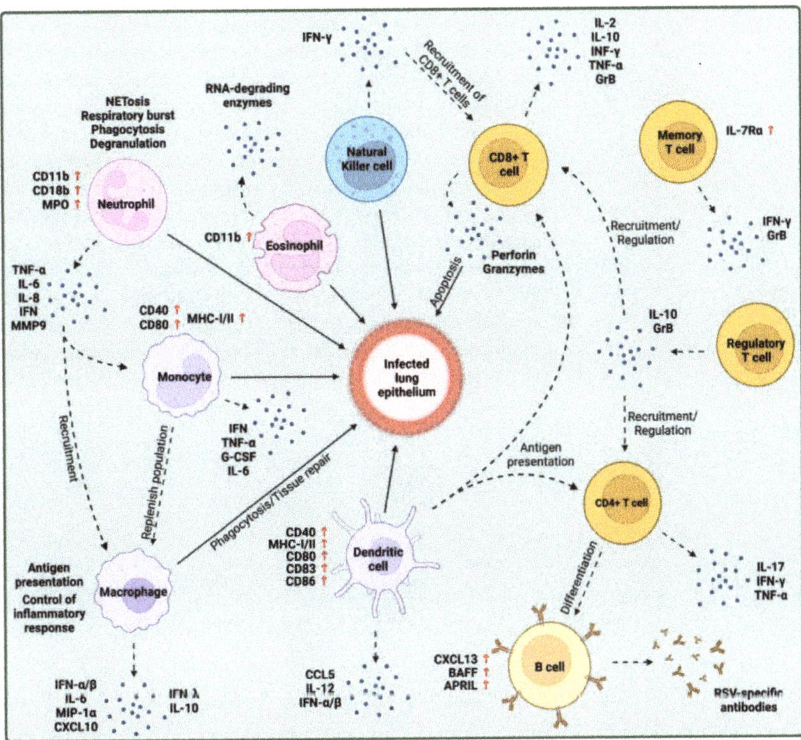

Figure 2. Schematic representation of the antiviral responses of innate and adaptive immune cells during RSV infection. Innate immune cells mediate the control of infection by direct interaction with the infected AECs, followed by their phagocytosis, or by direct interaction with virus particles, leading to the neutralization of the virus, e.g., by NETosis. Non-professional and professional antigen-presenting cells are responsible for the activation and recruitment of adaptive immune cells to the site of infection to further drive viral clearance. Regulation of immune responses is thereby crucial to avoid exaggerated, potentially immunopathogenic responses. IL: interleukin; IFN: interferon; TNF: tumor necrosis factor; GrB: granzyme B; BAFF: B cell-activating factor; APRIL: A proliferation-inducing ligand; MHC: major histocompatibility complex; MMP9: matrix metalloprotease 9. Created with BioRender.com.

Despite their potential role in virus elimination, in the context of RSV, eosinophils are mostly known for their alleged role in ERD in infants following FI-RSV vaccination. Early studies with the tissue of the two fatal cases (*ex vivo*) reported an increased infiltration of eosinophils into the lungs compared to control groups [29]. Eosinophilia was consequently considered a hallmark of ERD. In the following years, several studies indicated that eosinophils do not directly contribute to ERD [115]. Vaccination studies with FI-RSV followed by RSV challenge in cotton rats demonstrated that levels of neutrophils and lymphocytes were increased in vaccinated rats compared to unimmunized control rats [116]. In mice depleted of regulatory T cells (Tregs), RSV infection led to increased disease severity and increased levels of eosinophils, CD4$^+$, and CD8$^+$ T cells in the bronchoalveolar lavage (BAL) [117]. CD4$^+$ T cells found in the BALs of these mice had a Th2-type phenotype and secreted IL-13 during infection [117]. Therefore, excessive recruitment of eosinophilia may only result from an unbalanced Th2-biased T-cell response and the increased expression of Th2-type cytokines like IL-4, IL-10, and IL-13 [118].

Apart from their still tentative role in FI-RSV-associated ERD, several *in vitro* and *ex vivo* studies suggested a contribution of eosinophils in severe cases of RSV infection (Figure 3). In RSV-infected infants, increased levels of ECP were detected in nasopharyngeal secretions that correlated with increased disease severity compared to uninfected controls [110,119]. The direct interaction or infection of eosinophils by RSV may explain their altered immune response as activated eosinophils secrete proinflammatory cytokines like CCL5, IL-6, and MIP-1α after viral interaction resulting in a proinflammatory milieu and increased influx of proinflammatory cells [79,120–122]. In IL-5-deficient mice, eosinophilic infiltration into the lungs was abolished following RSV infection and restored after IL-5 transfer into the mice, indicating a crucial role of IL-5 in eosinophil recruitment [123]. However, eosinophilic recruitment to the site of infection led to airway hyperresponsiveness and increased disease severity compared to sham-infected controls [123]. A recent study showed that infection of neonatal mice with RSV resulted in airway hyperresponsiveness and increased numbers of eosinophils, macrophages, and CD4[+] T cells in the lungs compared to mock-infected mice, demonstrating that recruited eosinophils can contribute to disease severity [124].

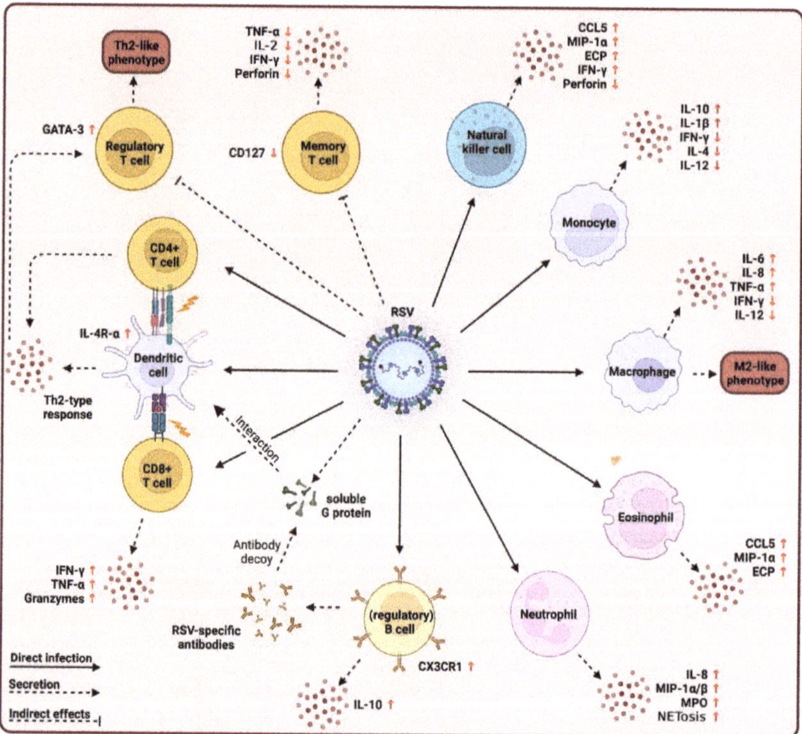

Figure 3. Overview of immunopathogenic responses mediated by innate and adaptive immune cells during RSV infection. Changes in the innate and adaptive immune response can be a consequence of microenvironmental changes, genetic predispositions, or interaction with the virus. RSV can directly infect a variety of immune cells or indirectly alter their immune response, e.g., by the secretion of soluble G, characterized by phenotype changes in immune cells, or the excessive secretion of immunosuppressive or proinflammatory cytokines. TNF-α: tumor necrosis factor α; IFN: interferon; IL: interleukin; ECP: eosinophil cationic protein; MIP-1α: macrophage inflammatory protein-1α; MPO: myeloperoxidase. Created with BioRender.com.

Together, these studies indicate that excessive eosinophil recruitment to the lungs and subsequent lung pathology may result from altered CD4$^+$ T cell signaling. Eosinophils usually contribute to protection against RSV infections. Th2-biased responses, however, lead to their increased recruitment and activation during infection and result in excessive and proinflammatory immune responses.

2.3. Neutrophils

Neutrophils belong to the polymorphonuclear leukocyte system and constitute the most abundant leukocyte in the blood [125]. Similar to eosinophils, neutrophils can migrate into tissue during a steady state and contribute to tissue homeostasis [125]. Under pathological conditions, like viral infections, neutrophils are involved in virus detection and elimination, cytokine production, recruitment of proinflammatory immune cells, and antigen presentation [126,127]. Neutrophil activity is mainly based on three central mechanisms: phagocytosis, degranulation, and NETosis [126].

During RSV infection, chemoattractants like ICAM-1 and IL-8 produced by epithelial cells lead to the recruitment of neutrophils to the site of infection [75,128]. Following recruitment, neutrophils interact with ICAM-1 on the surface of RSV-infected AECs leading to their activation characterized by increased expression of CD11b, CD18b, and myeloperoxidase (MPO) (Figure 2) [129,130]. Once activated, neutrophils can initiate several mechanisms to inhibit the further spread of infection. The formation of neutrophil extracellular traps (NETs) is a hallmark of neutrophil activation. NETs are extracellular structures of histones, elastase, MPO, and defensins [126]. Viral particles are captured by these structures and neutralized by MPO and defensin secretion [126]. It was shown that the interaction of RSV F protein with neutrophilic bactericidal/permeability-increasing protein (BPI) induces NET formation in a TLR4- and reactive oxygen species (ROS)-dependent manner, thus promoting viral neutralization [131,132]. Neutrophils can further limit viral infection and replication by producing ROS ('respiratory burst'), virus phagocytosis, and degranulation [129]. Neutrophils can release a subset of chemokines, cytokines, and antimicrobial substances like TNF-α, IL-6, IL-8, IFN, and matrix metalloprotease 9 (MMP9), resulting in the recruitment of monocytes and macrophages [133]. Neutrophils, therefore, can contribute to viral clearance during RSV infections by exhibiting an antiviral response and coordinating the immune signaling of innate and adaptive immune cells.

An immunopathogenic role during RSV infection has also been proposed for neutrophils since their numbers were increased in the airways of infants with severe disease [134,135]. Enhanced interaction of neutrophils with AECs may induce excessive cell damage characterized by lower ciliary activity, cilium loss, and detachment [133,136]. Recent studies further implicated pulmonary neutrophilia as contributing to ERD following FI-RSV vaccination [137]. Direct infection of neutrophils led to the secretion of IL-8, MIP-1α/β, and MPO, resulting in an increased proinflammatory milieu (Figure 3) [138,139]. The increased secretion of IL-8, a potent chemoattractant for neutrophils, by infected neutrophils may lead to further recruitment of neutrophils to the infection site, contributing to immune pathogenesis [139]. Although NETs generally have an antiviral role, their disproportional formation during infections might enhance disease. Increased generation and accumulation of NETs were shown to block the airways of infants with severe infections [132,140]. In a recent experimental human infection study, the inflammatory activity of neutrophils in the respiratory mucosa of volunteers before infection predisposed the latter to symptomatic RSV infection [141]. The presence and activity of neutrophils were linked with impaired antiviral immune responses. Asymptomatic outcomes were characterized by IL-17 expression in the early phase of infection [141]. Preexisting neutrophilic inflammation, therefore, may counteract the protective response during the early stages of infection and instead lead to a higher risk of symptomatic infection [141].

Similar to other immune cells, neutrophils exhibit a bivalent role during RSV infections where time and microenvironmental conditions are critical for neutrophilic activity. Neutrophils may contribute to viral clearance by inhibiting viral spread and coordinating innate and adaptive immune responses. However, excessive neutrophil recruitment and activation of neutrophils can lead to an exaggerated immune response, correlated with disease severity in infants.

2.4. Natural Killer Cells

Bone-marrow-derived NK cells are granular lymphocytes best known for their cytotoxic activity [142]. Since NK cells do not undergo somatic hypermutation like other lymphocytes (i.e., B and T cells), they are considered part of the innate immune response [143]. NK cells are involved in various processes, like maintaining immune homeostasis, clearing tumors or infections, and regulating immune responses by interaction with different immune cells like T cells, macrophages, or dendritic cells [142,144]. NK cells contribute to viral clearance through the secretion of IFN-γ and the lysis of infected cells by perforin- and granzyme-dependent antibody-dependent cellular cytotoxicity (ADCC) [144].

Following RSV infection, NK cells accumulate in the lung and are an important source of IFN-γ, especially during the early phase of infection [145]. Besides their cytotoxic role, NK cells are involved in the IFN-γ-mediated recruitment of CD8$^+$ T cells during infection (Figure 2) [145]. In RSV-infected mice, NK-derived IFN-γ prevented the development of lung eosinophilia in an IL-12-dependent manner [146]. Infection experiments in NK-deficient mice further showed an impaired IFN-γ production resulting in a Th2-skewed immune response and lung disease [147]. In secretions of RSV-infected infants with bronchiolitis, low numbers of NK cells and IFN-γ were detected, indicating that impaired NK activity is associated with more severe disease manifestations [148]. These studies thereby show a cytotoxic and immunoregulatory role of NK cells during RSV infection, mainly by the production of IFN-γ. In TLR4-deficient mice, the recruitment into the lungs and cytotoxic activity of NK cells was diminished upon infection compared to wild-type mice, indicating that RSV F protein activation of TLR4 is important for the recruitment and activity of NK cells [149].

Excessive recruitment of immune cells into the lungs during RSV infection, as seen for eosinophils and neutrophils, can induce immunopathogenic side effects. In RSV-infected mice, NK cells accumulated in the lungs during the early stage of infection and produced large amounts of IFN-γ (Figure 3) [150]. The depletion of NK cells reduced the disease severity and the influx of inflammatory cells like T cells [150]. In studies with nude and wild-type BALB/c mice, increased numbers of NK cells in the lung were associated with airway inflammation and hyperresponsiveness [151]. The depletion of NK cells reduced disease severity and decreased the levels of Th2-type cytokines IL-4, IL-5, and IL-13, indicating that NK cells might contribute to a Th2-skewed immune response at later stages [151]. Direct infection of NK cells might be an explanation for these immune responses. *In vitro* studies have shown that RSV infects neonatal and adult NK cells efficiently [152]. This infection of NK cells was further increased by sub-neutralizing antibodies, indicating the uptake of viral particles in an Fc-γ-receptor-dependent manner [152]. Infected NK cells were characterized by an increased expression of IFN-γ and a decrease in perforin secretion, suggesting a shift from the cytotoxic to the proinflammatory phenotype of RSV-infected NK cells [152].

The role of NK cells seems to depend on the stage of infection. In mice infected with recombinant RSV expressing IL-18, increased NK cell numbers were detected at early stages that correlated with increased lung inflammation and reduced viral replication [153]. Depleting NK cells during that infection stage reduced the disease severity and increased the viral load. However, the depletion of NK cells at later stages of infection was associated with increased disease severity mediated by the increased influx of CD8$^+$ T cells [153]. This demonstrates the biphasic role of NK cells during RSV infections.

The NK cell response strongly depends on environmental stimuli. At the early stages of infection, the cytotoxic and immune-recruiting activity of NK cells is required to restrict viral spread. However, the cytotoxic activity of NK cells may also affect uninfected cells and thereby contribute to increased lung pathology. At later stages of infection, as immune cells infiltrate the lung, NK activity must shift from a cytotoxic to a regulatory phenotype to limit immune-cell-mediated lung inflammation. An imbalance in this immune response, induced by direct RSV infection or an altered microenvironment, can lead to increased lung damage.

2.5. Monocytes

Monocytes originate in the bone marrow from a common myeloid progenitor and belong to the mononuclear phagocyte system [154]. Monocytes are classified into three groups based on surface markers and function [155]. Classical monocytes represent the largest population (~90%) and are $CD14^{++}CD16^{-}CCR2^{hi}CX3CR1^{low}$ with inflammatory and phagocytic activities [155]. Non-classical monocytes with a $CD14^{+}CD16^{+}CCR2^{low}CX3CR1^{hi}$ phenotype have antiviral and patrolling activities [155]. Intermediate monocytes represent a transitional population between classic and non-classic monocytes and have a $CD14^{+}CD16^{+}CX3CR1^{hi}CCR2^{low}$ phenotype. Intermediate monocytes are proinflammatory and show increased activity in antigen presentation [155]. Mature monocytes circulate in the blood and infiltrate the tissue in response to chemoattractants. In the steady state, monocytes can differentiate into macrophages and dendritic cells in the presence of granulocyte-macrophage colony-stimulating factor (GM-CSF) and macrophage colony-stimulating factor (M-CSF) [156,157]. During viral infections, monocytes and monocyte-derived phagocytes, i.e., macrophages and dendritic cells, contribute to viral clearance by the secretion of antiviral mediators and phagocytosis of infected cells [135,136], as shown for HIV-1 and the influenza virus [158–160].

Monocytes are recruited to the site of infection during the early phase by IL-6, IL-8, CCL2, CCL5, and MIP-1α derived from infected AECs and macrophages [108,161,162]. The kind of recruited monocytes depends strongly on the secreted cytokines. CCL2 mediates classical inflammatory cytokine recruitment while non-classical, anti-inflammatory monocytes are recruited by CX3CL1 [156]. Alterations in the secretion pattern, e.g., mediated by RSV, could lead to a modified and shifted immune response. Following recruitment and activation, monocytes produce CCL2 and MIP-1α/β, resulting in the recruitment of immune cells [108]. They directly contribute to viral clearance by phagocytosis of the virus in a surfactant A-enhanced manner [163]. In response to uptake, monocytes expressed type I IFNs, TNF-α, G-CSF, and IL-6 [109]. A byproduct of monocyte-mediated viral clearance is the decrease in CCL5 expression by AECs and, consequently, the limitation of immune cell infiltration and cell-mediated immunopathology [109]. Growth factors like G-CSF, however, are secreted continuously and mediate the survival and differentiation of recruited cells [109]. Direct interaction of the RSV F protein with TLR4 and CD14 on the surface of monocytes further initiated innate immune pathways that resulted in the secretion of IL-6, IL-8, TNF-α, and IL-1β, which in turn mediated the antiviral immune response [164]. In TLR4-deficient mice, recruitment of $CD14^{+}$ and NK cells was diminished upon infection while viral clearance was delayed [149,164].

The upregulation of co-stimulatory factors like CD40, CD80, and MHC-I/II on the surface of activated monocytes reflects their role in antigen presentation (Figure 2) [109,165]. Expression of HLA-DR on the surface of monocytes isolated from RSV-infected infants was reduced and correlated with increased disease severity, underlining their importance in antigen presentation and the initiation of adaptive immunity [165]. A previous study analyzed the immune responses in a CD14-deficient patient with recurrent RSV infections and discovered that this deficiency impaired the innate immune response, i.e., a decrease in IL-6 expression, mediated by monocytes to RSV, resulting in inadequate control of infection characterized by recurrent and severe RSV infections [166]. Monocytes, therefore, contribute to viral clearance by phagocytosis of viral particles and by supporting the proliferation, activation, and survival of other immune cells through the secretion of cytokines and upregulation of surface markers.

Several studies also described the involvement of monocytes in immunopathogenesis. *In vitro* restimulation of monocytes isolated from RSV-infected patients with bronchiolitis (*ex vivo*) showed an increase in IL-10, associated with recurrent wheezing, and decreased levels of IFN-γ, IL-12, and IL-4, indicating a direct contribution of monocytes to immunosuppression during severe infections (Figure 3) [167]. RSV can productively infect monocytes the same as other immune cells [168,169]. The permissiveness of monocytes to infection depended on their maturation and activation state, as *in vitro* studies of isolated cord blood monocytes were more permissive to infection than adult monocytes [168]. Infection of neonatal monocytes induced the expression and secretion of proinflammatory cytokine IL-1β [170]. The infection did not affect their cell viability, indicating that RSV inhibits the apoptosis of monocytes and instead alters their cytokine response [170].

The interaction of soluble G with CX3CR1 on the surface of monocytes may lead to altered gene and cytokine expression by inhibiting the nuclear translocation of NF-κB [171]. This change in cytokine response, notably the increased expression of IL-10, may skew for a Th2-typed immune response, a hallmark of severe infections [171]. Following their recruitment to the lungs, monocytes are exposed to different stimuli leading to their differentiation into macrophages with a proinflammatory (M1) or anti-inflammatory phenotype (M2) (described in Section 2.6). Based on this, it can be speculated that RSV alters the lung microenvironment resulting in increased recruitment of non-classical, anti-inflammatory monocytes. Microenvironmental stimuli and possibly the interaction of soluble G with CX3CR1 on their surface may result in alternatively activated, anti-inflammatory monocytes with an M2-like phenotype. This shift in the early immune responses would impair viral clearance.

Recruiting functional monocytes to the lungs in response to stimuli is pivotal for protection and immune regulation. In addition, monocyte recruitment is crucial to replenish the macrophage population during infection. Despite their protective effects, microenvironmental changes, especially due to G protein interactions, can affect their immune response. A modified cytokine expression following infection in the lung may shift monocyte differentiation toward an anti-inflammatory, M2-like phenotype, resulting in delayed viral clearance.

2.6. Macrophages

Macrophages belong to the mononuclear phagocyte system, like monocytes and dendritic cells, and are characterized by their phagocytic activity toward cell debris in different situations, like tissue repair or infections [154,172]. Macrophages circulate in the bloodstream but can migrate and reside in various tissues, like the lung, liver, and brain, in response to microenvironmental stimuli [172]. Two subsets of macrophages are present in the lung: alveolar macrophages, which are present on the surface of the alveolar cavity, and interstitial macrophages, which remain in the interstitium [173,174]. Lung-tissue-resident, alveolar macrophages show two distinct polarization states, M1 and M2, with different functions [156,173]. M1-like macrophages display a proinflammatory phenotype due to the secretion of proinflammatory cytokines, decreased expression of CX3CR1, and the increased expression of CD86, a co-stimulatory surface marker involved in activating T cells, demonstrating their role in antigen presentation [156,173]. M2-like macrophages, on the other hand, are characterized by the increased expression of CX3CR1, and mediate anti-inflammatory responses and contribute to tissue repair [156,173]. As mentioned above, monocytes are recruited to the lungs during viral infection to replenish the macrophage population. It should be noted that monocyte-derived macrophages are a distinct population formed during infection and do not have the role and function of tissue-resident, long-lived macrophages [156]. During the early stages of infection, monocytes differentiate into M1-like macrophages to mediate viral clearance and initiate adaptive immunity [156]. At later stages, monocyte differentiation switches from M1 to M2 polarization to limit the inflammatory response and promote tissue repair [156]. However, macrophage polarization is not irreversible and can shift between the different M-like subsets depending on endogenous stimuli [156].

Macrophages can detect RSV through MAVS-coupled RLRs and IFNAR signaling, producing type I IFNs, like IFN-α/β, IL-6, MIP-1α, and CXCL10 (Figure 2) [161,175,176]. It was shown that the macrophage-mediated release of IFNs is crucial for controlling RSV infection as they recruit immune cells to the infection site [161]. Several studies emphasized the role of macrophages in controlling and clearing viral infection. Depletion or impaired activity of macrophages led to increased viral replication, disease severity, inflammation, and excessive neutrophil/dendritic cell infiltration [177]. It is thought that the depletion of alveolar macrophages resulted in a proinflammatory environment characterized by increased expression of G-CSF, IL-17, TGF-β, MIP-1α, and IL-α/β, and decreased levels of TNF-α, IL-6, and type I IFNs [177]. It could thus be speculated that the functional immaturity of neonatal alveolar macrophages leads to an inefficient uptake of infected cells, recruitment of immune cells, and clearance of damaged cells [178]. The activity of different subsets of macrophages appears to be dependent on the stage of infection, similar to the activation pattern of NK cells. During the early stage of infection, M1-like, proinflammatory macrophages are essential for the restriction of viral spread, removal of infected cells, expression of type I IFNs for immune cell recruitment, and antigen presentation to initiate adaptive immune responses. Later during an infection, an M1 to M2 phenotypic switch can be observed, where M2 macrophages produce type III IFNs in a PPAR-γ-dependent manner and secrete IL-10 to control the immune-cell-mediated inflammatory response and mediate tissue repair [179]. M2 differentiation is thought to be mediated by TLR4 and IFN-β signaling [180]. This phenotypic switch is crucial for suppressing exaggerated, unspecific antiviral responses and preventing unnecessary tissue damage.

A possible contribution of macrophages to immunopathology remains inconclusive and requires further research. Several *in vitro* and *in vivo* studies show the infection of macrophages by RSV [181–183]. This infection is TLR4- and CX3CR1-dependent and polarizes the differentiation of neonatal macrophages toward an M2-like phenotype (Figure 3) [180,184,185]. It is unclear whether it is an abortive or productive infection, as various studies yielded different conclusions. While one study shows no loss of cell viability upon infection [181], another indicated that RSV induces necroptosis and suggests an enhancement of viral replication by M2-like macrophages, thus contributing to disease severity and lung pathology [183]. Regardless of that, *in vitro* infection of macrophages, isolated from adult or neonatal humans, has been shown to alter the immune response, characterized by the expression of proinflammatory cytokines like TNF-α, IL-6, and IL-8 [186,187], while the expression of IFN-γ and IL-12, a regulatory cytokine, was impaired [187]. Reduced expression of IFN-γ was associated with impaired activation of macrophages resulting in reduced phagocytosis of the virus and recruitment of immune cells like T cells [188]. These data suggest that the direct interaction with RSV impairs the activation and function of neonatal macrophages by skewing their differentiation toward an M2-like phenotype, thereby contributing to disease severity.

Macrophage responses during RSV infections follow a biphasic course similar to NK cell response. During the early stage of infection, proinflammatory M1-like macrophages are activated to restrict viral spread. At later stages, probably by the arrival of T cells, a phenotype switch towards an anti-inflammatory, immune regulatory M2-like type is necessary to reduce immunopathogenesis. It can be speculated that in infants with severe infections, the change in the macrophage response is impaired, leading to exaggerated T-cell responses and increased lung damage.

2.7. Dendritic Cells

Dendritic cells complete the mononuclear phagocyte system and are considered professional antigen-presenting cells as they bridge innate and adaptive immune responses during infections [189]. Compared to the other two members of the phagocyte system, DCs only display a low phagocytic activity [190]. Two DC subsets have been described based on the expression of surface markers, function, and origin. Conventional or myeloid DCs (cDCs) are involved in tissue damage sensing and the capture and presentation of antigens [191,192]. cDCs are further divided into cDC1s, located directly underneath the airway epithelium and responsible for MHC-I-mediated antigen presentation to CD8[+] T cells, and cDC2s, located in the lung parenchyma and presenting antigen to CD4[+] T cells [191,192]. Plasmacytoid DCs (pDCs) are found in blood and lymphoid tissues and are an essential source of IFN-α/β [191,192]. While DCs show reduced phagocytic activity and low antigen presentation under homeostatic conditions, their T cell-priming activity increases during viral infections as co-stimulatory markers, like CD40, CD80, and CD86, are upregulated on the surface [191].

Upon infection with RSV, DCs are recruited to the site of infection and sense the virus either through direct infection or by binding viral antigens through PRRs, like TLRs or RLRs [193,194]. As a result, co-stimulatory molecules CD40, CD80, CD83, CD86, and MHC-I/II were found to be upregulated on the surface of infected DCs [195], indicating their maturation (Figure 2). *Ex vivo* studies implied a 'two-step process' for maximal maturation of DCs since only the direct infection of pDCs led to increased expression of CD40 [196]. Once DCs are activated and matured, TLR7 and myeloid differentiation factor 88 (MyD88) mediate the production of antiviral and Th1-type cytokines (CCL5, IL-12, IFN-α, and IFN-β) [176,193,197,198]. Matured DCs then migrate to lung-draining lymph nodes to initiate the virus-specific T-cell response and have an essential role in initiating the virus-specific adaptive immune response and, consequently, in controlling RSV infection.

Despite their protective and crucial role in immunity, DC functions are heavily affected by RSV-induced immune modulatory mechanisms. Several studies showed that the direct infection of DCs impacts subsequent DC-mediated T-cell activation (Figure 3) [199,200]. Upon infection, the RSV N protein is expressed on the surface of infected DCs, which is thought to interfere with DC–T cell synapse formation [201]. This interference impairs or delays the formation of RSV-specific T-cell immunity [201]. Others found a decreased expression of co-stimulatory molecules on infected neonatal DCs that impaired their recruitment to the lymph nodes, ultimately reducing virus-specific T-cell responses compared to the control group [202,203]. The increased expression of IL-4R-α on the surface of neonatal cDC2s, in comparison to cDCs derived from adult mice, may also account for the altered DC responses and impaired recruitment of the cells into the lymph nodes as the deletion of the receptor resulted in increased recruitment to the lymph nodes, reduced Th2 polarization, and upregulation of IL-12 expression [204]. Interaction of the RSV G protein with DC-/L-SIGN inhibits the activation of DCs and interferes with the DC-induced production of cytokines, illustrating an immunomodulatory mechanism by RSV for interfering with the host's immune response [205].

DCs, as the bridge between innate and adaptive immunity, are of major importance for the induction of virus-specific immune responses. The direct infection of DCs and the associated functional changes may lead to severe consequences in initiating a functional adaptive immune response.

3. Adaptive Immune Responses to RSV Infection

The adaptive immune response represents the second line of defense against infections. Although innate immunity reacts to infection rapidly, the recognition of pathogen patterns is limited since the responsible receptors are germline-encoded [206]. Adaptive immune responses, on the other hand, are characterized by highly specific receptors on the surface of the lymphocytes allowing targeted, pathogen-specific responses. Activation of adaptive immune cells leads to their proliferation and clonal expansion, exerting their effector function, and ultimately to the generation of an 'immunological memory' that persists in the host and can be rapidly reactivated during reinfection [207,208]. The adaptive immune system comprises two cell types: T lymphocytes and antibody-secreting B cells [207]. Mature adaptive immune cells reside in secondary lymph nodes and can infiltrate the site of infection after stimulation with antigens presented by innate immune cells, like monocytes, macrophages, or DCs [207]. In the following part, the protective roles of B and

T cells and the potential impairment of their function by RSV during infections will be discussed in more detail.

3.1. B Cells

B cells originate in the bone marrow from hemopoietic stem cells [209]. In an immature state, immunoglobulin (Ig)-expressing B cells leave the bone marrow and migrate to secondary lymphoid organs for their final development and maturation [209]. Final B cell activation is mediated by two distinct processes: T-cell-independent and -dependent activation, both of which require the binding of an antigen to the B cell receptor (BCR) [209]. During T-cell-independent activation, antigens with multiple, repeating epitopes are necessary for the cross-linking of BCRs, leading to the proliferation and differentiation of B cells into antibody-secreting plasma cells [209]. For the majority of antigens, however, B cell activation requires the help of $CD4^+$ T helper (Th) cells. During this process, B cells internalize the antigen and present it on the cell surface through MHC-II. The antigen is then recognized by antigen-specific Th cells that initiate B cell proliferation and differentiation by transmitting activating signals [209]. Besides the generation of antibody-secreting plasma cells, the T cell-dependent activation of B cells also allows the development of B memory cells. B cell development is completed by somatic hypermutation, affinity maturation, and isotype switching, all of which result in the production of highly specific antibodies [209]. Upon viral infections, B cells secrete virus-specific antibodies, neutralizing the virus and preventing viral cell entrance by blocking the viral binding domain [210]. B cells and antibodies, therefore, are an important correlate of protection by eliciting virus-specific effector functions.

Following RSV infection in infants and adults, B cell numbers in the peripheral blood were increased compared to uninfected individuals. They displayed a differentiated phenotype by the expression of typical proliferation, differentiation, and survival markers, such as B cell-activating factor (BAFF) and A proliferation-inducing ligand (APRIL) (Figure 2) [211–215]. These observations are supported by studies in mice where RSV infection led to the upregulation of BAFF, APRIL, and CXCL13 on B cells [216,217]. These studies demonstrate the activation and circulation of B cells following RSV infection and indicate their migration to the site of infection. Although B cells are activated and recruited during RSV infection, studies suggest that these numbers rapidly decrease once the infection is cleared [218–220]. In blood samples of infected adults, low levels of RSV-specific B cells were detected in healthy 'non-healthcare' workers [219]. After experimental infection of the volunteers, the numbers of RSV-specific B cells rapidly decreased within months after exposure [219]. *Ex vivo* studies confirmed these observations after the experimental infection of adults in which RSV-specific B-cell responses generated in the volunteers were poorly maintained and returned to baseline levels one year after exposure, indicating a defective memory B-cell response in adults to a certain degree compared to samples from RSV-negative, hospitalized infants [220].

RSV-specific, neutralizing antibodies can be detected in humans as early as two days after infection [213]. They mainly target the RSV glyco- and fusion protein (preF and postF confirmation) and display a cross-reactivity for both RSV subtypes [212,221]. RSV-specific antibodies can mediate viral clearance through various mechanisms, like ADCC, antibody-dependent cellular phagocytosis, and the neutralization of viral particles [221]. Their protective role was confirmed in mice, where administering RSV-specific, neutralizing antibodies before RSV exposure reduced the morbidity and mortality of the animals compared to IgG-treated controls [222]. The presence of these neutralizing antibodies before the RSV challenge further restricted virus replication in the early stage of infection and thereby reduced the activation of $CD8^+$ memory T cells that displayed immunopathogenic activity in control mice through the increased secretion of IFN-γ [222]. In earlier studies with older adults, low RSV-neutralizing serum and mucosal antibody levels prior to infection correlated with disease severity [21,223–227]. Upon RSV infection, older adults develop stronger nasal and serum antibody responses than younger subjects, which is dependent on the duration of viral shedding [213,228]. In a recent study, older adults mounted poor nasal IgA responses, but robust serum IgG responses that correlated with recovery from the RSV infection [226]. Levels of RSV-specific, neutralizing antibodies, however, were relatively short-lived and titers dropped more than 4-fold within a year after infection in the majority of study subjects [218]. Despite the fact that older adults have comparable antibody levels to younger subjects prior to infection, they are more susceptible to severe RSV infections, which indicates that other immunological parameters, like dysfunctional T cells, may account for this [229,230].

Although antibodies mediate viral clearance, they may also contribute to immunopathogenesis during RSV infections. Isolated RSV-specific antibodies and B cells of infected infants show significant differences from those of adults [231]. F-specific antibodies of infants under 3 months of age showed restricted affinity, neutralization capacity, and recognition of antigenic sites compared to adult samples [231,232]. The RSV F protein contains six major antigenic sites (ø, I–V), of which only antigenic sites I and III were recognized by antibodies isolated from infants. Antibodies directed against antigenic site I do not display neutralizing activity and their preferred production in infants could result in lower levels of neutralizing antibodies. A high abundance of non-neutralizing antibodies in infants could ultimately result in the formation of immune complexes that may contribute to lung pathology if not cleared in time [231]. The defective generation of neutralizing antibodies seems to be age-dependent in childhood as the recognition of additional epitopes increases with age [231]. A genetic limitation in infants under three months of age may be responsible for this defect as infant B cells have limited use of antibody variable genes and lack somatic hypermutation [233]. Maternally derived antibodies may provide an explanation for this limitation since several studies indicated a suppressive effect of maternal antibodies on the infant's immune response in the first months of life during RSV infection *in vivo* [219,234,235]. Recent immunization experiments in infant mice indicate that maternal antibodies altered B cell differentiation and isotype switching, thereby limiting the B cell repertoire [236].

The virus itself further affects B-cell responses. Analysis of neonatal regulatory B cells (nBregs) isolated from cord blood demonstrated their infectability by RSV [237]. The interaction of the RSV F protein with the B cell receptor activated the regulatory B cells and increased the expression of CX3CR1, a potential viral receptor, via the CX3C motif of the RSV G protein. Therefore, the interaction of RSV F with BCRs on the surface of nBregs leads to increased expression of CX3CR1 on the surface of the cells, making them susceptible to direct infection. The infection of nBregs resulted in the secretion of IL-10, an immunoregulatory cytokine, which reduced the production of Th1-type cytokines [237]. This study, therefore, demonstrates the immune modulatory effects mediated by RSV F and G on nBregs, ultimately resulting in an immune response that favors viral survival (Figure 3) [237]. Besides the direct infection of nBregs, the secretion of soluble G may depict another immune modulatory effect on B cell activity. Soluble G acts as an antibody decoy during infections, thereby serving as an escape mechanism for viral progeny from G-specific antibodies and interfering with antibody-mediated viral clearance [238].

Antibodies are the major correlate of protection during RSV infection in infants, as shown by the prophylactic use of human polyclonal neutralizing antibodies [239–241] and the neutralizing monoclonal antibodies palivizumab and nirsevimab [242,243]. The presence of pre-existing, RSV-specific antibodies and, therefore, memory B cells during reinfections is crucial, as they are able to detect the virus before the infection of immune cells. Memory B cells thus have a relatively long-lasting effector function, while the effector functions of memory T cells are limited to the presence of antigen [244].

Direct modulation by RSV G and the genetic limitations of the immune system in infants under 3 months of age interfere with the generation of neutralizing antibodies, leaving the infant at risk of more severe infections. Direct infection of nBregs resulting in the secretion of immunoregulatory IL-10 demonstrates that interactions of viral proteins can shift the immune response, generating an environment that favors viral shedding.

Although RSV-specific immune responses have been studied extensively in infants, the induction of these responses in older adults has been underinvestigated and warrants more research, especially in the context of age-related immunosenescence [245].

3.2. T Cells

Unlike B cells, T cells arise from a common lymphoid progenitor in the thymus [208]. T cells migrate to secondary lymphoid tissues upon maturation to encounter antigens [207,208]. Two major subsets of T cells have been described, characterized by the expression of either CD4 or CD8 on their surface. Both subsets are activated by the formation of a so-called 'immunological synapse' which describes the interaction of different T cell markers, e.g., CD4/CD8, CD3, or T cell receptor, with antigen-bearing MHCs on the surface of professional (MHC-II; DCs, macrophages, monocytes, and B cells) and non-professional (MHC-I; epithelial cells) antigen-presenting cells [207,246]. Activated T cells can exert various effector functions, such as the elimination of infected cells, activation of other immune cells, like B cells, through the secretion of cytokines or direct interaction, or regulation of immune responses to prevent excessive damage [247]. In the following part, the protective role of the individual T cell subsets as well as their contribution to immune pathogenesis during an RSV infection will be illustrated in more detail.

3.2.1. CD4$^+$ T Cells

CD4$^+$ T cells are pivotal in the host's immune response as they orchestrate the activity of innate and adaptive immune cells through the secretion of several cytokines. They can be divided into different subsets (i.e., Th1, Th2, Th9, Th17, T-follicular helper, and Tregs), which mainly provide helper functions and are characterized by the secretion of specific cytokines [247]. For example, Th1-like cells are involved in activating mononuclear phagocytes or cytotoxic T cells by the secretion of IFN-γ, TNF-α, IL-2, IL-12, and IL-18. Th2-like cells, by contrast, are involved in eosinophil responses, and B cells maturation, and production of virus-specific antibodies by secreting IL-4, IL-5, IL-10, and IL-13 upon activation [247].

During RSV infection, CD4$^+$ T cells are recruited to the airways and orchestrate the immune response via secretion of IL-17, IFN-γ, and TNF-α (Figure 2). The adoptive transfer of CD4$^+$ T cells into nude mice demonstrated their beneficial role during infection and B cell differentiation as they were necessary to induce the production of RSV-specific antibodies [248]. Adoptive transfer of CD4$^+$ T cells from the airways of RSV-infected mice into naïve mice reduced disease severity by suppressing the secretion of TNF-α compared to control mice, further supporting their protective role during infection [249]. In human experimental infection studies, numbers of RSV F- and G-specific, CD4$^+$ T cells were increased in the airways of infected individuals [250]. By contrast, in infants with severe RSV infection with subsequent wheezing, numbers of CD4$^+$ T cells were reduced and produced lower levels of TNF-α after *in vitro* restimulation compared to infants without wheezing, indicating a partial impairment of the CD4$^+$ T-cell responses in these children [148,251,252]. The production of proinflammatory cytokines by CD4$^+$ T cells during viral infections is crucial for the activation and regulation of innate and adaptive immune responses. However, the balance between pro- and anti-inflammatory cytokine responses is of great importance, and any alteration in this balance may have serious consequences, as illustrated by the use of TNF-α. The studies mentioned above show that an imbalanced TNF-α response during RSV infections can lead to more severe disease and prolonged viral shedding.

A dysfunctional CD4$^+$ T-cell response during infection was linked to immunopathology after FI-RSV vaccination [253,254]. Studies in mice vaccinated with FI-RSV showed a Th2-skewed immune response characterized by the increased expression of Th2-type cytokines like IL-5 and IL-13, resulting in airway hyperresponsiveness (AHR) and increased mucus secretion compared to mock-immunized controls [255]. Depletion of CD4$^+$ T cells in the mice significantly improved the condition of the mice during infection, indicating that a biased Th2-type immune response contributes to disease severity in FI-RSV-vaccinated mice [255]. Similarly, RSV infection of neonatal mice and their reinfection in the adult phase led to an exaggerated Th2-type immune response resulting in AHR, mucus hypersecretion, and eosinophilia compared to mice that were infected during adulthood [256]. The increased expression of IL4-Rα on the CD4$^+$ Th cells of these mice indicates a preferential proliferation of CD4$^+$ T cells to a Th2-type [256]. In experimental infection studies of adults, increased levels of CD4$^+$ T-cell-derived IFN-γ, IL-2, IL-4, IL-10, and TNF-α and the number of CD8$^+$IFN-γ^+ T cells correlated with increased disease severity and hospitalization, emphasizing the potential immunopathogenic role of CD4$^+$ T cells during RSV infection [257]. Depletion of CD4$^+$ T cells, IL-10, or IL-4 before infection or vaccination ameliorated disease severity but consequently resulted in prolonged viral shedding [118,255,258–260]. It has been shown that RSV can productively infect CD4$^+$ and CD8$^+$ T cells, resulting in the decreased expression of IFN-γ and interfering with the T cell functions and development of Th1 cells, probably in an F-protein-dependent manner (Figure 3) [261,262]. This direct infection may hinder the generation of long-term protective immunity since RSV-specific lymphocyte responses only persist for about a year and do not boost virus-specific cell-mediated immunity upon reinfection [263].

A balanced CD4$^+$ T-cell response is pivotal for developing protective immune responses by coordinating the recruitment and activity of innate and adaptive immune cells through the secretion of cytokines. Direct infection by RSV and partial genetic dysfunction of the infant's CD4$^+$ T-cell response, however, may interfere with this balance, leading to a Th2-skewed immune response and more severe infections.

3.2.2. CD8$^+$ T Cells

CD8$^+$ T cells can recognize and eliminate virus-infected cells through their cytotoxic activity. During RSV infection, naïve CD8$^+$ T cells are activated through the interaction of the T cell receptor either with MHC-I on the surface of infected epithelial cells or with MHC-I on the surface of APCs in lung-draining lymph nodes (cross-presentation) [195]. Following activation, virus-specific effector CD8$^+$ T cells expand and migrate to the site of infection in response to chemoattractants CCL5 and CXCL10 [264,265]. Activated CTLs directly interact with antigen-bearing MHC-I on the surface of RSV-infected cells and induce the apoptosis of the target cell by the secretion of perforin and granzymes [266–268].

Several studies revealed a protective role of CD8$^+$ T cells during RSV infection. The depletion of CD8$^+$ T cells in RSV-infected mice resulted in delayed viral clearance compared to control mice. The adoptive transfer of CTLs, isolated from RSV-infected mice, into naïve mice during RSV infection reduced both the viral load and weight loss [249]. These results are supported by human studies analyzing the blood and tissue samples of infected infants, in whom reduced CD8$^+$ T cell counts were associated with delayed viral clearance and increased disease severity in comparison to uninfected infant samples, emphasizing their protective role during infection [148,252,269,270]. In an RSV-infected infant with severe immunodeficiency, bone marrow transplantation significantly reduced the nasal viral load [271]. This decrease in viral load began with the increasing activity of CTLs after transplantation [271]. CTLs mediate their effector functions during RSV infection by the secretion of several cytokines, e.g., IL-2, IL-10, INF-γ, TNF-α, and the serine protease granzyme B (Figure 2) [264,266,267,272–275]. The importance of CD8$^+$ T-cell-derived cytokines is demonstrated by studies in mice that were primed for an RSV-specific CD8$^+$ T-cell response prior to RSV challenge [276]. In the primed mice, CD8$^+$ T cells suppressed the exaggerated Th2-type cytokine response characterized by the decrease in IL-4 and IL-5, and consequently led to the suppression of excessive recruitment of eosinophils compared to controls. Thus, it can be assumed that CD8$^+$ T-cell-derived cytokines contribute to the regulation of CD4$^+$ T-cell responses [276].

Despite their crucial role in viral clearance, CTLs are also involved in immunopathology following RSV infection. RSV-infected mice displayed increased levels of CTL-secreted IFN-γ that correlated with disease severity in comparison to control mice [277]. Depletion of CTLs in mice markedly reduced disease severity following RSV infection, although viral shedding was prolonged, indicating that CTLs are involved in both protective and harmful immunity [258]. In BAL samples of hospitalized, RSV-infected infants, increased levels of granzymes A and B were detected compared to children without a pulmonary disease (Figure 3) [278,279]. The increased levels of granzymes correlated with IL-8 levels in the airways and disease severity [278,279]. In experimental infections of adults, the peak of CTL numbers in the respiratory mucosa was linked to the reduction in viral load and an increase in symptoms, emphasizing the dual role of CTLs during RSV infection [266]. Several studies have implicated CTL-derived IFN-γ and TNF-α as a driving force in CTL-mediated immunopathogenesis during RSV infection. Increased levels of IFN-γ in the BAL of RSV-infected mice correlated with disease severity, while its depletion with neutralizing antibodies or absence in IFN-γ-KO mice led to ameliorated disease [277,280]. A similar role in immunopathogenesis was shown for TNF-α. Its neutralization in mice before RSV infection reduced weight loss [281].

These data emphasize the role of CTLs in both virus elimination and immunopathogenesis. While CTLs mediate viral clearance through their cytotoxic activity and are involved in the regulation of a balanced CD4$^+$ T-cell response, their excessive activation and infiltration of the lung tissue may contribute to lung pathology and disease severity characterized by the increased secretion of granzymes.

3.2.3. Regulatory T Cells

Previously discussed immune cells, such as B and T cells, illustrate the importance of a balanced immune response during viral infection, like SARS-CoV-2 or hepatitis B virus [282,283]. Tregs are a subset of CD4$^+$ T cells characterized by the expression of CD25 and the transcription factor forkhead box protein 3 (FoxP3), which is pivotal for their development and regulatory function [284]. Tregs can control immune cell recruitment and proliferation and prevent exaggerated immune responses during infections by secreting TGF-β and IL-10 [285–288].

During RSV infection, Tregs accumulate early in the lungs and lymph nodes of mice [275,289,290]. The depletion of Tregs in mice resulted in the delayed recruitment of CTLs and CD4$^+$ T cells to the site of infection, followed by delayed viral clearance, increased weight loss, and slower recovery compared to naïve control mice, indicating that Tregs mediate the recruitment of T lymphocytes during RSV infection [117,290–293]. Despite the delayed recruitment of CTLs and CD4$^+$ T cells after Treg depletion in mice, their numbers were significantly increased later. They were associated with an excessive inflammatory response later during infection, characterized by the increased secretion of IFN-γ and TNF-α (Figure 2) [117,290]. Exaggerated and persistent Th2-like responses, characterized by IL-13-expressing CD4$^+$ T cells, were also noticed in the absence of Tregs, further illustrating the role of Tregs in the regulation and suppression of immunopathology during RSV infection [117]. *In vivo* studies showed an increased expression of granzyme B by lung-resident T cells compared to naïve mice, suggesting that the expression of granzyme B mediates the regulatory activity of Tregs and that Treg-derived granzyme B may eliminate activated lymphocytes, thereby regulating inflammatory responses [292]. Since Tregs are a major source of IL-10, their immunosuppressive function might be mediated through an IL-10-dependent mechanism. Studies in IL-10$^{-/-}$ mice show an increased accumulation of CTLs, increased levels of proinflammatory cytokines, like IFN-γ and TNF-α, and increased disease severity in the lungs of RSV-infected mice in comparison to control mice [294,295]. The absence of Tregs in mice further indirectly affected immunity to RSV as decreased concentrations of neutralizing and increased concentrations of non-neutralizing antibodies were measured compared to those in naïve mice [293]. These data underline the versatile role of Tregs during RSV infection.

In infants with severe infections, the numbers of activated Tregs were reduced in blood and nasal aspirates compared to uninfected children and remained at that level for several weeks after infection [296,297]. The decrease in IL-33 levels, a cytokine involved in the accumulation and homeostasis of Tregs in mucosal sites, in the nasal aspirates of RSV-infected infants, may explain the reduced numbers of Tregs in these samples (Figure 3) [296]. The reduced accumulation of Tregs and lower levels of IL-33 in these infants may be the consequence of a Th2-type cytokine milieu. The Th2-type environment, characterized by the increased secretion of IL-5 and IL-13, induced the Th2-like development of the Tregs effector phenotype and a loss of its immunosuppressive function [298]. Th2-like Tregs were characterized by the expression of GATA-3, a transcription factor that can inhibit the expression of FoxP3 in Tregs [298]. Since Tregs are involved in cell recruitment, especially during the early stage of infection, a decrease in their activity would mean a delayed infiltration of immune cells into the tissue, e.g., CTLs, and, consequently, a delay in viral clearance [290].

Tregs and their activity are crucial to limiting immunopathology during RSV infection, probably mediated by granzyme B and IL-10. Alteration in the host immune response to RSV infection towards a Th2-like phenotype indirectly interferes with Treg activity, resulting in delayed and increased recruitment of immune cells and exaggerated responses.

3.2.4. Memory T Cells

Memory T cells are essential for the rapid, antigen-specific immune response during virus reinfection. Following viral clearance, the virus-specific T cell population contracts, leaving only a small fraction of cells to survive, which differentiate into persisting memory T cells [299]. Memory T cells can be classified into four subsets based on surface marker expression [300]. Following infection, central memory T cells (CD45RA$^-$CCR7$^+$) home in on secondary lymphoid organs, while effector memory T cells (CD45RA$^-$CCR7$^-$) migrate into the peripheral tissue to mediate effector functions [300]. Late effector memory T cells are considered terminally differentiated and have a reduced functional capacity [300]. Tissue-resident memory T cells (Trm), characterized by the increased expression of CD69 and CD103, represent the last subset and reside within the peripheral tissue [266,301]. Lung Trms are further characterized by the expression of CD11a and CD49a [302].

During secondary infections with RSV, memory cell populations rapidly expand and induce a virus-specific immune response in a MAVS- and IFN-α-dependent manner [195,303,304]. Memory T cells in mice are recruited and activated in three waves: the initial activation of tissue-resident memory T cells in the lungs is followed by the secondary recruitment of circulating, RSV-specific memory cells from the periphery. The third wave consists of memory T cells activated by antigen-presenting cells in the lung-draining lymph nodes [305]. The fast response mediated by memory T cells leads to early virus recognition and elimination, reducing the disease severity and viral load [266]. In blood samples of young and elderly volunteers, resting CD8$^+$ memory T cells were identified that were characterized by an increased expression of CD27, CD28, and IL-7Rα (Figure 2) [303]. *In vitro* restimulation of these memory cells led to the production of IFN-γ and granzyme B, demonstrating that they can regain their effector functions and proliferate and expand during RSV infection [303]. Similar results were shown in RSV-infected mice [306]. The adoptive transfer of tissue-resident memory T cells into naïve mice before infection reduced weight loss and the viral load compared to control mice showing the protective effect of Trms during RSV infection [249].

Because repeated RSV infections are common, an impairment of RSV-specific memory T cell generation may be assumed (Figure 3). In mice, RSV infection induces RSV-specific Trms that correlate with viral clearance and reduced disease severity [306]. However, the number of RSV-specific Trms rapidly waned within five months, corresponding to the age of a young adult in humans [306]. Prime and challenge studies in mice showed similar results as RSV-specific Trms displayed an impaired IFN-γ response, and the numbers of Trms declined rapidly after the clearance of infection [307]. This defect in RSV-specific Trm response was restricted to the respiratory tract, indicating an immunosuppressive effect of RSV on the effector functions and the generation of memory of local T cells after infection [307]. Preexisting, RSV-specific CD8$^+$ memory T cells mediate viral clearance in the absence of RSV-specific CD4$^+$ T cells and antibodies in mice, emphasizing the protective role of CD8$^+$ memory T cells in the absence of other adaptive immune cells [308]. Besides their protective role, CD8$^+$ memory T cells were also involved in excessive weight loss and increased disease severity and immunopathology in infected mice mediated by the increased secretion of IFN-γ compared to control mice [308]. This study demonstrates that memory T cells mediate viral clearance upon reinfection but also lead to increased disease severity and immunopathogenesis in the absence of regulation factors like CD4$^+$ T cells and antibodies. In a recent study, peripheral blood mononuclear cells were isolated from healthy children who encountered RSV infection during infancy. Restimulation of these PBMCs showed an impaired memory T-cell response irrespective of the disease severity during RSV infection when compared to children that were not infected during infancy [309]. The altered memory T-cell response was indicated by reduced IFN-γ, TNF-α, and IL-2 secretion. This study demonstrates that RSV infections during infancy can attenuate the memory Th1- and Th17-cell responses to RSV later in life regardless of the disease severity of the primary infection or reinfection [309].

The reduced survival of memory T cells may be due to intrinsic cell programming, as studies in infant mice infected with influenza virus showed an increased expression of T-bet and a reduced expression of survival factor CD127 on the surface of T cells compared to infected adult mice. This altered phenotype persisted even after the adoptive transfer of infant Trms into adult mice, indicating that the host environment is not responsible for the inefficient establishment of memory and that infant Trms are intrinsically programmed for short-term immunity [310].

Generating memory T cells is essential for long-term protection against recurrent infections. Fully functional memory T cells are produced during RSV infections that can mediate protection upon reinfection. The rapid decrease in RSV-specific memory T cells and impaired functionality in infants and older adults indicate that this protection is not long-lasting and predisposes the host to recurrent infections. Further studies will be necessary to better understand the role of memory T cells during RSV infection and immunopathogenesis, and the role of RSV in regulating these processes.

4. Vaccine Development and Treatment Options with Novel Insights into Immune-Mediated Protection

After the first isolation of RSV from chimpanzees in 1956 and infants in 1957, the first formalin-inactivated vaccine, FI-RSV, was developed in the 1960s [311,312] and used in a cohort of infants in the USA. Vaccination with FI-RSV led to an increased frequency and severity of ALRIs in seronegative infants following their first natural infection and even led to the death of two vaccine recipients [29]. Possible explanations for ERD in children were discussed above (see Sections 2.2, 2.3 and 3.2.1). For over six decades, the search for an effective vaccine continued without success. RSV vaccine and human monoclonal antibody development changed drastically with the stabilization of the F protein in its prefusion conformation as antibodies directed against prefusion epitopes have a tenfold increase in neutralizing activity compared to antibodies directed to postfusion epitopes [313]. Recently, two stabilized, prefusion F-protein-based vaccines, Arexvy (GSK) and Abrysvo (Pfizer), were licensed for the prevention of ALRI in older adults. Abrysvo was recently approved for the vaccination of pregnant women in their third trimester for the prevention of RSV-induced ALRI in infants during their first RSV season. Since a broad range of RSV vaccines and treatment options are currently in clinical trials, we focus on the ones that are either licensed or are/were in phase III clinical trials (Figure 4).

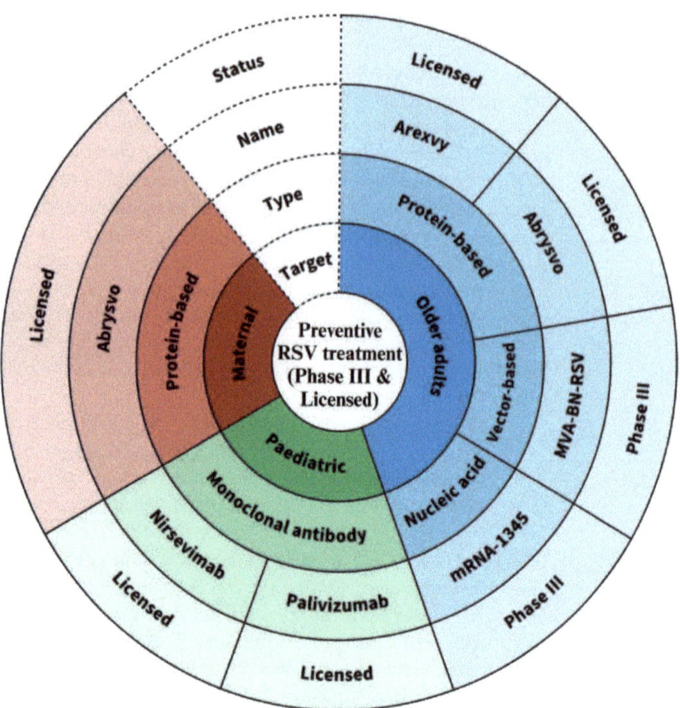

Figure 4. Schematic overview of current vaccines and monoclonal antibody-based treatment options available for RSV. The mentioned vaccines and monoclonal antibodies are either licensed or currently in phase III clinical trials for the treatment of patients at risk. Created with BioRender.com.

Vector-based vaccines have been used for decades for several pathogens, showing their potential to induce protective immunity in recipients after vaccination. They are generated by integrating a viral sequence encoding a target antigen into a non-pathogenic, often replication-deficient viral vector, like modified vaccinia virus or adenoviruses [314]. Vector-based vaccines generally aim to induce a robust antibody and CTL response and are characterized by high immunogenicity. The advantages of vector-based vaccines are the simple and safe production of a high number of vaccine doses [314]. Among the candidates for RSV, a modified vaccinia virus Ankara (MVA)-based vaccine expressing the RSV N, F, G, and M2 proteins (MVA-BN-RSV) is currently in phase III trials for Bavaria Nordic. MVA-BN-RSV would be recommended for use in older adults since vaccination induces stronger T-cell responses than antibody responses. In clinical trials, vaccination induced moderate neutralizing antibody responses and enhanced Th1-type immune responses indicated by the low levels of IL-4 in vaccinated individuals compared to placebo controls. A booster shot after 12 months further increased the antibody titers and T-cell responses in vaccine recipients [315,316]. Phase III clinical trials for MVA-BN-RSV are currently ongoing (NCT05238025). Although vector-based vaccines elicit strong T-cell responses, preexisting immunity to adenoviruses or poxviruses (in smallpox-vaccinated adults) may interfere with vaccine efficacy. Ad26.RSV.preF is an adenovirus-vectored vaccine candidate developed by Janssen Pharmaceuticals which expresses the preF protein. Similar to MVA-BN-RSV, it is designed for use in older adults and is additionally suitable for seropositive children. First trials with Ad26.RSV.preF resulted in increased antibody titers, reduced risk of ALRI, and reduced viral load in comparison to the unvaccinated group [317–319]. In March 2023, Janssen Pharmaceuticals discontinued its phase III vaccine development strategy based on Ad26.RSV.preF.

In contrast to whole virus vaccines, subunit vaccines are based on one or more purified viral antigens [320]. Therefore, the choice of antigen is critical in generating protective immune responses. Subunit vaccines are relatively safe as reversion to virulence is not possible, and a spread to unvaccinated individuals can be ruled out [320]. Compared to whole-virus vaccines, however, subunit vaccines must generally be delivered together with an adjuvant, and multiple doses may be necessary because of reduced immunogenicity [320]. A protein-based, bivalent vaccine directed against the prefusion F protein of both RSV subtypes (RSVPreF) was designed by Pfizer for the vaccination of pregnant women and older adults. Intramuscular vaccination of healthy adult men and nonpregnant women (18–49 years) as well as older adults (50–85 years) with RSVPreF induced robust neutralizing antibody responses against both RSV subtypes, with elevated titers through 12 months post-vaccination, while being safe and well tolerated [321–323]. In phase III trials of RSVPreF in older adults, vaccination efficiently prevented RSV-associated ALRI in the vaccine recipients compared to the placebo group [324]. Consistent with previous studies, RSVPreF was safe and well-tolerated [324]. In May 2023, RSVPreF, marketed as Abrysvo, was approved by the FDA for the vaccination of older adults ($\geq$60 years) for the prevention of RSV-associated ALRI. Phase III trials for RSVPreF in older adults are currently ongoing to evaluate the efficacy and safety of a second dose of RSVPreF throughout two RSV seasons (NCT05035212). Single-dose vaccination of pregnant women with RSVPreF, administered by the beginning of their third trimester, showed the induction of neutralizing antibodies and efficient transplacental migration of these antibodies to the fetus. Maternal vaccination effectively reduced the risk of ALRI and protected infants from RSV infection during the first six months of life compared to the placebo group [325,326]. The vaccine, marketed as Abrysvo, was recently approved by the FDA for the vaccination of pregnant women during their third trimester representing the first vaccine targeting the infant population. Phase III trials for maternal vaccination are currently ongoing to further evaluate the efficacy and safety of the vaccine in pregnant women and infants (NCT04424316).

A similar subunit vaccine, RSVPreF3, is based on a prefusion F protein antigen and was developed by GSK. Intramuscular vaccination of healthy pregnant women during their late second or third trimester with unadjuvanted RSVPreF3 was safe and induced neutralizing antibodies that were transferred to the fetus [327]. Transplacentally derived antibodies in newborns were high and had waned after approximately six months [327]. In February 2022, GSK stopped recruitment for their phase III trials due to safety concerns, as premature birth was more likely in vaccinated women compared to the placebo group (NCT04605159, NCT04980391, and NCT05229068). Single-dose vaccination of older adults with RSVPreF3 (RSVPreF3 OA) adjuvanted with AS01$_E$, administered intramuscularly, significantly reduced the risk of ALRI in vaccine recipients compared to the placebo group [328]. The generated antibodies were detectable for at least six months. Vaccination was safe and displayed a similar efficacy regardless of RSV strain, age group, or comorbidities [328]. In May 2023, RSVPreF3 OA, marketed as Arexvy, was approved by the FDA for the vaccination of older adults ($\geq$60 years) to reduce the risk of ALRI during an RSV season. Phase III clinical trials for RSVPreF OA are currently ongoing to monitor the effects of repeated vaccination and the long-term effects of vaccination in volunteers (NCT04886596).

During the SARS-CoV-2 pandemic, mRNA vaccines against COVID-19 attracted much interest. These mRNA-based vaccines, coding for a specific viral antigen, are administered intramuscularly using lipid nanoparticles [329]. After translation, mRNAs are degraded, reducing toxicity and the risk of side effects. The successful vaccination campaign during the pandemic also showed that this vaccine system could be safe and immunogenic in different age groups [329]. Although mRNA vaccines can be produced fast and on a large scale, they tend to be relatively unstable and degrade quickly if not stored properly [329]. mRNA-1345, which was developed by Moderna, is a nucleic-acid-based vaccine encoding a membrane-anchored RSV F protein stabilized in the prefusion conformation. Single-dose vaccination of older adults with mRNA-1345 resulted in increased antibody titers compared to the unvaccinated control group, which persisted for six months and effectively prevented

RSV-associated ALRI in vaccine recipients [330]. Vaccinations were well-tolerated in older adults without safety concerns. Phase III studies for mRNA-1345 are currently ongoing to further evaluate the efficacy and safety of the vaccine as primary efficacy endpoints in older adults were recently met (NCT05127434).

Prophylactic treatment with monoclonal antibodies represents an alternative to vaccination. Palivizumab, a humanized IgG1-type monoclonal antibody, was approved in 1998 to treat high-risk infants during their first RSV season [243,331]. Palivizumab targets the antigenic site II of the F protein and prevents viral infection and subsequent replication. Although antigenic site II is present on both preF and postF confirmation, antibodies directed against this epitope display a reduced neutralizing activity compared to antigenic site Ø, which is only present in the preF confirmation [332,333]. Trials in high-risk infants, to whom palivizumab was administered every 30 days for five months, showed reduced RSV-associated hospitalization and disease severity following intramuscular injection while being well tolerated compared to the placebo group [243,331]. Due to palivizumab's relatively short half-life and the necessity of monthly administrations during RSV season associated with high costs, the American Academy of Pediatrics recommended using palivizumab for high-risk infants during their first RSV season [334], advice that was followed in several other countries [335]. Recently, a new monoclonal antibody, nirsevimab, was approved. This recombinant human IgG1-type monoclonal antibody is directed against the antigenic site Ø in the RSV F prefusion confirmation. A triple amino acid substitution in the Fc domain increased nirsevimab's half-life, giving it a significant advantage over palivizumab [336]. A single-dose administration of nirsevimab to preterm infants before the RSV season significantly reduced the risk of ALRI and hospitalization compared to placebo-treated infants [337,338]. This effect lasted the whole season and was protective against both RSV serotypes without any safety issues [337,338]. Nirsevimab thus has two critical advantages over palivizumab: the higher neutralization activity of the antibodies and the increased half-life, which means that a single dose before the start of the RSV season is sufficient for the protection of high-risk infants. This makes nirsevimab more cost-effective. The recent approval of nirsevimab offers at least a cost-effective and efficient alternative to vaccination for the infant population. However, it is unclear if RSV can accumulate resistance mutations towards the neutralizing activity of nirsevimab, which has been the case for palivizumab [339].

The stabilization of the prefusion F protein is a critical factor for the development of vaccines and human monoclonal antibodies for the prevention of severe RSV infections in infants, older adults, and the immunocompromised. The progress in vaccine development is best illustrated by the recent approval of Arexvy and Abrysvo, which offer protection for the older population from RSV-associated ALRI. For infants, the vaccination of pregnant women seems an attractive option as Pfizer's PreF-protein vaccine was approved recently by the FDA based on promising data from phase III clinical trials.

5. Concluding Remarks

Despite decades of research, RSV remains a substantial global burden, mainly affecting infants, older adults, and immunocompromised individuals. After the early failure of FI-RSV-based vaccines in the 1960s, research has focused on understanding the host's immune response and possible immunomodulatory effects of RSV that may counteract a protective response and may even enhance immunopathological effects. This review emphasizes that effective and well-balanced immune responses are fundamental for reducing RSV-associated disease. As summarized, RSV-altered immune cell activity and differentiation robustly impact the cellular and humoral immune response. This dysregulation is mainly characterized by a Th2-skewed immune response leading to more severe disease and immunopathology. In the end, altered RSV-mediated responses of the innate and adaptive immune system may affect and suppress the generation of a proper immunological memory leading to impaired and short-lived immunity. Comprehending these alterations is thereby of great interest, although further studies will be necessary to address this in more detail.

Solving the prefusion conformation of the RSV F protein and its stabilization demonstrated the immense progress made in vaccine development after decades of basic research. As a consequence of this breakthrough, two preF-based vaccines have recently received licensure for use in older adults as both safety and efficacy were demonstrated. Although the recently developed RSV vaccines seem highly efficacious in older adults, their transmission-reducing potential and durability of protection in this age group require further investigation. The risk of waning antibody levels, also seen after RSV infection, may call for annual vaccination to protect this vulnerable age group.

A preF-protein-based vaccine for the vaccination of pregnant women was licensed recently by the FDA. This way, infants can be protected by vaccine-induced maternally derived antibodies during their first RSV season. Furthermore, several vaccine candidates are currently in phase III clinical trials showing promising results for the vaccination of older adults and pregnant women. Maternal vaccination seems to be a promising approach for the infant population, at least during their first RSV season, by the transplacental transfer of neutralizing antibodies, as direct vaccination of newborns may result in a suboptimal immune response considering their immature immune system. Maternal vaccination, however, may bear certain risks, taking into account the current developments in relation to the RSVPreF3 vaccine and the higher abundance of premature births after vaccination with one candidate vaccine.

Although a significant amount of data has been collected recently, our understanding of the immunological mechanisms during a natural RSV infection is still incomplete. Considering the phenomena of immunosenescence and preexisting immunity of older adults on one side and the immaturity of the infant's immune system on the other, analyzing the differences between the immune responses of the two main age groups at risk is critical for vaccine development. Further studies in infants and older adults will be necessary to fully comprehend the protective and harmful immune responses upon RSV infection and vaccination and to pave the way toward a better understanding of cell- and antibody-mediated immunity to RSV infection.

Author Contributions: A.A., M.L., R.M. and G.F.R. conceptualized and composed the manuscript; S.M.K., A.D.M.E.O. and M.L. reviewed the final manuscript; R.M. and G.F.R. oversaw all aspects of the manuscript preparation. All authors have read and agreed to the published version of the manuscript.

Funding: This work was financially supported by the Innovative Medicines Initiative 2 of the European Union's Horizon 2020 research and innovation program (EU IMI101007799, Inno4Vac) and the Deutsche Forschungsgemeinschaft (DFG, German Research Foundation)—398066876/GRK 2485/2, Research Training Group VIPER. Open Access publication was funded by the DFG—491094227 "Open Access Publication Funding" and the University of Veterinary Medicine Hannover Foundation.

References

1. Collins, P.L.; Fearns, R.; Graham, B.S. Respiratory syncytial virus: Virology, reverse genetics, and pathogenesis of disease. *Curr. Top. Microbiol. Immunol.* **2013**, *372*, 3–38. [CrossRef] [PubMed]
2. Mufson, M.A.; Orvell, C.; Rafnar, B.; Norrby, E. Two distinct subtypes of human respiratory syncytial virus. *J. Gen. Virol.* **1985**, *66*, 2111–2124. [CrossRef] [PubMed]
3. Yu, J.M.; Fu, Y.H.; Peng, X.L.; Zheng, Y.P.; He, J.S. Genetic diversity and molecular evolution of human respiratory syncytial virus A and B. *Sci. Rep.* **2021**, *11*, 12941. [CrossRef] [PubMed]
4. Cantu-Flores, K.; Rivera-Alfaro, G.; Munoz-Escalante, J.C.; Noyola, D.E. Global distribution of respiratory syncytial virus A and B infections: A systematic review. *Pathog. Glob. Health* **2022**, *116*, 398–409. [CrossRef] [PubMed]
5. Hall, C.B.; Weinberg, G.A.; Iwane, M.K.; Blumkin, A.K.; Edwards, K.M.; Staat, M.A.; Auinger, P.; Griffin, M.R.; Poehling, K.A.; Erdman, D.; et al. The burden of respiratory syncytial virus infection in young children. *N. Engl. J. Med.* **2009**, *360*, 588–598. [CrossRef]
6. Falsey, A.R.; Hennessey, P.A.; Formica, M.A.; Cox, C.; Walsh, E.E. Respiratory Syncytial Virus Infection in Elderly and High-Risk Adults. *N. Engl. J. Med.* **2005**, *352*, 1749–1759. [CrossRef]
7. Falsey, A.R.; Walsh, E.E. Respiratory syncytial virus infection in elderly adults. *Drugs Aging* **2005**, *22*, 577–587. [CrossRef]

8.	Nair, H.; Nokes, D.J.; Gessner, B.D.; Dherani, M.; Madhi, S.A.; Singleton, R.J.; O'Brien, K.L.; Roca, A.; Wright, P.F.; Bruce, N.; et al. Global burden of acute lower respiratory infections due to respiratory syncytial virus in young children: A systematic review and meta-analysis. *Lancet* **2010**, *375*, 1545–1555. [CrossRef]

9.	Li, Y.; Wang, X.; Blau, D.M.; Caballero, M.T.; Feikin, D.R.; Gill, C.J.; Madhi, S.A.; Omer, S.B.; Simoes, E.A.F.; Campbell, H.; et al. Global, regional, and national disease burden estimates of acute lower respiratory infections due to respiratory syncytial virus in children younger than 5 years in 2019: A systematic analysis. *Lancet* **2022**, *399*, 2047–2064. [CrossRef]

10.	Savic, M.; Penders, Y.; Shi, T.; Branche, A.; Pircon, J.Y. Respiratory syncytial virus disease burden in adults aged 60 years and older in high-income countries: A systematic literature review and meta-analysis. *Influenza Other Respir. Viruses* **2023**, *17*, e13031. [CrossRef]

11.	Chow, E.J.; Uyeki, T.M.; Chu, H.Y. The effects of the COVID-19 pandemic on community respiratory virus activity. *Nat. Rev. Microbiol.* **2023**, *21*, 195–210. [CrossRef] [PubMed]

12.	Chuang, Y.C.; Lin, K.P.; Wang, L.A.; Yeh, T.K.; Liu, P.Y. The Impact of the COVID-19 Pandemic on Respiratory Syncytial Virus Infection: A Narrative Review. *Infect. Drug Resist.* **2023**, *16*, 661–675. [CrossRef] [PubMed]

13.	Di Mattia, G.; Nenna, R.; Mancino, E.; Rizzo, V.; Pierangeli, A.; Villani, A.; Midulla, F. During the COVID-19 pandemic where has respiratory syncytial virus gone? *Pediatr. Pulmonol.* **2021**, *56*, 3106–3109. [CrossRef]

14.	Smyth, R.L.; Openshaw, P.J. Bronchiolitis. *Lancet* **2006**, *368*, 312–322. [CrossRef]

15.	Henderson, F.W.; Collier, A.M.; Clyde, W.A., Jr.; Denny, F.W. Respiratory-syncytial-virus infections, reinfections and immunity. A prospective, longitudinal study in young children. *N. Engl. J. Med.* **1979**, *300*, 530–534. [CrossRef] [PubMed]

16.	Glezen, W.P.; Taber, L.H.; Frank, A.L.; Kasel, J.A. Risk of primary infection and reinfection with respiratory syncytial virus. *Am. J. Dis. Child.* **1986**, *140*, 543–546. [CrossRef]

17.	Collins, P.L.; Graham, B.S. Viral and host factors in human respiratory syncytial virus pathogenesis. *J. Virol.* **2008**, *82*, 2040–2055. [CrossRef]

18.	Pickles, R.J.; DeVincenzo, J.P. Respiratory syncytial virus (RSV) and its propensity for causing bronchiolitis. *J. Pathol.* **2015**, *235*, 266–276. [CrossRef]

19.	McNamara, P.S.; Smyth, R.L. The pathogenesis of respiratory syncytial virus disease in childhood. *Br. Med. Bull.* **2002**, *61*, 13–28. [CrossRef]

20.	Hall, C.B. Respiratory syncytial virus and parainfluenza virus. *N. Engl. J. Med.* **2001**, *344*, 1917–1928. [CrossRef]

21.	Walsh, E.E.; Falsey, A.R. Respiratory syncytial virus infection in adult populations. *Infect. Disord. Drug Targets* **2012**, *12*, 98–102. [CrossRef] [PubMed]

22.	Shi, T.; Ooi, Y.; Zaw, E.M.; Utjesanovic, N.; Campbell, H.; Cunningham, S.; Bont, L.; Nair, H.; Investigators, R. Association Between Respiratory Syncytial Virus-Associated Acute Lower Respiratory Infection in Early Life and Recurrent Wheeze and Asthma in Later Childhood. *J. Infect. Dis.* **2020**, *222*, S628–S633. [CrossRef]

23.	Coutts, J.; Fullarton, J.; Morris, C.; Grubb, E.; Buchan, S.; Rodgers-Gray, B.; Thwaites, R. Association between respiratory syncytial virus hospitalization in infancy and childhood asthma. *Pediatr. Pulmonol.* **2020**, *55*, 1104–1110. [CrossRef] [PubMed]

24.	van Wijhe, M.; Johannesen, C.K.; Simonsen, L.; Jorgensen, I.M.; Investigators, R.; Fischer, T.K. A Retrospective Cohort Study on Infant Respiratory Tract Infection Hospitalizations and Recurrent Wheeze and Asthma Risk: Impact of Respiratory Syncytial Virus. *J. Infect. Dis.* **2022**, *226*, S55–S62. [CrossRef] [PubMed]

25.	McGinley, J.P.; Lin, G.L.; Oner, D.; Golubchik, T.; O'Connor, D.; Snape, M.D.; Gruselle, O.; Langedijk, A.C.; Wildenbeest, J.; Openshaw, P.; et al. Clinical and Viral Factors Associated With Disease Severity and Subsequent Wheezing in Infants With Respiratory Syncytial Virus Infection. *J. Infect. Dis.* **2022**, *226*, S45–S54. [CrossRef]

26.	Rosas-Salazar, C.; Chirkova, T.; Gebretsadik, T.; Chappell, J.D.; Peebles, R.S., Jr.; Dupont, W.D.; Jadhao, S.J.; Gergen, P.J.; Anderson, L.J.; Hartert, T.V. Respiratory syncytial virus infection during infancy and asthma during childhood in the USA (INSPIRE): A population-based, prospective birth cohort study. *Lancet* **2023**, *401*, 1669–1680. [CrossRef]

27.	Glezen, W.P.; Greenberg, S.B.; Atmar, R.L.; Piedra, P.A.; Couch, R.B. Impact of respiratory virus infections on persons with chronic underlying conditions. *JAMA* **2000**, *283*, 499–505. [CrossRef]

28.	Fixler, D.E. Respiratory syncytial virus infection in children with congenital heart disease: A review. *Pediatr. Cardiol.* **1996**, *17*, 163–168. [CrossRef] [PubMed]

29.	Kim, H.W.; Canchola, J.G.; Brandt, C.D.; Pyles, G.; Chanock, R.M.; Jensen, K.; Parrott, R.H. Respiratory syncytial virus disease in infants despite prior administration of antigenic inactivated vaccine. *Am. J. Epidemiol.* **1969**, *89*, 422–434. [CrossRef] [PubMed]

30.	Palacios-Pedrero, M.A.; Osterhaus, A.; Becker, T.; Elbahesh, H.; Rimmelzwaan, G.F.; Saletti, G. Aging and Options to Halt Declining Immunity to Virus Infections. *Front. Immunol.* **2021**, *12*, 681449. [CrossRef]

31.	Andrade, C.A.; Pacheco, G.A.; Galvez, N.M.S.; Soto, J.A.; Bueno, S.M.; Kalergis, A.M. Innate Immune Components that Regulate the Pathogenesis and Resolution of hRSV and hMPV Infections. *Viruses* **2020**, *12*, 637. [CrossRef] [PubMed]

32.	Ballegeer, M.; Saelens, X. Cell-Mediated Responses to Human Metapneumovirus Infection. *Viruses* **2020**, *12*, 542. [CrossRef] [PubMed]

33.	Farrag, M.A.; Almajhdi, F.N. Human Respiratory Syncytial Virus: Role of Innate Immunity in Clearance and Disease Progression. *Viral Immunol.* **2016**, *29*, 11–26. [CrossRef] [PubMed]

34.	Sun, Y.; Lopez, C.B. The innate immune response to RSV: Advances in our understanding of critical viral and host factors. *Vaccine* **2017**, *35*, 481–488. [CrossRef]

35. Turvey, S.E.; Broide, D.H. Innate immunity. *J. Allergy Clin. Immunol.* **2010**, *125*, S24–S32. [CrossRef]
36. Mogensen, T.H. Pathogen recognition and inflammatory signaling in innate immune defenses. *Clin. Microbiol. Rev.* **2009**, *22*, 240–273. [CrossRef]
37. Openshaw, P.J.M.; Chiu, C.; Culley, F.J.; Johansson, C. Protective and Harmful Immunity to RSV Infection. *Annu. Rev. Immunol.* **2017**, *35*, 501–532. [CrossRef]
38. Vareille, M.; Kieninger, E.; Edwards, M.R.; Regamey, N. The airway epithelium: Soldier in the fight against respiratory viruses. *Clin. Microbiol. Rev.* **2011**, *24*, 210–229. [CrossRef]
39. Kuek, L.E.; Lee, R.J. First contact: The role of respiratory cilia in host-pathogen interactions in the airways. *Am. J. Physiol. Lung Cell Mol. Physiol.* **2020**, *319*, L603–L619. [CrossRef]
40. Wallace, L.E.; Liu, M.; van Kuppeveld, F.J.M.; de Vries, E.; de Haan, C.A.M. Respiratory mucus as a virus-host range determinant. *Trends Microbiol.* **2021**, *29*, 983–992. [CrossRef]
41. Ridley, C.; Thornton, D.J. Mucins: The frontline defence of the lung. *Biochem. Soc. Trans.* **2018**, *46*, 1099–1106. [CrossRef] [PubMed]
42. Chatterjee, M.; van Putten, J.P.M.; Strijbis, K. Defensive Properties of Mucin Glycoproteins during Respiratory Infections-Relevance for SARS-CoV-2. *mBio* **2020**, *11*, e02374-20. [CrossRef] [PubMed]
43. LeMessurier, K.S.; Tiwary, M.; Morin, N.P.; Samarasinghe, A.E. Respiratory Barrier as a Safeguard and Regulator of Defense Against Influenza A Virus and Streptococcus pneumoniae. *Front. Immunol.* **2020**, *11*, 3. [CrossRef] [PubMed]
44. Sanders, C.J.; Doherty, P.C.; Thomas, P.G. Respiratory epithelial cells in innate immunity to influenza virus infection. *Cell Tissue Res.* **2011**, *343*, 13–21. [CrossRef]
45. Zanin, M.; Baviskar, P.; Webster, R.; Webby, R. The Interaction between Respiratory Pathogens and Mucus. *Cell Host Microbe* **2016**, *19*, 159–168. [CrossRef]
46. Bailey, K.L. Aging Diminishes Mucociliary Clearance of the Lung. *Adv. Geriatr. Med. Res.* **2022**, *4*, e220005. [CrossRef]
47. Ioannidis, I.; McNally, B.; Willette, M.; Peeples, M.E.; Chaussabel, D.; Durbin, J.E.; Ramilo, O.; Mejias, A.; Flano, E. Plasticity and virus specificity of the airway epithelial cell immune response during respiratory virus infection. *J. Virol.* **2012**, *86*, 5422–5436. [CrossRef]
48. Zhang, L.; Peeples, M.E.; Boucher, R.C.; Collins, P.L.; Pickles, R.J. Respiratory syncytial virus infection of human airway epithelial cells is polarized, specific to ciliated cells, and without obvious cytopathology. *J. Virol.* **2002**, *76*, 5654–5666. [CrossRef]
49. Barnes, M.V.C.; Openshaw, P.J.M.; Thwaites, R.S. Mucosal Immune Responses to Respiratory Syncytial Virus. *Cells* **2022**, *11*, 1153. [CrossRef]
50. Feng, Z.; Xu, L.; Xie, Z. Receptors for Respiratory Syncytial Virus Infection and Host Factors Regulating the Life Cycle of Respiratory Syncytial Virus. *Front. Cell Infect. Microbiol.* **2022**, *12*, 858629. [CrossRef]
51. Tayyari, F.; Marchant, D.; Moraes, T.J.; Duan, W.; Mastrangelo, P.; Hegele, R.G. Identification of nucleolin as a cellular receptor for human respiratory syncytial virus. *Nat. Med.* **2011**, *17*, 1132–1135. [CrossRef] [PubMed]
52. Mastrangelo, P.; Chin, A.A.; Tan, S.; Jeon, A.H.; Ackerley, C.A.; Siu, K.K.; Lee, J.E.; Hegele, R.G. Identification of RSV Fusion Protein Interaction Domains on the Virus Receptor, Nucleolin. *Viruses* **2021**, *13*, 261. [CrossRef] [PubMed]
53. Green, G.; Johnson, S.M.; Costello, H.; Brakel, K.; Harder, O.; Oomens, A.G.; Peeples, M.E.; Moulton, H.M.; Niewiesk, S. CX3CR1 Is a Receptor for Human Respiratory Syncytial Virus in Cotton Rats. *J. Virol.* **2021**, *95*, e0001021. [CrossRef] [PubMed]
54. Johnson, S.M.; McNally, B.A.; Ioannidis, I.; Flano, E.; Teng, M.N.; Oomens, A.G.; Walsh, E.E.; Peeples, M.E. Respiratory Syncytial Virus Uses CX3CR1 as a Receptor on Primary Human Airway Epithelial Cultures. *PLoS Pathog.* **2015**, *11*, e1005318. [CrossRef]
55. Chirkova, T.; Lin, S.; Oomens, A.G.P.; Gaston, K.A.; Boyoglu-Barnum, S.; Meng, J.; Stobart, C.C.; Cotton, C.U.; Hartert, T.V.; Moore, M.L.; et al. CX3CR1 is an important surface molecule for respiratory syncytial virus infection in human airway epithelial cells. *J. Gen. Virol.* **2015**, *96*, 2543–2556. [CrossRef]
56. Anderson, C.S.; Chu, C.Y.; Wang, Q.; Mereness, J.A.; Ren, Y.; Donlon, K.; Bhattacharya, S.; Misra, R.S.; Walsh, E.E.; Pryhuber, G.S.; et al. CX3CR1 as a respiratory syncytial virus receptor in pediatric human lung. *Pediatr. Res.* **2020**, *87*, 862–867. [CrossRef]
57. Krusat, T.; Streckert, H.J. Heparin-dependent attachment of respiratory syncytial virus (RSV) to host cells. *Arch. Virol.* **1997**, *142*, 1247–1254. [CrossRef]
58. Bourgeois, C.; Bour, J.B.; Lidholt, K.; Gauthray, C.; Pothier, P. Heparin-like structures on respiratory syncytial virus are involved in its infectivity in vitro. *J. Virol.* **1998**, *72*, 7221–7227. [CrossRef]
59. Feldman, S.A.; Audet, S.; Beeler, J.A. The Fusion Glycoprotein of Human Respiratory Syncytial Virus Facilitates Virus Attachment and Infectivity via an Interaction with Cellular Heparan Sulfate. *J. Virol.* **2000**, *74*, 6442–6447. [CrossRef]
60. Behera, A.K.; Matsuse, H.; Kumar, M.; Kong, X.; Lockey, R.F.; Mohapatra, S.S. Blocking intercellular adhesion molecule-1 on human epithelial cells decreases respiratory syncytial virus infection. *Biochem. Biophys. Res. Commun.* **2001**, *280*, 188–195. [CrossRef]
61. Imai, T.; Hieshima, K.; Haskell, C.; Baba, M.; Nagira, M.; Nishimura, M.; Kakizaki, M.; Takagi, S.; Nomiyama, H.; Schall, T.J.; et al. Identification and molecular characterization of fractalkine receptor CX3CR1, which mediates both leukocyte migration and adhesion. *Cell* **1997**, *91*, 521–530. [CrossRef] [PubMed]
62. Kim, K.W.; Vallon-Eberhard, A.; Zigmond, E.; Farache, J.; Shezen, E.; Shakhar, G.; Ludwig, A.; Lira, S.A.; Jung, S. In vivo structure/function and expression analysis of the CX3C chemokine fractalkine. *Blood* **2011**, *118*, e156–e167. [CrossRef] [PubMed]
63. Jones, B.A.; Beamer, M.; Ahmed, S. Fractalkine/CX3CL1: A Potential New Target for Inflammatory Diseases. *Mol. Interv.* **2010**, *10*, 263–270. [CrossRef] [PubMed]

64. Bazan, J.F.; Bacon, K.B.; Hardiman, G.; Wang, W.; Soo, K.; Rossi, D.; Greaves, D.R.; Zlotnik, A.; Schall, T.J. A new class of membrane-bound chemokine with a CX3C motif. *Nature* **1997**, *385*, 640–644. [CrossRef]
65. Tripp, R.A.; Jones, L.P.; Haynes, L.M.; Zheng, H.; Murphy, P.M.; Anderson, L.J. CX3C chemokine mimicry by respiratory syncytial virus G glycoprotein. *Nat. Immunol.* **2001**, *2*, 732–738. [CrossRef]
66. Ha, B.; Chirkova, T.; Boukhvalova, M.S.; Sun, H.Y.; Walsh, E.E.; Anderson, C.S.; Mariani, T.J.; Anderson, L.J. Mutation of Respiratory Syncytial Virus G Protein's CX3C Motif Attenuates Infection in Cotton Rats and Primary Human Airway Epithelial Cells. *Vaccines* **2019**, *7*, 69. [CrossRef]
67. Malik, G.; Zhou, Y. Innate Immune Sensing of Influenza A Virus. *Viruses* **2020**, *12*, 755. [CrossRef]
68. Islamuddin, M.; Mustfa, S.A.; Ullah, S.; Omer, U.; Kato, K.; Parveen, S. Innate Immune Response and Inflammasome Activation During SARS-CoV-2 Infection. *Inflammation* **2022**, *45*, 1849–1863. [CrossRef]
69. O'Connor, C.M.; Sen, G.C. Innate Immune Responses to Herpesvirus Infection. *Cells* **2021**, *10*, 2122. [CrossRef]
70. Loo, Y.M.; Fornek, J.; Crochet, N.; Bajwa, G.; Perwitasari, O.; Martinez-Sobrido, L.; Akira, S.; Gill, M.A.; Garcia-Sastre, A.; Katze, M.G.; et al. Distinct RIG-I and MDA5 signaling by RNA viruses in innate immunity. *J. Virol.* **2008**, *82*, 335–345. [CrossRef]
71. Meineke, R.; Rimmelzwaan, G.F.; Elbahesh, H. Influenza Virus Infections and Cellular Kinases. *Viruses* **2019**, *11*, 171. [CrossRef]
72. Carty, M.; Guy, C.; Bowie, A.G. Detection of Viral Infections by Innate Immunity. *Biochem. Pharmacol.* **2021**, *183*, 114316. [CrossRef]
73. Liu, P.; Jamaluddin, M.; Li, K.; Garofalo, R.P.; Casola, A.; Brasier, A.R. Retinoic acid-inducible gene I mediates early antiviral response and Toll-like receptor 3 expression in respiratory syncytial virus-infected airway epithelial cells. *J. Virol.* **2007**, *81*, 1401–1411. [CrossRef] [PubMed]
74. Rudd, B.D.; Burstein, E.; Duckett, C.S.; Li, X.; Lukacs, N.W. Differential role for TLR3 in respiratory syncytial virus-induced chemokine expression. *J. Virol.* **2005**, *79*, 3350–3357. [CrossRef] [PubMed]
75. Bueno, S.M.; Gonzalez, P.A.; Riedel, C.A.; Carreno, L.J.; Vasquez, A.E.; Kalergis, A.M. Local cytokine response upon respiratory syncytial virus infection. *Immunol. Lett.* **2011**, *136*, 122–129. [CrossRef] [PubMed]
76. Noah, T.L.; Becker, S. Respiratory syncytial virus-induced cytokine production by a human bronchial epithelial cell line. *Am. J. Physiol.* **1993**, *265*, 472–478. [CrossRef]
77. Noah, T.L.; Henderson, F.W.; Wortman, I.A.; Devlin, R.B.; Handy, J.; Koren, H.S.; Becker, S. Nasal cytokine production in viral acute upper respiratory infection of childhood. *J. Infect. Dis.* **1995**, *171*, 584–592. [CrossRef]
78. Saito, T.; Deskin, R.W.; Casola, A.; Häeberle, H.; Olszewska, B.; Ernst, P.B.; Alam, R.; Ogra, P.L.; Garofalo, R. Respiratory Syncytial Virus Induces Selective Production of the Chemokine RANTES by Upper Airway Epithelial Cells. *J. Infect. Dis.* **1996**, *175*, 497–504. [CrossRef]
79. Olszewska-Pazdrak, B.; Casola, A.; Saito, T.; Alam, R.; Crowe, S.E.; Mei, F.; Ogra, P.L.; Garofalo, R.P. Cell-Specific Expression of RANTES, MCP-1, and MIP-1a by Lower Airway Epithelial Cells and Eosinophils Infected with Respiratory Syncytial Virus. *J. Virol.* **1998**, *72*, 4756–4764. [CrossRef]
80. Wang, S.-Z.; Hallsworth, P.G.; Dowling, K.D.; Alpers, J.H.; Bowden, J.J.; Forsyth, K.D. Adhesion molecule expression on epithelial cells infected with respiratory syncytial virus. *Eur. Respir. J.* **2000**, *15*, 358–366. [CrossRef]
81. Guo, X.; Liu, T.; Shi, H.; Wang, J.; Ji, P.; Wang, H.; Hou, Y.; Tan, R.X.; Li, E. Respiratory Syncytial Virus Infection Upregulates NLRC5 and Major Histocompatibility Complex Class I Expression through RIG-I Induction in Airway Epithelial Cells. *J. Virol.* **2015**, *89*, 7636–7645. [CrossRef]
82. Levitz, R.; Wattier, R.; Phillips, P.; Solomon, A.; Lawler, J.; Lazar, I.; Weibel, C.; Kahn, J.S. Induction of IL-6 and CCL5 (RANTES) in human respiratory epithelial (A549) cells by clinical isolates of respiratory syncytial virus is strain specific. *Virol. J.* **2012**, *9*, 190. [CrossRef]
83. Touzelet, O.; Broadbent, L.; Armstrong, S.D.; Aljabr, W.; Cloutman-Green, E.; Power, U.F.; Hiscox, J.A. The Secretome Profiling of a Pediatric Airway Epithelium Infected with hRSV Identified Aberrant Apical/Basolateral Trafficking and Novel Immune Modulating (CXCL6, CXCL16, CSF3) and Antiviral (CEACAM1) Proteins. *Mol. Cell Proteom.* **2020**, *19*, 793–807. [CrossRef]
84. Van Royen, T.; Rossey, I.; Sedeyn, K.; Schepens, B.; Saelens, X. How RSV Proteins Join Forces to Overcome the Host Innate Immune Response. *Viruses* **2022**, *14*, 419. [CrossRef]
85. Moore, E.C.; Barber, J.; Tripp, R.A. Respiratory syncytial virus (RSV) attachment and nonstructural proteins modify the type I interferon response associated with suppressor of cytokine signaling (SOCS) proteins and IFN-stimulated gene-15 (ISG15). *Virol. J.* **2008**, *5*, 116. [CrossRef]
86. Xu, X.; Zheng, J.; Zheng, K.; Hou, Y.; Zhao, F.; Zhao, D. Respiratory syncytial virus NS1 protein degrades STAT2 by inducing SOCS1 expression. *Intervirology* **2014**, *57*, 65–73. [CrossRef]
87. Pei, J.; Beri, N.R.; Zou, A.J.; Hubel, P.; Dorando, H.K.; Bergant, V.; Andrews, R.D.; Pan, J.; Andrews, J.M.; Sheehan, K.C.F.; et al. Nuclear-localized human respiratory syncytial virus NS1 protein modulates host gene transcription. *Cell Rep.* **2021**, *37*, 109803. [CrossRef]
88. Boyapalle, S.; Wong, T.; Garay, J.; Teng, M.; San Juan-Vergara, H.; Mohapatra, S.; Mohapatra, S. Respiratory syncytial virus NS1 protein colocalizes with mitochondrial antiviral signaling protein MAVS following infection. *PLoS ONE* **2012**, *7*, e29386. [CrossRef]
89. Ling, Z.; Tran, K.C.; Teng, M.N. Human respiratory syncytial virus nonstructural protein NS2 antagonizes the activation of beta interferon transcription by interacting with RIG-I. *J. Virol.* **2009**, *83*, 3734–3742. [CrossRef]

90. Spann, K.M.; Tran, K.C.; Collins, P.L. Effects of nonstructural proteins NS1 and NS2 of human respiratory syncytial virus on interferon regulatory factor 3, NF-kappaB, and proinflammatory cytokines. *J. Virol.* **2005**, *79*, 5353–5362. [CrossRef]
91. Ren, J.; Liu, T.; Pang, L.; Li, K.; Garofalo, R.P.; Casola, A.; Bao, X. A novel mechanism for the inhibition of interferon regulatory factor-3-dependent gene expression by human respiratory syncytial virus NS1 protein. *J. Gen. Virol.* **2011**, *92*, 2153–2159. [CrossRef]
92. Bitko, V.; Shulyayeva, O.; Mazumder, B.; Musiyenko, A.; Ramaswamy, M.; Look, D.C.; Barik, S. Nonstructural proteins of respiratory syncytial virus suppress premature apoptosis by an NF-kappaB-dependent, interferon-independent mechanism and facilitate virus growth. *J. Virol.* **2007**, *81*, 1786–1795. [CrossRef]
93. Takeuchi, R.; Tsutsumi, H.; Osaki, M.; Haseyama, K.; Mizue, N.; Chiba, C. Respiratory syncytial virus infection of human alveolar epithelial cells enhances interferon regulatory factor 1 and interleukin-1beta-converting enzyme gene expression but does not cause apoptosis. *J. Virol.* **1998**, *72*, 4498–4502. [CrossRef]
94. Simpson, J.; Loh, Z.; Ullah, M.A.; Lynch, J.P.; Werder, R.B.; Collinson, N.; Zhang, V.; Dondelinger, Y.; Bertrand, M.J.M.; Everard, M.L.; et al. Respiratory Syncytial Virus Infection Promotes Necroptosis and HMGB1 Release by Airway Epithelial Cells. *Am. J. Respir. Crit. Care Med.* **2020**, *201*, 1358–1371. [CrossRef]
95. Lifland, A.W.; Jung, J.; Alonas, E.; Zurla, C.; Crowe, J.E., Jr.; Santangelo, P.J. Human respiratory syncytial virus nucleoprotein and inclusion bodies antagonize the innate immune response mediated by MDA5 and MAVS. *J. Virol.* **2012**, *86*, 8245–8258. [CrossRef]
96. Rincheval, V.; Lelek, M.; Gault, E.; Bouillier, C.; Sitterlin, D.; Blouquit-Laye, S.; Galloux, M.; Zimmer, C.; Eleouet, J.F.; Rameix-Welti, M.A. Functional organization of cytoplasmic inclusion bodies in cells infected by respiratory syncytial virus. *Nat. Commun.* **2017**, *8*, 563. [CrossRef]
97. Groskreutz, D.J.; Babor, E.C.; Monick, M.M.; Varga, S.M.; Hunninghake, G.W. Respiratory syncytial virus limits alpha subunit of eukaryotic translation initiation factor 2 (eIF2alpha) phosphorylation to maintain translation and viral replication. *J. Biol. Chem.* **2010**, *285*, 24023–24031. [CrossRef]
98. Meineke, R.; Stelz, S.; Busch, M.; Werlein, C.; Kühnel, M.; Jonigk, D.; Rimmelzwaan, G.F.; Elbahesh, H. FDA-approved Abl/EGFR/PDGFR kinase inhibitors show potent efficacy against pandemic and seasonal influenza A virus infections of human lung explants. *iScience* **2023**, *26*, 106309. [CrossRef]
99. Harcourt, J.L.; Haynes, L.M. Establishing a liquid-covered culture of polarized human airway epithelial Calu-3 cells to study host cell response to respiratory pathogens in vitro. *J. Vis. Exp.* **2013**, *72*, e50157. [CrossRef]
100. Ito, K.; Daly, L.; Coates, M. An impact of age on respiratory syncytial virus infection in air-liquid-interface culture bronchial epithelium. *Front Med.* **2023**, *10*, 1144050. [CrossRef]
101. Rajan, A.; Weaver, A.M.; Aloisio, G.M.; Jelinski, J.; Johnson, H.L.; Venable, S.F.; McBride, T.; Aideyan, L.; Piedra, F.A.; Ye, X.; et al. The human nose organoid respiratory virus model: An ex-vivo human challenge model to study RSV and SARS-CoV-2 pathogenesis and evaluate therapeutics. *bioRxiv* **2021**. [CrossRef]
102. Rothenberg, M.E.; Hogan, S.P. The eosinophil. *Annu. Rev. Immunol.* **2006**, *24*, 147–174. [CrossRef] [PubMed]
103. Flores-Torres, A.S.; Salinas-Carmona, M.C.; Salinas, E.; Rosas-Taraco, A.G. Eosinophils and Respiratory Viruses. *Viral Immunol.* **2019**, *32*, 198–207. [CrossRef] [PubMed]
104. Samarasinghe, A.E.; Melo, R.C.; Duan, S.; LeMessurier, K.S.; Liedmann, S.; Surman, S.L.; Lee, J.J.; Hurwitz, J.L.; Thomas, P.G.; McCullers, J.A. Eosinophils Promote Antiviral Immunity in Mice Infected with Influenza A Virus. *J. Immunol.* **2017**, *198*, 3214–3226. [CrossRef]
105. Drake, M.G.; Bivins-Smith, E.R.; Proskocil, B.J.; Nie, Z.; Scott, G.D.; Lee, J.J.; Lee, N.A.; Fryer, A.D.; Jacoby, D.B. Human and Mouse Eosinophils Have Antiviral Activity against Parainfluenza Virus. *Am. J. Respir. Cell Mol. Biol.* **2016**, *55*, 387–394. [CrossRef]
106. Handzel, Z.T.; Busse, W.W.; Sedgwick, J.B.; Vrtis, R.; Lee, W.M.; Kelly, E.A.B.; Gern, J.E. Eosinophils Bind Rhinovirus and Activate Virus-Specific T Cells. *J. Immunol.* **1998**, *160*, 1279–1284. [CrossRef]
107. Ho, K.S.; Howell, D.; Rogers, L.; Narasimhan, B.; Verma, H.; Steiger, D. The relationship between asthma, eosinophilia, and outcomes in coronavirus disease 2019 infection. *Ann. Allergy Asthma Immunol.* **2021**, *127*, 42–48. [CrossRef]
108. Becker, S.; Soukup, J.M. Airway epithelial cell-induced activation of monocytes and eosinophils in respiratory syncytial viral infection. *Immunobiology* **1999**, *201*, 88–106. [CrossRef]
109. Soukup, J.M.; Becker, S. Role of monocytes and eosinophils in human respiratory syncytial virus infection in vitro. *Clin. Immunol.* **2003**, *107*, 178–185. [CrossRef]
110. Harrison, A.M.; Bonville, C.A.; Rosenberg, H.F.; Domachowske, J.B. Respiratory syncytical virus-induced chemokine expression in the lower airways: Eosinophil recruitment and degranulation. *Am. J. Respir. Crit. Care Med.* **1999**, *159*, 1918–1924. [CrossRef]
111. Phipps, S.; Lam, C.E.; Mahalingam, S.; Newhouse, M.; Ramirez, R.; Rosenberg, H.F.; Foster, P.S.; Matthaei, K.I. Eosinophils contribute to innate antiviral immunity and promote clearance of respiratory syncytial virus. *Blood* **2007**, *110*, 1578–1586. [CrossRef] [PubMed]
112. Lindemans, C.A.; Kimpen, J.L.; Luijk, B.; Heidema, J.; Kanters, D.; van der Ent, C.K.; Koenderman, L. Systemic eosinophil response induced by respiratory syncytial virus. *Clin. Exp. Immunol.* **2006**, *144*, 409–417. [CrossRef] [PubMed]
113. Stark, J.M.; Godding, V.; Sedgwick, J.B.; Busse, W.W. Respiratory syncytial virus infection enhances neutrophil and eosinophil adhesion to cultured respiratory epithelial cells. Roles of CD18 and intercellular adhesion molecule-1. *J. Immunol.* **1996**, *156*, 4774–4782. [CrossRef] [PubMed]

114. Olszewska-Pazdrak, B.; Pazdrak, K.; Ogra, P.L.; Garofalo, R.P. Respiratory Syncytial Virus-Infected Pulmonary Epithelial Cells Induce Eosinophil Degranulation by a CD18-Mediated Mechanism. *J. Immunol.* **1998**, *160*, 4889–4895. [CrossRef]

115. Castilow, E.M.; Legge, K.L.; Varga, S.M. Cutting edge: Eosinophils do not contribute to respiratory syncytial virus vaccine-enhanced disease. *J. Immunol.* **2008**, *181*, 6692–6696. [CrossRef]

116. Prince, G.A.; Jenson, A.B.; Hemming, V.; Murphy, B.; Walsh, E.; Horswood, R.; Chanock, R. Enhancement of respiratory syncytial virus pulmonary pathology in cotton rats by prior intramuscular inoculation of formalin-inactiva ted virus. *J. Virol.* **1986**, *57*, 721–728. [CrossRef]

117. Durant, L.R.; Makris, S.; Voorburg, C.M.; Loebbermann, J.; Johansson, C.; Openshaw, P.J. Regulatory T cells prevent Th2 immune responses and pulmonary eosinophilia during respiratory syncytial virus infection in mice. *J. Virol.* **2013**, *87*, 10946–10954. [CrossRef]

118. Conners, M.; Giese, N.A.; Kulkarni, A.B.; Firestone, C.Y.; Morse, H.C.r.; Murphy, B.R. Enhanced pulmonary histopathology induced by respiratory syncytial virus (RSV) challenge of formalin-inactivated RSV-immunized BALB/c mice is abrogated by depletion of interleukin-4 (IL-4) and IL-10. *J. Virol.* **1994**, *68*, 5321–5325. [CrossRef]

119. Garofalo, R.; Kimpen, J.L.L.; Welliver, R.C.; Ogra, P.L. Eosinophil degranulation in the respiratory tract during naturally acquired respiratory syncytial virus infection. *J. Pediatr.* **1992**, *120*, 28–32. [CrossRef]

120. Dyer, K.D.; Percopo, C.M.; Fischer, E.R.; Gabryszewski, S.J.; Rosenberg, H.F. Pneumoviruses infect eosinophils and elicit MyD88-dependent release of chemoattractant cytokines and interleukin-6. *Blood* **2009**, *114*, 2649–2656. [CrossRef]

121. Kimpen, J.L.; Garofalo, R.; Welliver, R.C.; Fujihara, K.; Ogra, P.L. An ultrastructural study of the interaction of human eosinophils with respiratory syncytial virus. *Pediatr. Allergy Immunol.* **1996**, *7*, 48–53. [CrossRef] [PubMed]

122. Kimpen, J.L.L.; Garofalo, R.; Welliver, R.C.; Ogra, P.L. Activation of Human Eosinophils In vitro by Respiratory Syncytial Virus. *Pediatr. Res.* **1992**, *32*, 160–164. [CrossRef] [PubMed]

123. Schwarze, J.; Cieslewicz, G.; Hamelmann, E.; Joetham, A.; Shultz, L.D.; Lamers, M.C.; Gelfand, E.W. IL-5 and Eosinophils Are Essential for the Development of Airway Hyperresponsiveness Following Acute Respiratory Syncytial Virus Infection. *J. Immunol.* **1999**, *162*, 2997–3004. [CrossRef]

124. Zhang, D.; Yang, J.; Zhao, Y.; Shan, J.; Wang, L.; Yang, G.; He, S.; Li, E. RSV Infection in Neonatal Mice Induces Pulmonary Eosinophilia Responsible for Asthmatic Reaction. *Front. Immunol.* **2022**, *13*, 817113. [CrossRef]

125. Ng, L.G.; Ostuni, R.; Hidalgo, A. Heterogeneity of neutrophils. *Nat. Rev. Immunol.* **2019**, *19*, 255–265. [CrossRef] [PubMed]

126. Kruger, P.; Saffarzadeh, M.; Weber, A.N.; Rieber, N.; Radsak, M.; von Bernuth, H.; Benarafa, C.; Roos, D.; Skokowa, J.; Hartl, D. Neutrophils: Between host defence, immune modulation, and tissue injury. *PLoS Pathog.* **2015**, *11*, e1004651. [CrossRef]

127. Ma, Y.; Zhang, Y.; Zhu, L. Role of neutrophils in acute viral infection. *Immun. Inflamm. Dis.* **2021**, *9*, 1186–1196. [CrossRef]

128. McNamara, P.S.; Flanagan, B.F.; Hart, C.A.; Smyth, R.L. Production of chemokines in the lungs of infants with severe respiratory syncytial virus bronchiolitis. *J. Infect. Dis.* **2005**, *191*, 1225–1232. [CrossRef]

129. Bataki, E.L.; Evans, G.S.; Everard, M.L. Respiratory syncytial virus and neutrophil activation. *Clin. Exp. Immunol.* **2005**, *140*, 470–477. [CrossRef]

130. Deng, Y.; Herbert, J.A.; Smith, C.M.; Smyth, R.L. An in vitro transepithelial migration assay to evaluate the role of neutrophils in Respiratory Syncytial Virus (RSV) induced epithelial damage. *Sci. Rep.* **2018**, *8*, 6777. [CrossRef]

131. Funchal, G.A.; Jaeger, N.; Czepielewski, R.S.; Machado, M.S.; Muraro, S.P.; Stein, R.T.; Bonorino, C.B.; Porto, B.N. Respiratory syncytial virus fusion protein promotes TLR-4-dependent neutrophil extracellular trap formation by human neutrophils. *PLoS ONE* **2015**, *10*, e0124082. [CrossRef] [PubMed]

132. Souza, P.S.S.; Barbosa, L.V.; Diniz, L.F.A.; da Silva, G.S.; Lopes, B.R.P.; Souza, P.M.R.; de Araujo, G.C.; Pessoa, D.; de Oliveira, J.; Souza, F.P.; et al. Neutrophil extracellular traps possess anti-human respiratory syncytial virus activity: Possible interaction with the viral F protein. *Virus Res.* **2018**, *251*, 68–77. [CrossRef] [PubMed]

133. Deng, Y.; Herbert, J.A.; Robinson, E.; Ren, L.; Smyth, R.L.; Smith, C.M. Neutrophil-Airway Epithelial Interactions Result in Increased Epithelial Damage and Viral Clearance during Respiratory Syncytial Virus Infection. *J. Virol.* **2020**, *94*, e02161-19. [CrossRef] [PubMed]

134. McNamara, P.S.; Ritson, P.; Selby, A.; Hart, C.A.; Smyth, R.L. Bronchoalveolar lavage cellularity in infants with severe respiratory syncytial virus bronchiolitis. *Arch. Dis. Child.* **2003**, *88*, 922–926. [CrossRef] [PubMed]

135. Everard, M.L.; Swarbrick, A.; Wrightham, M.; McIntyre, J.; Dunkley, C.; James, P.D.; Sewell, H.F.; Milner, A.D. Analysis of cells obtained by bronchial lavage of infants with respiratory syncytial virus infection. *Arch. Dis. Child.* **1994**, *71*, 428–432. [CrossRef]

136. Wang, S.Z.; Xu, H.; Wraith, A.; Bowden, J.J.; Alpers, J.H.; Forsyth, K.D. Neutrophils induce damage to respiratory epithelial cells infected with respiratory syncytial virus. *Eur. Respir. J.* **1998**, *12*, 612–618. [CrossRef] [PubMed]

137. Acosta, P.L.; Caballero, M.T.; Polack, F.P. Brief History and Characterization of Enhanced Respiratory Syncytial Virus Disease. *Clin. Vaccine Immunol.* **2015**, *23*, 189–195. [CrossRef] [PubMed]

138. Halfhide, C.P.; Flanagan, B.F.; Brearey, S.P.; Hunt, J.A.; Fonceca, A.M.; McNamara, P.S.; Howarth, D.; Edwards, S.; Smyth, R.L. Respiratory syncytial virus binds and undergoes transcription in neutrophils from the blood and airways of infants with severe bronchiolitis. *J. Infect. Dis.* **2011**, *204*, 451–458. [CrossRef]

139. Jaovisidha, P.; Peeples, M.E.; Brees, A.A.; Carpenter, L.R.; Moy, J.N. Respiratory syncytial virus stimulates neutrophil degranulation and chemokine release. *J. Immunol.* **1999**, *163*, 2816–2820. [CrossRef]

140. Cortjens, B.; de Boer, O.J.; de Jong, R.; Antonis, A.F.; Sabogal Pineros, Y.S.; Lutter, R.; van Woensel, J.B.; Bem, R.A. Neutrophil extracellular traps cause airway obstruction during respiratory syncytial virus disease. *J. Pathol.* **2016**, *238*, 401–411. [CrossRef]
141. Habibi, M.S.; Thwaites, R.S.; Chang, M.; Jozwik, A.; Paras, A.; Kirsebom, F.; Varese, A.; Owen, A.; Cuthbertson, L.; James, P.; et al. Neutrophilic inflammation in the respiratory mucosa predisposes to RSV infection. *Science* **2020**, *370*, eaba9301. [CrossRef] [PubMed]
142. Vivier, E.; Tomasello, E.; Baratin, M.; Walzer, T.; Ugolini, S. Functions of natural killer cells. *Nat. Immunol.* **2008**, *9*, 503–510. [CrossRef] [PubMed]
143. Lodoen, M.B.; Lanier, L.L. Natural killer cells as an initial defense against pathogens. *Curr. Opin. Immunol.* **2006**, *18*, 391–398. [CrossRef] [PubMed]
144. Bhat, R.; Farrag, M.A.; Almajhdi, F.N. Double-edged role of natural killer cells during RSV infection. *Int. Rev. Immunol.* **2020**, *39*, 233–244. [CrossRef] [PubMed]
145. Hussell, T.; Openshaw, P.J. Intracellular IFN-? expression in natural killer cells precedes lung CD8 T cell recruitment during respiratory syncytial virus infectio. *J. Gen. Virol.* **1998**, *79*, 2593–2601. [CrossRef]
146. Hussell, T.; Openshaw, P.J. IL-12-activated NK cells reduce lung eosinophilia to the attachment protein of respiratory syncytial virus but do not enhance the severity of illness in CD8 T cell-immunodeficient conditions. *J. Immunol.* **2000**, *165*, 7109–7115. [CrossRef]
147. Kaiko, G.E.; Phipps, S.; Angkasekwinai, P.; Dong, C.; Foster, P.S. NK cell deficiency predisposes to viral-induced Th2-type allergic inflammation via epithelial-derived IL-25. *J. Immunol.* **2010**, *185*, 4681–4690. [CrossRef]
148. Welliver, T.P.; Garofalo, R.P.; Hosakote, Y.; Hintz, K.H.; Avendano, L.; Sanchez, K.; Velozo, L.; Jafri, H.; Chavez-Bueno, S.; Ogra, P.L.; et al. Severe human lower respiratory tract illness caused by respiratory syncytial virus and influenza virus is characterized by the absence of pulmonary cytotoxic lymphocyte responses. *J. Infect. Dis.* **2007**, *195*, 1126–1136. [CrossRef]
149. Haynes, L.M.; Moore, D.D.; Kurt-Jones, E.A.; Finberg, R.W.; Anderson, L.J.; Tripp, R.A. Involvement of toll-like receptor 4 in innate immunity to respiratory syncytial virus. *J. Virol.* **2001**, *75*, 10730–10737. [CrossRef]
150. Li, F.; Zhu, H.; Sun, R.; Wei, H.; Tian, Z. Natural killer cells are involved in acute lung immune injury caused by respiratory syncytial virus infection. *J. Virol.* **2012**, *86*, 2251–2258. [CrossRef]
151. Long, X.; Xie, J.; Zhao, K.; Li, W.; Tang, W.; Chen, S.; Zang, N.; Ren, L.; Deng, Y.; Xie, X.; et al. NK cells contribute to persistent airway inflammation and AHR during the later stage of RSV infection in mice. *Med. Microbiol. Immunol.* **2016**, *205*, 459–470. [CrossRef] [PubMed]
152. van Erp, E.A.; Feyaerts, D.; Duijst, M.; Mulder, H.L.; Wicht, O.; Luytjes, W.; Ferwerda, G.; van Kasteren, P.B. Respiratory Syncytial Virus Infects Primary Neonatal and Adult Natural Killer Cells and Affects Their Antiviral Effector Function. *J. Infect. Dis.* **2019**, *219*, 723–733. [CrossRef] [PubMed]
153. Harker, J.A.; Godlee, A.; Wahlsten, J.L.; Lee, D.C.; Thorne, L.G.; Sawant, D.; Tregoning, J.S.; Caspi, R.R.; Bukreyev, A.; Collins, P.L.; et al. Interleukin 18 coexpression during respiratory syncytial virus infection results in enhanced disease mediated by natural killer cells. *J. Virol.* **2010**, *84*, 4073–4082. [CrossRef] [PubMed]
154. Hettinger, J.; Richards, D.M.; Hansson, J.; Barra, M.M.; Joschko, A.C.; Krijgsveld, J.; Feuerer, M. Origin of monocytes and macrophages in a committed progenitor. *Nat. Immunol.* **2013**, *14*, 821–830. [CrossRef] [PubMed]
155. Ziegler-Heitbrock, L. Monocyte subsets in man and other species. *Cell Immunol.* **2014**, *289*, 135–139. [CrossRef]
156. Italiani, P.; Boraschi, D. From Monocytes to M1/M2 Macrophages: Phenotypical vs. Functional Differentiation. *Front. Immunol.* **2014**, *5*, 514. [CrossRef]
157. Shi, C.; Pamer, E.G. Monocyte recruitment during infection and inflammation. *Nat. Rev. Immunol.* **2011**, *11*, 762–774. [CrossRef]
158. Naidoo, K.K.; Ndumnego, O.C.; Ismail, N.; Dong, K.L.; Ndung'u, T. Antigen Presenting Cells Contribute to Persistent Immune Activation Despite Antiretroviral Therapy Initiation During Hyperacute HIV-1 Infection. *Front. Immunol.* **2021**, *12*, 738743. [CrossRef]
159. Cao, W.; Taylor, A.K.; Biber, R.E.; Davis, W.G.; Kim, J.H.; Reber, A.J.; Chirkova, T.; De La Cruz, J.A.; Pandey, A.; Ranjan, P.; et al. Rapid differentiation of monocytes into type I IFN-producing myeloid dendritic cells as an antiviral strategy against influenza virus infection. *J. Immunol.* **2012**, *189*, 2257–2265. [CrossRef]
160. Gill, M.A.; Long, K.; Kwon, T.; Muniz, L.; Mejias, A.; Connolly, J.; Roy, L.; Banchereau, J.; Ramilo, O. Differential recruitment of dendritic cells and monocytes to respiratory mucosal sites in children with influenza virus or respiratory syncytial virus infection. *J. Infect. Dis.* **2008**, *198*, 1667–1676. [CrossRef]
161. Goritzka, M.; Makris, S.; Kausar, F.; Durant, L.R.; Pereira, C.; Kumagai, Y.; Culley, F.J.; Mack, M.; Akira, S.; Johansson, C. Alveolar macrophage-derived type I interferons orchestrate innate immunity to RSV through recruitment of antiviral monocytes. *J. Exp. Med.* **2015**, *212*, 699–714. [CrossRef]
162. Kim, T.H.; Kim, C.W.; Oh, D.S.; Jung, H.E.; Lee, H.K. Monocytes Contribute to IFN-beta Production via the MyD88-Dependent Pathway and Cytotoxic T-Cell Responses against Mucosal Respiratory Syncytial Virus Infection. *Immune Netw.* **2021**, *21*, e27. [CrossRef] [PubMed]
163. Barr, F.E.; Pedigo, H.; Johnson, T.R.; Shepherd, V.L. Surfactant protein-A enhances uptake of respiratory syncytial virus by monocytes and U937 macrophages. *Am. J. Respir. Cell Mol. Biol.* **2000**, *23*, 586–592. [CrossRef] [PubMed]

164. Kurt-Jones, E.A.; Popova, L.; Kwinn, L.; Haynes, L.M.; Jones, L.P.; Tripp, R.A.; Walsh, E.E.; Freeman, M.W.; Golenbock, D.T.; Anderson, L.J.; et al. Pattern recognition receptors TLR4 and CD14 mediate response to respiratory syncytial virus. *Nat. Immunol.* **2000**, *1*, 398–401. [CrossRef] [PubMed]

165. Ahout, I.M.; Jans, J.; Haroutiounian, L.; Simonetti, E.R.; van der Gaast-de Jongh, C.; Diavatopoulos, D.A.; de Jonge, M.I.; de Groot, R.; Ferwerda, G. Reduced Expression of HLA-DR on Monocytes During Severe Respiratory Syncytial Virus Infections. *Pediatr. Infect. Dis. J.* **2016**, *35*, e89–e96. [CrossRef]

166. Besteman, S.B.; Phung, E.; Raeven, H.H.M.; Amatngalim, G.D.; Rumpret, M.; Crabtree, J.; Schepp, R.M.; Rodenburg, L.W.; Siemonsma, S.G.; Verleur, N.; et al. Recurrent Respiratory Syncytial Virus Infection in a CD14-Deficient Patient. *J. Infect. Dis.* **2022**, *226*, 258–269. [CrossRef]

167. Bont, L.; Heijnen, C.J.; Kavelaars, A.; van Aalderen, W.M.; Brus, F.; Draaisma, J.T.; Geelen, S.M.; Kimpen, J.L. Monocyte IL-10 production during respiratory syncytial virus bronchiolitis is associated with recurrent wheezing in a one-year follow-up study. *Am. J. Respir. Crit. Care Med.* **2000**, *161*, 1518–1523. [CrossRef]

168. Midulla, F.; Huang, Y.T.; Gilbert, I.A.; Cirino, N.M.; McFadden, E.R., Jr.; Panuska, J.R. Respiratory syncytial virus infection of human cord and adult blood monocytes and alveolar macrophages. *Am. Rev. Respir. Dis.* **1989**, *140*, 771–777. [CrossRef]

169. Krilov, L.R.; Hendry, R.M.; Godfrey, E.; McIntosh, K. Respiratory virus infection of peripheral blood monocytes: Correlation with ageing of cells and interferon production in vitro. *J. Gen. Virol.* **1987**, *68*, 1749–1753. [CrossRef]

170. Takeuchi, R.; Tsutsumi, H.; Osaki, M.; Sone, S.; Imai, S.; Chiba, S. Respiratory syncytial virus infection of neonatal monocytes stimulates synthesis of interferon regulatory factor 1 and interleukin-1beta (IL-1beta)-converting enzyme and secretion of IL-1beta. *J. Virol.* **1998**, *72*, 837–840. [CrossRef]

171. Polack, F.P.; Irusta, P.M.; Hoffman, S.J.; Schiatti, M.P.; Melendi, G.A.; Delgado, M.F.; Laham, F.R.; Thumar, B.; Hendry, R.M.; Melero, J.A.; et al. The cysteine-rich region of respiratory syncytial virus attachment protein inhibits innate immunity elicited by the virus and endotoxin. *Proc. Natl. Acad. Sci. USA* **2005**, *102*, 8996–9001. [CrossRef] [PubMed]

172. Mosser, D.M.; Edwards, J.P. Exploring the full spectrum of macrophage activation. *Nat. Rev. Immunol.* **2008**, *8*, 958–969. [CrossRef] [PubMed]

173. Hou, F.; Xiao, K.; Tang, L.; Xie, L. Diversity of Macrophages in Lung Homeostasis and Diseases. *Front. Immunol.* **2021**, *12*, 753940. [CrossRef] [PubMed]

174. Wang, Y.; Zheng, J.; Wang, X.; Yang, P.; Zhao, D. Alveolar macrophages and airway hyperresponsiveness associated with respiratory syncytial virus infection. *Front. Immunol.* **2022**, *13*, 1012048. [CrossRef]

175. Makris, S.; Bajorek, M.; Culley, F.J.; Goritzka, M.; Johansson, C. Alveolar Macrophages Can Control Respiratory Syncytial Virus Infection in the Absence of Type I Interferons. *J. Innate Immun.* **2016**, *8*, 452–463. [CrossRef]

176. Oh, D.S.; Kim, T.H.; Lee, H.K. Differential Role of Anti-Viral Sensing Pathway for the Production of Type I Interferon beta in Dendritic Cells and Macrophages Against Respiratory Syncytial Virus A2 Strain Infection. *Viruses* **2019**, *11*, 62. [CrossRef]

177. Kolli, D.; Gupta, M.R.; Sbrana, E.; Velayutham, T.S.; Chao, H.; Casola, A.; Garofalo, R.P. Alveolar macrophages contribute to the pathogenesis of human metapneumovirus infection while protecting against respiratory syncytial virus infection. *Am. J. Respir. Cell Mol. Biol.* **2014**, *51*, 502–515. [CrossRef]

178. Reed, J.L.; Brewah, Y.A.; Delaney, T.; Welliver, T.; Burwell, T.; Benjamin, E.; Kuta, E.; Kozhich, A.; McKinney, L.; Suzich, J.; et al. Macrophage impairment underlies airway occlusion in primary respiratory syncytial virus bronchiolitis. *J. Infect. Dis.* **2008**, *198*, 1783–1793. [CrossRef]

179. Huang, S.; Zhu, B.; Cheon, I.S.; Goplen, N.P.; Jiang, L.; Zhang, R.; Peebles, R.S.; Mack, M.; Kaplan, M.H.; Limper, A.H.; et al. PPAR-gamma in Macrophages Limits Pulmonary Inflammation and Promotes Host Recovery following Respiratory Viral Infection. *J. Virol.* **2019**, *93*, e00030-19. [CrossRef]

180. Shirey, K.A.; Pletneva, L.M.; Puche, A.C.; Keegan, A.D.; Prince, G.A.; Blanco, J.C.; Vogel, S.N. Control of RSV-induced lung injury by alternatively activated macrophages is IL-4R alpha-, TLR4-, and IFN-beta-dependent. *Mucosal Immunol.* **2010**, *3*, 291–300. [CrossRef]

181. Franke-Ullmann, G.; Pförtner, C.; Walter, P.; Steinmüller, C.; Lohmann-Matthes, M.L.; Kobzik, L.; Freihorst, J. Alteration of pulmonary macrophage function by respiratory syncytial virus infection in vitro. *J. Immunol.* **1995**, *154*, 268–280. [CrossRef] [PubMed]

182. Panuska, J.R.; Cirino, N.M.; Midulla, F.; Despot, J.E.; McFadden, E.R., Jr.; Huang, Y.T. Productive infection of isolated human alveolar macrophages by respiratory syncytial virus. *J. Clin. Investig.* **1990**, *86*, 113–119. [CrossRef] [PubMed]

183. Santos, L.D.; Antunes, K.H.; Muraro, S.P.; de Souza, G.F.; da Silva, A.G.; Felipe, J.S.; Zanetti, L.C.; Czepielewski, R.S.; Magnus, K.; Scotta, M.; et al. TNF-mediated alveolar macrophage necroptosis drives disease pathogenesis during respiratory syncytial virus infection. *Eur. Respir. J.* **2021**, *57*, 2003764. [CrossRef] [PubMed]

184. Haeberle, H.A.; Takizawa, R.; Casola, A.; Brasie, A.R.; Dieterich, H.-J.; van Rooijen, N.; Gatalica, Z.; Garofalo, R.P. Respiratory syncytial virus-induced activation of nuclear factor-kappaB in the lung involves alveolar macrophages and toll-like receptor 4-dependent pathways. *J. Infect. Dis.* **2002**, *186*, 1199–1206. [CrossRef] [PubMed]

185. Harris, J.; Werling, D. Binding and entry of respiratory syncytial virus into host cells and initiation of the innate immune response. *Cell Microbiol.* **2003**, *5*, 671–680. [CrossRef]

186. Becker, S.; Quay, J.; Soukup, J. Cytokine (tumor necrosis factor, IL-6, and IL-8) production by respiratory syncytial virus-infected human alveolar macrophages. *J. Immunol.* **1991**, *147*, 4307–4312. [CrossRef]

187. Tsutsumi, H.; Matsuda, K.; Sone, S.; Takeuchi, R.; Chiba, S. Respiratory syncytial virus-induced cytokine production by neonatal macrophages. *Clin. Exp. Immunol.* **1996**, *106*, 442–446. [CrossRef]
188. Eichinger, K.M.; Egana, L.; Orend, J.G.; Resetar, E.; Anderson, K.B.; Patel, R.; Empey, K.M. Alveolar macrophages support interferon gamma-mediated viral clearance in RSV-infected neonatal mice. *Respir. Res.* **2015**, *16*, 122. [CrossRef]
189. Stockwin, L.H.; McGonagle, D.; Martin, I.G.; Blair, G.E. Dendritic cells: Immunological sentinels with a central role in health and disease. *Immunol. Cell Biol.* **2000**, *78*, 91–102. [CrossRef]
190. Zaslona, Z.; Wilhelm, J.; Cakarova, L.; Marsh, L.M.; Seeger, W.; Lohmeyer, J.; von Wulffen, W. Transcriptome profiling of primary murine monocytes, lung macrophages and lung dendritic cells reveals a distinct expression of genes involved in cell trafficking. *Respir. Res.* **2009**, *10*, 2. [CrossRef]
191. Boltjes, A.; van Wijk, F. Human dendritic cell functional specialization in steady-state and inflammation. *Front. Immunol.* **2014**, *5*, 131. [CrossRef] [PubMed]
192. Geissmann, F.; Manz, M.G.; Jung, S.; Sieweke, M.H.; Merad, M.; Ley, K. Development of Monocytes, Macrophages, and Dendritic Cells. *Science* **2010**, *327*, 656–661. [CrossRef] [PubMed]
193. Kim, T.H.; Oh, D.S.; Jung, H.E.; Chang, J.; Lee, H.K. Plasmacytoid Dendritic Cells Contribute to the Production of IFN-beta via TLR7-MyD88-Dependent Pathway and CTL Priming during Respiratory Syncytial Virus Infection. *Viruses* **2019**, *11*, 730. [CrossRef] [PubMed]
194. Klouwenberg, P.K.; Tan, L.; Werkman, W.; van Bleek, G.M.; Coenjaerts, F. The role of Toll-like receptors in regulating the immune response against respiratory syncytial virus. *Crit. Rev. Immunol.* **2009**, *29*, 531–550. [CrossRef] [PubMed]
195. Lukens, M.V.; Kruijsen, D.; Coenjaerts, F.E.; Kimpen, J.L.; van Bleek, G.M. Respiratory syncytial virus-induced activation and migration of respiratory dendritic cells and subsequent antigen presentation in the lung-draining lymph node. *J. Virol.* **2009**, *83*, 7235–7243. [CrossRef] [PubMed]
196. Johnson, T.R.; Johnson, C.N.; Corbett, K.S.; Edwards, G.C.; Graham, B.S. Primary human mDC1, mDC2, and pDC dendritic cells are differentially infected and activated by respiratory syncytial virus. *PLoS ONE* **2011**, *6*, e16458. [CrossRef] [PubMed]
197. Rudd, B.D.; Schaller, M.A.; Smit, J.J.; Kunkel, S.L.; Neupane, R.; Kelley, L.; Berlin, A.A.; Lukacs, N.W. MyD88-mediated instructive signals in dendritic cells regulate pulmonary immune responses during respiratory virus infection. *J. Immunol.* **2007**, *178*, 5820–5827. [CrossRef]
198. Schijf, M.A.; Lukens, M.V.; Kruijsen, D.; van Uden, N.O.; Garssen, J.; Coenjaerts, F.E.; Van't Land, B.; van Bleek, G.M. Respiratory syncytial virus induced type I IFN production by pDC is regulated by RSV-infected airway epithelial cells, RSV-exposed monocytes and virus specific antibodies. *PLoS ONE* **2013**, *8*, e81695. [CrossRef]
199. Bartz, H.; Türkel, O.; Hoffjan, S.; Rothoeft, T.; Gonschorek, A.; Schauer, U. Respiratory syncytial virus decreases the capacity of myeloid dendritic cells to induce interferon-gamma in naïve T cells. *Immunology* **2003**, *109*, 49–57. [CrossRef]
200. de Graaff, P.M.A.; de Jong, E.C.; van Capel, T.M.; van Dijk, M.E.A.; Roholl, P.J.M.; Boes, J.; Luytjes, W.; Kimpen, J.L.L.; van Bleek, G.M. Respiratory syncytial virus infection of monocyte-derived dendritic cells decreases their capacity to activate CD4 T cells. *J. Immunol.* **2005**, *175*, 5904–5911. [CrossRef]
201. Cespedes, P.F.; Bueno, S.M.; Ramirez, B.A.; Gomez, R.S.; Riquelme, S.A.; Palavecino, C.E.; Mackern-Oberti, J.P.; Mora, J.E.; Depoil, D.; Sacristan, C.; et al. Surface expression of the hRSV nucleoprotein impairs immunological synapse formation with T cells. *Proc. Natl. Acad. Sci. USA* **2014**, *111*, E3214–E3223. [CrossRef] [PubMed]
202. Lau-Kilby, A.W.; Turfkruyer, M.; Kehl, M.; Yang, L.; Buchholz, U.J.; Hickey, K.; Malloy, A.M.W. Type I IFN ineffectively activates neonatal dendritic cells limiting respiratory antiviral T-cell responses. *Mucosal Immunol.* **2020**, *13*, 371–380. [CrossRef] [PubMed]
203. Malloy, A.M.; Ruckwardt, T.J.; Morabito, K.M.; Lau-Kilby, A.W.; Graham, B.S. Pulmonary Dendritic Cell Subsets Shape the Respiratory Syncytial Virus-Specific CD8+ T Cell Immunodominance Hierarchy in Neonates. *J. Immunol.* **2017**, *198*, 394–403. [CrossRef] [PubMed]
204. Shrestha, B.; You, D.; Saravia, J.; Siefker, D.T.; Jaligama, S.; Lee, G.I.; Sallam, A.A.; Harding, J.N.; Cormier, S.A. IL-4Ralpha on dendritic cells in neonates and Th2 immunopathology in respiratory syncytial virus infection. *J. Leukoc. Biol.* **2017**, *102*, 153–161. [CrossRef] [PubMed]
205. Johnson, T.R.; McLellan, J.S.; Graham, B.S. Respiratory syncytial virus glycoprotein G interacts with DC-SIGN and L-SIGN to activate ERK1 and ERK2. *J. Virol.* **2012**, *86*, 1339–1347. [CrossRef] [PubMed]
206. Stambas, J.; Lu, C.; Tripp, R.A. Innate and adaptive immune responses in respiratory virus infection: Implications for the clinic. *Expert. Rev. Respir. Med.* **2020**, *14*, 1141–1147. [CrossRef]
207. Bonilla, F.A.; Oettgen, H.C. Adaptive immunity. *J. Allergy Clin. Immunol.* **2010**, *125*, S33–S40. [CrossRef]
208. Boehm, T.; Swann, J.B. Origin and evolution of adaptive immunity. *Annu. Rev. Anim. Biosci.* **2014**, *2*, 259–283. [CrossRef]
209. Wang, Y.; Liu, J.; Burrows, P.D.; Wang, J.Y. B Cell Development and Maturation. *Adv. Exp. Med. Biol.* **2020**, *1254*, 1–22. [CrossRef]
210. Lam, J.H.; Smith, F.L.; Baumgarth, N. B Cell Activation and Response Regulation During Viral Infections. *Viral Immunol.* **2020**, *33*, 294–306. [CrossRef]
211. Roman, M.; Calhoun, W.J.; Hinton, K.L.; Avendano, L.F.; Simon, V.; Escobar, A.M.; Gaggero, A.; Diaz, P.V. Respiratory syncytial virus infection in infants is associated with predominant Th-2-like response. *Am. J. Respir. Crit. Care Med.* **1997**, *156*, 190–195. [CrossRef] [PubMed]

212. Andreano, E.; Paciello, I.; Bardelli, M.; Tavarini, S.; Sammicheli, C.; Frigimelica, E.; Guidotti, S.; Torricelli, G.; Biancucci, M.; D'Oro, U.; et al. The respiratory syncytial virus (RSV) prefusion F-protein functional antibody repertoire in adult healthy donors. *EMBO Mol. Med.* **2021**, *13*, e14035. [CrossRef] [PubMed]

213. Lee, F.E.; Falsey, A.R.; Halliley, J.L.; Sanz, I.; Walsh, E.E. Circulating antibody-secreting cells during acute respiratory syncytial virus infection in adults. *J. Infect. Dis.* **2010**, *202*, 1659–1666. [CrossRef]

214. Raes, M.; Alliet, P.; Gillis, P.; Kortleven, J.; Magerman, K.; Rummens, J.L. Peripheral blood T and B lymphocyte subpopulations in infants with acute respiratory syncytial virus brochiolitis. *Pediatr. Allergy Immunol.* **1997**, *8*, 97–102. [CrossRef]

215. Xiao, X.; Tang, A.; Cox, K.S.; Wen, Z.; Callahan, C.; Sullivan, N.L.; Nahas, D.D.; Cosmi, S.; Galli, J.D.; Minnier, M.; et al. Characterization of potent RSV neutralizing antibodies isolated from human memory B cells and identification of diverse RSV/hMPV cross-neutralizing epitopes. *MAbs* **2019**, *11*, 1415–1427. [CrossRef] [PubMed]

216. Alturaiki, W.; McFarlane, A.J.; Rose, K.; Corkhill, R.; McNamara, P.S.; Schwarze, J.; Flanagan, B.F. Expression of the B cell differentiation factor BAFF and chemokine CXCL13 in a murine model of Respiratory Syncytial Virus infection. *Cytokine* **2018**, *110*, 267–271. [CrossRef]

217. Reed, J.L.; Welliver, T.P.; Sims, G.P.; McKinney, L.; Velozo, L.; Avendano, L.; Hintz, K.; Luma, J.; Coyle, A.J.; Welliver, R.C., Sr. Innate Immune Signals Modulate Antiviral and Polyreactive Antibody Responses during Severe Respiratory Syncytial Virus Infection. *J. Infect. Dis.* **2009**, *199*, 1128–1138. [CrossRef]

218. Falsey, A.R.; Singh, H.K.; Walsh, E.E. Serum antibody decay in adults following natural respiratory syncytial virus infection. *J. Med. Virol.* **2006**, *78*, 1493–1497. [CrossRef]

219. Green, C.A.; Sande, C.J.; de Lara, C.; Thompson, A.J.; Silva-Reyes, L.; Napolitano, F.; Pierantoni, A.; Capone, S.; Vitelli, A.; Klenerman, P.; et al. Humoral and cellular immunity to RSV in infants, children and adults. *Vaccine* **2018**, *36*, 6183–6190. [CrossRef]

220. Habibi, M.S.; Jozwik, A.; Makris, S.; Dunning, J.; Paras, A.; DeVincenzo, J.P.; de Haan, C.A.; Wrammert, J.; Openshaw, P.J.; Chiu, C.; et al. Impaired Antibody-mediated Protection and Defective IgA B-Cell Memory in Experimental Infection of Adults with Respiratory Syncytial Virus. *Am. J. Respir. Crit. Care Med.* **2015**, *191*, 1040–1049. [CrossRef]

221. Cortjens, B.; Yasuda, E.; Yu, X.; Wagner, K.; Claassen, Y.B.; Bakker, A.Q.; van Woensel, J.B.M.; Beaumont, T. Broadly Reactive Anti-Respiratory Syncytial Virus G Antibodies from Exposed Individuals Effectively Inhibit Infection of Primary Airway Epithelial Cells. *J. Virol.* **2017**, *91*, e02357-16. [CrossRef] [PubMed]

222. Schmidt, M.E.; Meyerholz, D.K.; Varga, S.M. Pre-existing neutralizing antibodies prevent CD8 T cell-mediated immunopathology following respiratory syncytial virus infection. *Mucosal Immunol.* **2020**, *13*, 507–517. [CrossRef] [PubMed]

223. Falsey, A.R.; Walsh, E.E. Relationship of serum antibody to risk of respiratory syncytial virus infection in elderly adults. *J. Infect. Dis.* **1998**, *177*, 463–466. [CrossRef] [PubMed]

224. Walsh, E.E.; Falsey, A.R. Humoral and mucosal immunity in protection from natural respiratory syncytial virus infection in adults. *J. Infect. Dis.* **2004**, *190*, 373–378. [CrossRef]

225. Walsh, E.E.; Peterson, D.R.; Falsey, A.R. Risk factors for severe respiratory syncytial virus infection in elderly persons. *J. Infect. Dis.* **2004**, *189*, 233–238. [CrossRef]

226. Ascough, S.; Dayananda, P.; Kalyan, M.; Kuong, S.U.; Gardener, Z.; Bergstrom, E.; Paterson, S.; Kar, S.; Avadhan, V.; Thwaites, R.; et al. Divergent age-related humoral correlates of protection against respiratory syncytial virus infection in older and young adults: A pilot, controlled, human infection challenge model. *Lancet Healthy Longev.* **2022**, *3*, e405–e416. [CrossRef]

227. Falsey, A.R.; Walsh, E.E. Humoral immunity to respiratory syncytial virus infection in the elderly. *J. Med. Virol.* **1992**, *36*, 39–43. [CrossRef]

228. Walsh, E.E.; Peterson, D.R.; Kalkanoglu, A.E.; Lee, F.E.; Falsey, A.R. Viral shedding and immune responses to respiratory syncytial virus infection in older adults. *J. Infect. Dis.* **2013**, *207*, 1424–1432. [CrossRef]

229. Falsey, A.R.; Walsh, E.E.; Looney, R.J.; Kolassa, J.E.; Formica, M.A.; Criddle, M.C.; Hall, W.J. Comparison of respiratory syncytial virus humoral immunity and response to infection in young and elderly adults. *J. Med. Virol.* **1999**, *59*, 221–226. [CrossRef]

230. Walsh, E.E.; Falsey, A.R. Age related differences in humoral immune response to respiratory syncytial virus infection in adults. *J. Med. Virol.* **2004**, *73*, 295–299. [CrossRef]

231. Goodwin, E.; Gilman, M.S.A.; Wrapp, D.; Chen, M.; Ngwuta, J.O.; Moin, S.M.; Bai, P.; Sivasubramanian, A.; Connor, R.I.; Wright, P.F.; et al. Infants Infected with Respiratory Syncytial Virus Generate Potent Neutralizing Antibodies that Lack Somatic Hypermutation. *Immunity* **2018**, *48*, 339–349 e335. [CrossRef]

232. van Erp, E.A.; Lakerveld, A.J.; de Graaf, E.; Larsen, M.D.; Schepp, R.M.; Hipgrave Ederveen, A.L.; Ahout, I.M.; de Haan, C.A.; Wuhrer, M.; Luytjes, W.; et al. Natural killer cell activation by respiratory syncytial virus-specific antibodies is decreased in infants with severe respiratory infections and correlates with Fc-glycosylation. *Clin. Transl. Immunol.* **2020**, *9*, e1112. [CrossRef] [PubMed]

233. Williams, J.V.; Weitkamp, J.H.; Blum, D.L.; LaFleur, B.J.; Crowe, J.E., Jr. The human neonatal B cell response to respiratory syncytial virus uses a biased antibody variable gene repertoire that lacks somatic mutations. *Mol. Immunol.* **2009**, *47*, 407–414. [CrossRef]

234. Murphy, B.R.; Graham, B.S.; Prince, G.A.; Walsh, E.E.; Chanock, R.M.; Karzon, D.T.; Wright, P.F. Serum and nasal-wash immunoglobulin G and A antibody response of infants and children to respiratory syncytial virus F and G glycoproteins following primary infection. *J. Clin. Microbiol.* **1986**, *23*, 1009–1014. [CrossRef] [PubMed]

235. Kruijsen, D.; Bakkers, M.J.; van Uden, N.O.; Viveen, M.C.; van der Sluis, T.C.; Kimpen, J.L.; Leusen, J.H.; Coenjaerts, F.E.; van Bleek, G.M. Serum antibodies critically affect virus-specific CD4+/CD8+ T cell balance during respiratory syncytial virus infections. *J. Immunol.* **2010**, *185*, 6489–6498. [CrossRef] [PubMed]

236. Vono, M.; Eberhardt, C.S.; Auderset, F.; Mastelic-Gavillet, B.; Lemeille, S.; Christensen, D.; Andersen, P.; Lambert, P.H.; Siegrist, C.A. Maternal Antibodies Inhibit Neonatal and Infant Responses to Vaccination by Shaping the Early-Life B Cell Repertoire within Germinal Centers. *Cell Rep.* **2019**, *28*, 1773–1784 e1775. [CrossRef]

237. Zhivaki, D.; Lemoine, S.; Lim, A.; Morva, A.; Vidalain, P.O.; Schandene, L.; Casartelli, N.; Rameix-Welti, M.A.; Herve, P.L.; Deriaud, E.; et al. Respiratory Syncytial Virus Infects Regulatory B Cells in Human Neonates via Chemokine Receptor CX3CR1 and Promotes Lung Disease Severity. *Immunity* **2017**, *46*, 301–314. [CrossRef]

238. Bukreyev, A.; Yang, L.; Fricke, J.; Cheng, L.; Ward, J.M.; Murphy, B.R.; Collins, P.L. The secreted form of respiratory syncytial virus G glycoprotein helps the virus evade antibody-mediated restriction of replication by acting as an antigen decoy and through effects on Fc receptor-bearing leukocytes. *J. Virol.* **2008**, *82*, 12191–12204. [CrossRef]

239. Groothuis, J.R.; Levin, M.J.; Rodriguez, W.; Hall, C.B.; Long, C.E.; Kim, H.W.; Lauer, B.A.; Hemming, V.G. Use of intravenous gamma globulin to passively immunize high-risk children against respiratory syncytial virus: Safety and pharmacokinetics. The RSVIG Study Group. *Antimicrob. Agents Chemother.* **1991**, *35*, 1469–1473. [CrossRef]

240. Groothuis, J.R.; Simoes, E.A.; Hemming, V.G. Respiratory syncytial virus (RSV) infection in preterm infants and the protective effects of RSV immune globulin (RSVIG). Respiratory Syncytial Virus Immune Globulin Study Group. *Pediatrics* **1995**, *95*, 463–467.

241. Rodriguez, W.J.; Gruber, W.C.; Groothuis, J.R.; Simoes, E.A.; Rosas, A.J.; Lepow, M.; Kramer, A.; Hemming, V. Respiratory syncytial virus immune globulin treatment of RSV lower respiratory tract infection in previously healthy children. *Pediatrics* **1997**, *100*, 937–942. [CrossRef] [PubMed]

242. Keam, S.J. Nirsevimab: First Approval. *Drugs* **2023**, *83*, 181–187. [CrossRef] [PubMed]

243. Group, I.-R.S. Palivizumab, a Humanized Respiratory Syncytial Virus Monoclonal Antibody, Reduces Hospitalization From Respiratory Syncytial Virus Infection in High-risk Infants. *Pediatrics* **1998**, *102*, 531–537. [CrossRef]

244. Kalia, V.; Sarkar, S.; Gourley, T.S.; Rouse, B.T.; Ahmed, R. Differentiation of memory B and T cells. *Curr. Opin. Immunol.* **2006**, *18*, 255–264. [CrossRef]

245. Palacios-Pedrero, M.A.; Jansen, J.M.; Blume, C.; Stanislawski, N.; Jonczyk, R.; Molle, A.; Hernandez, M.G.; Kaiser, F.K.; Jung, K.; Osterhaus, A.; et al. Signs of immunosenescence correlate with poor outcome of mRNA COVID-19 vaccination in older adults. *Nat. Aging* **2022**, *2*, 896–905. [CrossRef]

246. Nicholson, L.B. The immune system. *Essays Biochem.* **2016**, *60*, 275–301. [CrossRef]

247. Benova, K.; Hanckova, M.; Koci, K.; Kudelova, M.; Betakova, T. T cells and their function in the immune response to viruses. *Acta Virol.* **2020**, *64*, 131–143. [CrossRef]

248. Sealy, R.E.; Surman, S.L.; Hurwitz, J.L. CD4(+) T cells support establishment of RSV-specific IgG and IgA antibody secreting cells in the upper and lower murine respiratory tract following RSV infection. *Vaccine* **2017**, *35*, 2617–2621. [CrossRef]

249. Kinnear, E.; Lambert, L.; McDonald, J.U.; Cheeseman, H.M.; Caproni, L.J.; Tregoning, J.S. Airway T cells protect against RSV infection in the absence of antibody. *Mucosal Immunol.* **2018**, *11*, 249–256. [CrossRef]

250. Guvenel, A.; Jozwik, A.; Ascough, S.; Ung, S.K.; Paterson, S.; Kalyan, M.; Gardener, Z.; Bergstrom, E.; Kar, S.; Habibi, M.S.; et al. Epitope-specific airway-resident CD4+ T cell dynamics during experimental human RSV infection. *J. Clin. Investig.* **2020**, *130*, 523–538. [CrossRef]

251. Kitcharoensakkul, M.; Bacharier, L.B.; Yin-Declue, H.; Boomer, J.S.; Sajol, G.; Leung, M.K.; Wilson, B.; Schechtman, K.B.; Atkinson, J.P.; Green, J.M.; et al. Impaired tumor necrosis factor-alpha secretion by CD4 T cells during respiratory syncytial virus bronchiolitis associated with recurrent wheeze. *Immun. Inflamm. Dis.* **2020**, *8*, 30–39. [CrossRef] [PubMed]

252. Welliver, T.P.; Reed, J.L.; Welliver, R.C., Sr. Respiratory syncytial virus and influenza virus infections: Observations from tissues of fatal infant cases. *Pediatr. Infect. Dis. J.* **2008**, *27*, S92–S96. [CrossRef] [PubMed]

253. De Swart, R.L.; Kuiken, T.; Timmerman, H.H.; van Amerongen, G.; Van Den Hoogen, B.G.; Vos, H.W.; Neijens, H.J.; Andeweg, A.C.; Osterhaus, A.D. Immunization of macaques with formalin-inactivated respiratory syncytial virus (RSV) induces interleukin-13-associated hypersensitivity to subsequent RSV infection. *J. Virol.* **2002**, *76*, 11561–11569. [CrossRef]

254. Waris, M.E.; Tsou, C.; Erdman, D.D.; Zaki, S.R.; Anderson, L.J. Respiratory synctial virus infection in BALB/c mice previously immunized with formalin-inactivated virus induces enhanced pulmonary inflammatory response with a predominant Th2-like cytokine pattern. *J. Virol.* **1996**, *70*, 2852–2860. [CrossRef] [PubMed]

255. Knudson, C.J.; Hartwig, S.M.; Meyerholz, D.K.; Varga, S.M. RSV vaccine-enhanced disease is orchestrated by the combined actions of distinct CD4 T cell subsets. *PLoS Pathog.* **2015**, *11*, e1004757. [CrossRef]

256. You, D.; Marr, N.; Saravia, J.; Shrestha, B.; Lee, G.I.; Turvey, S.E.; Brombacher, F.; Herbert, D.R.; Cormier, S.A. IL-4Ralpha on CD4+ T cells plays a pathogenic role in respiratory syncytial virus reinfection in mice infected initially as neonates. *J. Leukoc. Biol.* **2013**, *93*, 933–942. [CrossRef] [PubMed]

257. Roumanes, D.; Falsey, A.R.; Quataert, S.; Secor-Socha, S.; Lee, F.E.; Yang, H.; Bandyopadhyay, S.; Holden-Wiltse, J.; Topham, D.J.; Walsh, E.E. T-Cell Responses in Adults During Natural Respiratory Syncytial Virus Infection. *J. Infect. Dis.* **2018**, *218*, 418–428. [CrossRef]

258. Graham, B.S.; Bunton, L.A.; Wright, P.F.; Karzon, D.T. Role of T lymphocyte subsets in the pathogenesis of primary infection and rechallenge with respiratory syncytial virus in mice. *J. Clin. Investig.* **1991**, *88*, 1026–1033. [CrossRef]

259. Conners, M.; Kulkarni, A.B.; Firestone, C.Y.; Holmes, K.L.; Morse, H.C.r.; Sotnikov, A.V.; Murphy, B.R. Pulmonary histopathology induced by respiratory syncytial virus (RSV) challenge of formalin-inactivated RSV-immunized BALB/c mice is abrogated by depletion of CD4+ T cells. *J. Virol.* **1992**, *66*, 7444–7451. [CrossRef]

260. Schneider-Ohrum, K.; Snell Bennett, A.; Rajani, G.M.; Hostetler, L.; Maynard, S.K.; Lazzaro, M.; Cheng, L.I.; O'Day, T.; Cayatte, C. CD4(+) T Cells Drive Lung Disease Enhancement Induced by Immunization with Suboptimal Doses of Respiratory Syncytial Virus Fusion Protein in the Mouse Model. *J. Virol.* **2019**, *93*, e00695-19. [CrossRef]

261. Raiden, S.; Sananez, I.; Remes-Lenicov, F.; Pandolfi, J.; Romero, C.; De Lillo, L.; Ceballos, A.; Geffner, J.; Arruvito, L. Respiratory Syncytial Virus (RSV) Infects CD4+ T Cells: Frequency of Circulating CD4+ RSV+ T Cells as a Marker of Disease Severity in Young Children. *J. Infect. Dis.* **2017**, *215*, 1049–1058. [CrossRef] [PubMed]

262. Schlender, J.; Walliser, G.; Fricke, J.; Conzelmann, K.K. Respiratory syncytial virus fusion protein mediates inhibition of mitogen-induced T-cell proliferation by contact. *J. Virol.* **2002**, *76*, 1163–1170. [CrossRef] [PubMed]

263. Bont, L. Natural Reinfection with Respiratory Syncytial Virus Does Not Boost Virus-Specific T-Cell Immunity. *Pediatr. Res.* **2002**, *52*, 363–367. [CrossRef]

264. Heidema, J.; Lukens, M.V.; van Maren, W.W.C.; van Dijk, M.E.A.; Otten, H.G.; van Vught, A.J.; van der Werff, D.B.M.; van Gestel, S.J.P.; Semple, M.G.; Smyth, R.L.; et al. CD8+ T cell responses in bronchoalveolar lavage fluid and peripheral blood mononuclear cells of infants with severe primary respiratory syncytial virus infections. *J. Immunol.* **2007**, *179*, 8410–8417. [CrossRef] [PubMed]

265. Knudson, C.J.; Weiss, K.A.; Hartwig, S.M.; Varga, S.M. The pulmonary localization of virus-specific T lymphocytes is governed by the tissue tropism of infection. *J. Virol.* **2014**, *88*, 9010–9016. [CrossRef]

266. Jozwik, A.; Habibi, M.S.; Paras, A.; Zhu, J.; Guvenel, A.; Dhariwal, J.; Almond, M.; Wong, E.H.C.; Sykes, A.; Maybeno, M.; et al. RSV-specific airway resident memory CD8+ T cells and differential disease severity after experimental human infection. *Nat. Commun.* **2015**, *6*, 10224. [CrossRef]

267. Ruckwardt, T.J.; Luongo, C.; Malloy, A.M.; Liu, J.; Chen, M.; Collins, P.L.; Graham, B.S. Responses against a subdominant CD8+ T cell epitope protect against immunopathology caused by a dominant epitope. *J. Immunol.* **2010**, *185*, 4673–4680. [CrossRef]

268. Taylor, G.; Stott, E.J.; Hayle, A.J. Cytotoxic lymphocytes in the lungs of mice infected with respiratory syncytial virus. *J. Gen. Virol.* **1985**, *66*, 2533–2538. [CrossRef]

269. Aberle, J.H.; Aberle, S.W.; Dworzak, M.N.; Mandl, C.W.; Rebhandl, W.; Vollnhofer, G.; Kundi, M.; Popow-Kraupp, T. Reduced interferon-gamma expression in peripheral blood mononuclear cells of infants with severe respiratory syncytial virus disease. *Am. J. Respir. Crit. Care Med.* **1999**, *160*, 1263–1268. [CrossRef]

270. De Weerd, W.; Twilhaar, W.N.; Kimpen, J.L. T cell subset analysis in peripheral blood of children with RSV bronchiolitis. *Scand. J. Infect. Dis.* **1998**, *30*, 77–80. [CrossRef]

271. El Saleeby, C.M.; Suzich, J.; Conley, M.E.; DeVincenzo, J.P. Quantitative effects of palivizumab and donor-derived T cells on chronic respiratory syncytial virus infection, lung disease, and fusion glycoprotein amino acid sequences in a patient before and after bone marrow transplantation. *Clin. Infect. Dis.* **2004**, *39*, 17–20. [CrossRef]

272. Heidema, J.; de Bree, G.J.; de Graaff, P.M.A.; van Maren, W.W.C.; Hoogerhout, P.; Out, T.A.; Kimpen, J.L.L.; van Bleek, G.M. Human CD8(+) T cell responses against five newly identified respiratory syncytial virus-derived epitopes. *J. Gen. Virol.* **2004**, *85*, 2365–2374. [CrossRef] [PubMed]

273. Lukens, M.V.; Claassen, E.A.; de Graaff, P.M.; van Dijk, M.E.; Hoogerhout, P.; Toebes, M.; Schumacher, T.N.; van der Most, R.G.; Kimpen, J.L.; van Bleek, G.M. Characterization of the CD8+ T cell responses directed against respiratory syncytial virus during primary and secondary infection in C57BL/6 mice. *Virology* **2006**, *352*, 157–168. [CrossRef] [PubMed]

274. Ostler, T.; Ehl, S. Pulmonary T cells induced by respiratory syncytial virus are functional and can make an important contribution to long-lived protective immunity. *Eur. J. Immunol.* **2002**, *32*, 2562–2569. [CrossRef] [PubMed]

275. Ruckwardt, T.J.; Bonaparte, K.L.; Nason, M.C.; Graham, B.S. Regulatory T cells promote early influx of CD8+ T cells in the lungs of respiratory syncytial virus-infected mice and diminish immunodominance disparities. *J. Virol.* **2009**, *83*, 3019–3028. [CrossRef]

276. Srikiatkhachorn, A.; Braciale, T.J. Virus-specific CD8+ T lymphocytes downregulate T helper cell type 2 cytokine secretion and pulmonary eosinophilia during experimental murine respiratory syncytial virus infection. *J. Exp. Med.* **1997**, *186*, 421–432. [CrossRef]

277. Ostler, T.; Davidson, W.; Ehl, S. Virus clearance and immunopathology by CD8(+) T cells during infection with respiratory syncytial virus are mediated by IFN-gamma. *Eur. J. Immunol.* **2002**, *32*, 2117–2123. [CrossRef]

278. Bem, R.A.; Bos, A.P.; Bots, M.; Wolbink, A.M.; van Ham, S.M.; Medema, J.P.; Lutter, R.; van Woensel, J.B.M. Activation of the granzyme pathway in children with severe respiratory syncytial virus infection. *Pediatr. Res.* **2008**, *63*, 650–655. [CrossRef]

279. Siefker, D.T.; Vu, L.; You, D.; McBride, A.; Taylor, R.; Jones, T.L.; DeVincenzo, J.; Cormier, S.A. Respiratory Syncytial Virus Disease Severity Is Associated with Distinct CD8(+) T-Cell Profiles. *Am. J. Respir. Crit. Care Med.* **2020**, *201*, 325–334. [CrossRef]

280. van Schaik, S.M.; Obot, N.; Enhorning, G.; Hintz, K.; Gross, K.; Hancock, G.E.; Stack, A.M.; Welliver, R.C. Role of interferon gamma in the pathogenesis of primary respiratory syncytial virus infection in BALB/c mice. *J. Med. Virol.* **2000**, *62*, 257–266. [CrossRef]

281. Rutigliano, J.A.; Graham, B.S. Prolonged production of TNF-alpha exacerbates illness during respiratory syncytial virus infection. *J. Immunol.* **2004**, *173*, 3408–3417. [CrossRef] [PubMed]

282. Meckiff, B.J.; Ramirez-Suastegui, C.; Fajardo, V.; Chee, S.J.; Kusnadi, A.; Simon, H.; Eschweiler, S.; Grifoni, A.; Pelosi, E.; Weiskopf, D.; et al. Imbalance of Regulatory and Cytotoxic SARS-CoV-2-Reactive CD4(+) T Cells in COVID-19. *Cell* **2020**, *183*, 1340–1353 e1316. [CrossRef] [PubMed]

283. Lan, Y.T.; Wang, Z.L.; Tian, P.; Gong, X.N.; Fan, Y.C.; Wang, K. Treg/Th17 imbalance and its clinical significance in patients with hepatitis B-associated liver cirrhosis. *Diagn. Pathol.* **2019**, *14*, 114. [CrossRef] [PubMed]

284. Chatila, T.A. Role of regulatory T cells in human diseases. *J. Allergy Clin. Immunol.* **2005**, *116*, 949–959, quiz 960. [CrossRef] [PubMed]

285. Ghiringhelli, F.; Menard, C.; Terme, M.; Flament, C.; Taieb, J.; Chaput, N.; Puig, P.E.; Novault, S.; Escudier, B.; Vivier, E.; et al. CD4+CD25+ regulatory T cells inhibit natural killer cell functions in a transforming growth factor-beta-dependent manner. *J. Exp. Med.* **2005**, *202*, 1075–1085. [CrossRef] [PubMed]

286. Lewkowicz, P.; Lewkowicz, N.; Sasiak, A.; Tchorzewski, H. Lipopolysaccharide-activated CD4+CD25+ T regulatory cells inhibit neutrophil function and promote their apoptosis and death. *J. Immunol.* **2006**, *177*, 7155–7163. [CrossRef]

287. Rudensky, A.V. Regulatory T cells and Foxp3. *Immunol. Rev.* **2011**, *241*, 260–268. [CrossRef]

288. Weiss, K.A.; Christiaansen, A.F.; Fulton, R.B.; Meyerholz, D.K.; Varga, S.M. Multiple CD4+ T cell subsets produce immunomodulatory IL-10 during respiratory syncytial virus infection. *J. Immunol.* **2011**, *187*, 3145–3154. [CrossRef]

289. Demoulins, T.; Brugger, M.; Zumkehr, B.; Oliveira Esteves, B.I.; Mehinagic, K.; Fahmi, A.; Borcard, L.; Taddeo, A.; Jandrasits, D.; Posthaus, H.; et al. The specific features of the developing T cell compartment of the neonatal lung are a determinant of respiratory syncytial virus immunopathogenesis. *PLoS Pathog.* **2021**, *17*, e1009529. [CrossRef]

290. Fulton, R.B.; Meyerholz, D.K.; Varga, S.M. Foxp3+ CD4 regulatory T cells limit pulmonary immunopathology by modulating the CD8 T cell response during respiratory syncytial virus infection. *J. Immunol.* **2010**, *185*, 2382–2392. [CrossRef]

291. Lee, D.C.; Harker, J.A.; Tregoning, J.S.; Atabani, S.F.; Johansson, C.; Schwarze, J.; Openshaw, P.J. CD25+ natural regulatory T cells are critical in limiting innate and adaptive immunity and resolving disease following respiratory syncytial virus infection. *J. Virol.* **2010**, *84*, 8790–8798. [CrossRef] [PubMed]

292. Loebbermann, J.; Thornton, H.; Durant, L.; Sparwasser, T.; Webster, K.E.; Sprent, J.; Culley, F.J.; Johansson, C.; Openshaw, P.J. Regulatory T cells expressing granzyme B play a critical role in controlling lung inflammation during acute viral infection. *Mucosal Immunol.* **2012**, *5*, 161–172. [CrossRef] [PubMed]

293. Shao, H.Y.; Huang, J.Y.; Lin, Y.W.; Yu, S.L.; Chitra, E.; Chang, C.K.; Sung, W.C.; Chong, P.; Chow, Y.H. Depletion of regulatory T-cells leads to moderate B-cell antigenicity in respiratory syncytial virus infection. *Int. J. Infect. Dis.* **2015**, *41*, 56–64. [CrossRef]

294. Loebbermann, J.; Schnoeller, C.; Thornton, H.; Durant, L.; Sweeney, N.P.; Schuijs, M.; O'Garra, A.; Johansson, C.; Openshaw, P.J. IL-10 regulates viral lung immunopathology during acute respiratory syncytial virus infection in mice. *PLoS ONE* **2012**, *7*, e32371. [CrossRef] [PubMed]

295. Sun, J.; Cardani, A.; Sharma, A.K.; Laubach, V.E.; Jack, R.S.; Muller, W.; Braciale, T.J. Autocrine regulation of pulmonary inflammation by effector T-cell derived IL-10 during infection with respiratory syncytial virus. *PLoS Pathog.* **2011**, *7*, e1002173. [CrossRef]

296. Christiaansen, A.F.; Syed, M.A.; Ten Eyck, P.P.; Hartwig, S.M.; Durairaj, L.; Kamath, S.S.; Varga, S.M. Altered Treg and cytokine responses in RSV-infected infants. *Pediatr. Res.* **2016**, *80*, 702–709. [CrossRef]

297. Raiden, S.; Pandolfi, J.; Payaslian, F.; Anderson, M.; Rivarola, N.; Ferrero, F.; Urtasun, M.; Fainboim, L.; Geffner, J.; Arruvito, L. Depletion of circulating regulatory T cells during severe respiratory syncytial virus infection in young children. *Am. J. Respir. Crit. Care Med.* **2014**, *189*, 865–868. [CrossRef]

298. Krishnamoorthy, N.; Khare, A.; Oriss, T.B.; Raundhal, M.; Morse, C.; Yarlagadda, M.; Wenzel, S.E.; Moore, M.L.; Peebles, R.S., Jr.; Ray, A.; et al. Early infection with respiratory syncytial virus impairs regulatory T cell function and increases susceptibility to allergic asthma. *Nat. Med.* **2012**, *18*, 1525–1530. [CrossRef]

299. Badovinac, V.P.; Porter, B.B.; Harty, J.T. CD8+ T cell contraction is controlled by early inflammation. *Nat. Immunol.* **2004**, *5*, 809–817. [CrossRef]

300. Mahnke, Y.D.; Brodie, T.M.; Sallusto, F.; Roederer, M.; Lugli, E. The who's who of T-cell differentiation: Human memory T-cell subsets. *Eur. J. Immunol.* **2013**, *43*, 2797–2809. [CrossRef]

301. Turner, D.L.; Bickham, K.L.; Thome, J.J.; Kim, C.Y.; D'Ovidio, F.; Wherry, E.J.; Farber, D.L. Lung niches for the generation and maintenance of tissue-resident memory T cells. *Mucosal Immunol.* **2014**, *7*, 501–510. [CrossRef]

302. Zheng, M.Z.M.; Wakim, L.M. Tissue resident memory T cells in the respiratory tract. *Mucosal Immunol.* **2022**, *15*, 379–388. [CrossRef] [PubMed]

303. De Bree, G.J.; Heidema, J.; van Leeuwen, E.M.M.; van Bleek, G.M.; Jonkers, R.E.; Jansen, H.M.; van Lier, R.A.W.; Out, T.A. Respiratory syncytial virus-specific CD8+ memory T cell responses in elderly persons. *J. Infect. Dis.* **2005**, *191*, 1710–1718. [CrossRef] [PubMed]

304. Varese, A.; Nakawesi, J.; Farias, A.; Kirsebom, F.C.M.; Paulsen, M.; Nuriev, R.; Johansson, C. Type I interferons and MAVS signaling are necessary for tissue resident memory CD8+ T cell responses to RSV infection. *PLoS Pathog.* **2022**, *18*, e1010272. [CrossRef] [PubMed]

305. Heidema, J.; Rossen, J.W.A.; Lukens, M.V.; Ketel, M.S.; Scheltens, E.; Kranendonk, M.E.G.; van Maren, W.W.C.; van Loon, A.M.; Otten, H.G.; Kimpen, J.L.L.; et al. Dynamics of human respiratory virus-specific CD8+ T cell responses in blood and airways during episodes of common cold. *J. Immunol.* **2008**, *181*, 5551–5559. [CrossRef] [PubMed]

306. Luangrath, M.A.; Schmidt, M.E.; Hartwig, S.M.; Varga, S.M. Tissue-Resident Memory T Cells in the Lungs Protect against Acute Respiratory Syncytial Virus Infection. *Immunohorizons* **2021**, *5*, 59–69. [CrossRef] [PubMed]

307. Chang, J.; Braciale, T.J. Respiratory syncytial virus infection suppresses lung CD8+ Tcell effector activity and peripheral CD8+ T-cell memory in the respiratory tract. *Nat. Med.* **2002**, *8*, 54–60. [CrossRef]

308. Schmidt, M.E.; Knudson, C.J.; Hartwig, S.M.; Pewe, L.L.; Meyerholz, D.K.; Langlois, R.A.; Harty, J.T.; Varga, S.M. Memory CD8 T cells mediate severe immunopathology following respiratory syncytial virus infection. *PLoS Pathog.* **2018**, *14*, e1006810. [CrossRef]

309. Chirkova, T.; Rosas-Salazar, C.; Gebretsadik, T.; Jadhao, S.J.; Chappell, J.D.; Peebles, R.S., Jr.; Dupont, W.D.; Newcomb, D.C.; Berdnikovs, S.; Gergen, P.J.; et al. Effect of Infant RSV Infection on Memory T Cell Responses at Age 2-3 Years. *Front. Immunol.* **2022**, *13*, 826666. [CrossRef]

310. Zens, K.D.; Chen, J.K.; Guyer, R.S.; Wu, F.L.; Cvetkovski, F.; Miron, M.; Farber, D.L. Reduced generation of lung tissue-resident memory T cells during infancy. *J. Exp. Med.* **2017**, *214*, 2915–2932. [CrossRef]

311. Chanock, R.; Roizman, B.; Myers, R. Recovery from infants with respiratory illness of a virus related to chimpanzee coryza agent (CCA). I. Isolation, properties and characterization. *Am. J. Hyg.* **1957**, *66*, 281–290. [CrossRef] [PubMed]

312. Blount, R.E.J.; Morris, J.A.; Savage, R.E. Recovery of cytopathogenic agent from chimpanzees with coryza. *Proc. Soc. Exp. Biol. Med.* **1956**, *92*, 544–594. [CrossRef] [PubMed]

313. Crank, M.C.; Ruckwardt, T.J.; Chen, M.; Morabito, K.M.; Phung, E.; Costner, P.J.; Holman, L.A.; Hickman, S.P.; Berkowitz, N.M.; Gordon, I.J.; et al. A proof of concept for structure-based vaccine design targeting RSV in humans. *Science* **2019**, *365*, 505–509. [CrossRef] [PubMed]

314. Ura, T.; Okuda, K.; Shimada, M. Developments in Viral Vector-Based Vaccines. *Vaccines* **2014**, *2*, 624–641. [CrossRef]

315. Jordan, E.; Lawrence, S.J.; Meyer, T.P.H.; Schmidt, D.; Schultz, S.; Mueller, J.; Stroukova, D.; Koenen, B.; Gruenert, R.; Silbernagl, G.; et al. Broad Antibody and Cellular Immune Response From a Phase 2 Clinical Trial With a Novel Multivalent Poxvirus-Based Respiratory Syncytial Virus Vaccine. *J. Infect. Dis.* **2021**, *223*, 1062–1072. [CrossRef]

316. Samy, N.; Reichhardt, D.; Schmidt, D.; Chen, L.M.; Silbernagl, G.; Vidojkovic, S.; Meyer, T.P.; Jordan, E.; Adams, T.; Weidenthaler, H.; et al. Safety and immunogenicity of novel modified vaccinia Ankara-vectored RSV vaccine: A randomized phase I clinical trial. *Vaccine* **2020**, *38*, 2608–2619. [CrossRef]

317. Falsey, A.R.; Williams, K.; Gymnopoulou, E.; Bart, S.; Ervin, J.; Bastian, A.R.; Menten, J.; De Paepe, E.; Vandenberghe, S.; Chan, E.K.H.; et al. Efficacy and Safety of an Ad26.RSV.preF-RSV preF Protein Vaccine in Older Adults. *N. Engl. J. Med.* **2023**, *388*, 609–620. [CrossRef]

318. Stuart, A.S.V.; Virta, M.; Williams, K.; Seppa, I.; Hartvickson, R.; Greenland, M.; Omoruyi, E.; Bastian, A.R.; Haazen, W.; Salisch, N.; et al. Phase 1/2a Safety and Immunogenicity of an Adenovirus 26 Vector Respiratory Syncytial Virus (RSV) Vaccine Encoding Prefusion F in Adults 18-50 Years and RSV-Seropositive Children 12-24 Months. *J. Infect. Dis.* **2022**, *227*, 71–82. [CrossRef]

319. Sadoff, J.; De Paepe, E.; DeVincenzo, J.; Gymnopoulou, E.; Menten, J.; Murray, B.; Rosemary Bastian, A.; Vandebosch, A.; Haazen, W.; Noulin, N.; et al. Prevention of Respiratory Syncytial Virus Infection in Healthy Adults by a Single Immunization of Ad26.RSV.preF in a Human Challenge Study. *J. Infect. Dis.* **2022**, *226*, 396–406. [CrossRef]

320. Baxter, D. Active and passive immunity, vaccine types, excipients and licensing. *Occup. Med.* **2007**, *57*, 552–556. [CrossRef]

321. Walsh, E.E.; Falsey, A.R.; Scott, D.A.; Gurtman, A.; Zareba, A.M.; Jansen, K.U.; Gruber, W.C.; Dormitzer, P.R.; Swanson, K.A.; Radley, D.; et al. A Randomized Phase 1/2 Study of a Respiratory Syncytial Virus Prefusion F Vaccine. *J. Infect. Dis.* **2022**, *225*, 1357–1366. [CrossRef] [PubMed]

322. Falsey, A.R.; Walsh, E.E.; Scott, D.A.; Gurtman, A.; Zareba, A.; Jansen, K.U.; Gruber, W.C.; Dormitzer, P.R.; Swanson, K.A.; Jiang, Q.; et al. Phase 1/2 Randomized Study of the Immunogenicity, Safety, and Tolerability of a Respiratory Syncytial Virus Prefusion F Vaccine in Adults With Concomitant Inactivated Influenza Vaccine. *J. Infect. Dis.* **2022**, *225*, 2056–2066. [CrossRef] [PubMed]

323. Schmoele-Thoma, B.; Zareba, A.M.; Jiang, Q.; Maddur, M.S.; Danaf, R.; Mann, A.; Eze, K.; Fok-Seang, J.; Kabir, G.; Catchpole, A.; et al. Vaccine Efficacy in Adults in a Respiratory Syncytial Virus Challenge Study. *N. Engl. J. Med.* **2022**, *386*, 2377–2386. [CrossRef] [PubMed]

324. Walsh, E.E.; Pérez Marc, G.; Zareba, A.M.; Falsey, A.R.; Jiang, Q.; Patton, M.; Polack, F.P.; Llapur, C.; Doreski, P.A.; Ilangovan, K.; et al. Efficacy and Safety of a Bivalent RSV Prefusion F Vaccine in Older Adults. *N. Engl. J. Med.* **2023**, *388*, 1465–1477. [CrossRef]

325. Kampmann, B.; Madhi, S.A.; Munjal, I.; Simões, E.A.F.; Pahud, B.A.; Llapur, C.; Baker, J.; Pérez Marc, G.; Radley, D.; Shittu, E.; et al. Bivalent Prefusion F Vaccine in Pregnancy to Prevent RSV Illness in Infants. *N. Engl. J. Med.* **2023**, *388*, 1451–1464. [CrossRef]

326. Simoes, E.A.F. Respiratory Syncytial Virus Disease in Young Children and Older Adults in Europe: A Burden and Economic Perspective. *J. Infect. Dis.* **2022**, *226*, S1–S9. [CrossRef]

327. Bebia, Z.; Reyes, O.; Jeanfreau, R.; Kantele, A.; De Leon, R.G.; Garcia Sanchez, M.; Banooni, P.; Gardener, G.J.; Bartha Rasero, J.L.; Encinas Pardilla, M.B.; et al. Safety and immunogenicity of an investigational respiratory syncytial virus vaccine (RSVPreF3) in mothers and their infants: A phase 2 randomized trial. *J. Infect. Dis.* **2023**, *228*, 299–310. [CrossRef]

328. Papi, A.; Ison, M.G.; Langley, J.M.; Lee, D.G.; Leroux-Roels, I.; Martinon-Torres, F.; Schwarz, T.F.; van Zyl-Smit, R.N.; Campora, L.; Dezutter, N.; et al. Respiratory Syncytial Virus Prefusion F Protein Vaccine in Older Adults. *N. Engl. J. Med.* **2023**, *388*, 595–608. [CrossRef]

329. Gebre, M.S.; Brito, L.A.; Tostanoski, L.H.; Edwards, D.K.; Carfi, A.; Barouch, D.H. Novel approaches for vaccine development. *Cell* **2021**, *184*, 1589–1603. [CrossRef]

330. Chen, G.L.; Mithani, R.; Kapoor, A.; Lu, S.; El Asmar, L.; Panozzo, C.A.; Shaw, C.A.; Stoszek, S.K.; August, A. 234. Safety and Immunogenicity of mRNA-1345, an mRNA-Based RSV Vaccine in Younger and Older Adult Cohorts: Results from a Phase 1, Randomized Clinical Trial. *Open Forum Infect. Dis.* **2022**, *9*, ofac492.312. [CrossRef]
331. Rodriguez-Fernandez, R.; Mejias, A.; Ramilo, O. Monoclonal Antibodies for Prevention of Respiratory Syncytial Virus Infection. *Pediatr. Infect. Dis. J.* **2021**, *40*, S35–S39. [CrossRef] [PubMed]
332. Gilman, M.S.; Castellanos, C.A.; Chen, M.; Ngwuta, J.O.; Goodwin, E.; Moin, S.M.; Mas, V.; Melero, J.A.; Wright, P.F.; Graham, B.S.; et al. Rapid profiling of RSV antibody repertoires from the memory B cells of naturally infected adult donors. *Sci. Immunol.* **2016**, *1*, eaaj1879. [CrossRef] [PubMed]
333. McLellan, J.S.; Chen, M.; Joyce, M.G.; Sastry, M.; Stewart-Jones, G.B.; Yang, Y.; Zhang, B.; Chen, L.; Srivatsan, S.; Zheng, A.; et al. Structure-based design of a fusion glycoprotein vaccine for respiratory syncytial virus. *Science* **2013**, *342*, 592–598. [CrossRef] [PubMed]
334. American Academy of Pediatrics Committee on Infectious Diseases; American Academy of Pediatrics Bronchiolitis Guidelines Committee. Updated Guidance for Palivizumab Prophylaxis Among Infants and Young Children at Increased Risk of Hospitalization for Respiratory Syncytial Virus Infection. *Pediatrics* **2014**, *134*, 415–420. [CrossRef]
335. Tulloh, R.M.; Feltes, T.F. The European Forum for Clinical Management: Prophylaxis against the respiratory syncytial virus in infants and young children with congenital cardiac disease. *Cardiol. Young* **2005**, *15*, 274–278. [CrossRef]
336. Zhu, Q.; McLellan, J.S.; Kallewaard, N.L.; Ulbrandt, N.D.; Palaszynski, S.; Zhang, J.; Moldt, B.; Khan, A.; Svabek, C.; McAuliffe, J.M.; et al. A highly potent extended half-life antibody as a potential RSV vaccine surrogate for all infants. *Sci. Transl. Med.* **2017**, *9*, eaaj1928. [CrossRef]
337. Hammitt, L.L.; Dagan, R.; Yuan, Y.; Baca Cots, M.; Bosheva, M.; Madhi, S.A.; Muller, W.J.; Zar, H.J.; Brooks, D.; Grenham, A.; et al. Nirsevimab for Prevention of RSV in Healthy Late-Preterm and Term Infants. *N. Engl. J. Med.* **2022**, *386*, 837–846. [CrossRef]
338. Griffin, M.P.; Yuan, Y.; Takas, T.; Domachowske, J.B.; Madhi, S.A.; Manzoni, P.; Simoes, E.A.F.; Esser, M.T.; Khan, A.A.; Dubovsky, F.; et al. Single-Dose Nirsevimab for Prevention of RSV in Preterm Infants. *N. Engl. J. Med.* **2020**, *383*, 415–425. [CrossRef]
339. Lin, G.L.; Drysdale, S.B.; Snape, M.D.; O'Connor, D.; Brown, A.; MacIntyre-Cockett, G.; Mellado-Gomez, E.; de Cesare, M.; Bonsall, D.; Ansari, M.A.; et al. Distinct patterns of within-host virus populations between two subgroups of human respiratory syncytial virus. *Nat. Commun.* **2021**, *12*, 5125. [CrossRef]

Review

Molecular Interaction of Nonsense-Mediated mRNA Decay with Viruses

Md Robel Ahmed and Zhiyou Du *

College of Life Sciences and Medicine, Zhejiang Sci-Tech University, Hangzhou 310018, China
* Correspondence: duzy@zstu.edu.cn; Tel.: +86-571-86843195

Abstract: The virus–host interaction is dynamic and evolutionary. Viruses have to fight with hosts to establish successful infection. Eukaryotic hosts are equipped with multiple defenses against incoming viruses. One of the host antiviral defenses is the nonsense-mediated mRNA decay (NMD), an evolutionarily conserved mechanism for RNA quality control in eukaryotic cells. NMD ensures the accuracy of mRNA translation by removing the abnormal mRNAs harboring pre-matured stop codons. Many RNA viruses have a genome that contains internal stop codon(s) (iTC). Akin to the premature termination codon in aberrant RNA transcripts, the presence of iTC would activate NMD to degrade iTC-containing viral genomes. A couple of viruses have been reported to be sensitive to the NMD-mediated antiviral defense, while some viruses have evolved with specific *cis*-acting RNA features or *trans*-acting viral proteins to overcome or escape from NMD. Recently, increasing light has been shed on the NMD–virus interaction. This review summarizes the current scenario of NMD-mediated viral RNA degradation and classifies various molecular means by which viruses compromise the NMD-mediated antiviral defense for better infection in their hosts.

Keywords: nonsense-mediated mRNA decay (NMD); RNA virus; host–virus interaction

1. Introduction

Citation: Ahmed, M.R.; Du, Z. Molecular Interaction of Nonsense-Mediated mRNA Decay with Viruses. *Viruses* **2023**, *15*, 816. https://doi.org/10.3390/v15040816

Academic Editors: Yiping Li and Yuliang Liu

Received: 14 December 2022
Revised: 14 February 2023
Accepted: 28 February 2023
Published: 23 March 2023

Nonsense-mediated mRNA decay (NMD) is an intrinsic mechanism for eukaryotic cells to remove aberrant cellular transcripts harboring a premature termination codon (PTC) [1–3] or a long 3′ untranslated region (3′ UTR) [4–7]. RNA transcripts with PTC created by alternative splicing or mutation produce a non-functional toxic protein that is detrimental to the cells. To maintain stability and functionality, the cellular NMD pathway removes at least part, if not all, of such transcripts and protects the cell from the accumulation of the non-functional mRNA and toxic protein [8–11]. The degradation and regulation of cellular mRNA by NMD have been identified in various eukaryotes, including *Saccharomyces cerevisiae*, *Drosophila melanogaster*, *Arabidopsis thaliana*, and humans [12]. By targeting non-mutant transcripts, such as mRNAs holding an unusually long 3′ UTR [13], NMD plays a significant role in the growth and development of eukaryotes [14–17]. Although primarily conceived as a quality control pathway, its regulation of gene expression goes far beyond quality control.

Nascent mRNA transcripts synthesized by RNA polymerase II undergo typical translation, stopping at the correct termination codon near the 3′ end (Figure 1) [18]. Once an unexpected stop codon is introduced in the coding frame, translation terminates ahead, resulting in incomplete translation. Premature termination accelerates the destruction of concerned mRNA through the NMD pathway. The NMD pathway is executed by the sequentially combined action of universally conserved up frameshift protein factors (UPF), eukaryotic release factors (eRF), and suppressor with morphological effect on genitalia (SMG) proteins, in spite of varied functions in different organisms [18]. In the exon junction complex (EJC)-dependent NMD, PTC can trigger the NMD response efficiently when EJC locates at 50–55 nucleotides (nt) upstream of the exon–exon junction [18–21] (Figure 1).

EJC is a dynamic structure installed on pre-mRNAs during splicing and is associated with matured mRNAs until the first round of translation. Usually, the translating ribosome removes EJC from the coding sequences [22,23]. If mRNA translation terminates early due to the presence of PTC, EJC keeps an association with mRNAs, which activates NMD effectors to decay the abnormal mRNA in an EJC-dependent way [24].

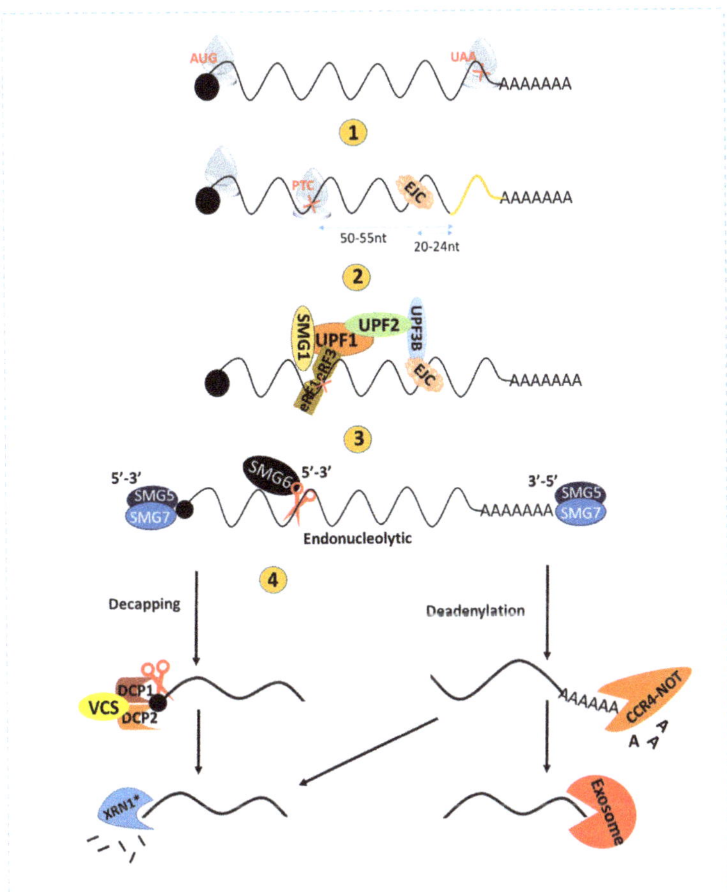

Figure 1. PTC-based NMD activation and RNA decay. When PTC is located >50-55 nucleotides upstream of the exon–exon junction. EJC deposits 20–24 nucleotides upstream of the exon–exon junction and cannot be removed by translating ribosomes. Subsequent deposition of eRF1 and eRF2 dissociates ribosome, thereby binding eRF3 and UPF1. SMG1-mediated phosphorylation of UPF1 initiates a conformational change of UPF1 to recruit UPF2 and UPF3, which precedes the reaction further. SMG1, UPF1, eRF1, and eRF3 together form the SURF complex. Afterward, UPF2 connects EJC and UPF1 via UPF3, which recruits relevant decay factors. Target transcripts can be degraded by two major cellular cleavage pathways, endo-nucleolytic and exo-nucleolytic. SMG6 is an endonuclease recruited by phosphorylated UPF1 at the PTC, cleaving the NMD substrate. SMG5/7 promote the exo nucleolytic pathway by recruiting decapping complex DCP1a/DCP2 at the 5′ end and recruiting deadenylation complex CCR4-NOT at the 3′ end. Nascent decapped transcripts were further digested in a 5′→3′ manner via XRN1*. Deadenylated RNA transcripts can be degraded in two ways: exosome-mediated digestion in a 3′→5′ manner and XRN1*-mediated digestion in a 5′→3′ manner. [XRN1*: XRN1 functions in animal cells, while XRN4 does in plant cells].

Usually, eRF components bind to ribosomes for dynamic translation termination and dissociate the ribosomes from translating templates. In the PTC-containing transcript, EJC is placed 20–24 nt upstream of the exon–exon junction and acts as the anchoring point for the assembly of several NMD components to form the SURF complex (SMG-1-Upf1-eRF1-eRF3) [20], playing a critical role in initiating the NMD pathway [25]. SMG1-mediated phosphorylation of UPF1 triggers the formation of SURF and promotes further steps of NMD activation [26,27], where UPF2 acts as a bridge structure for the connection and assembly of the SURF complex with EJC components [28] (Figure 1). Furthermore, EJC-independent NMD activation is subjected to the 3′ UTR length of a translating mRNA [6,29], which is present in *Drosophila melanogaster, Caenorhabditis elegans, Tetrahymena thermophile, Schizosaccarhomyces pombe,* and so on [23,30–32]. The long 3′ UTR (>300 nt) delays the ribosome dissociation circumstances and efficient translation termination, such as potential interaction between eRF and PolyA Binding Protein C1 (PABPC1), which leads to the assembly of essential NMD components and corresponding RNA decay [24,33,34]. NMD-mediated degradation of RNA transcripts is initiated by SMG5 and SMG7 in plants or other SMG6-depleted cells [35] or SMG6 in mammals [36]. Subsequently, SMG7 recruits CCR4-NOT by binding with one of its catalytic subunits, which then promotes the deadenylation-dependent decapping [35]. It also recruits decapping complexes such as DCP1, DCP2, and Varicose (VCS), which removes the cap structure of RNA transcripts. The decapped transcripts are further degraded in a 5′→3′ manner through the activity of XRN4 in plants or XRN1 and XRN2 in some other organisms [34,35,37]. A post-transcriptional QC system, NMD is unique and evolutionarily conserved in eukaryotes [38–40]. Interestingly, NMD regulates endogenous mRNA turnover but also engages in host defense against virus infection. A broad range of positive-sense RNA viruses are susceptible to and compromised by host NMD.

NMD is a translation-dependent molecular event in eukaryotic cells. Viruses rely on host translation machinery to synthesize viral proteins using viral RNAs as templates. Many RNA viruses have a polycistronic genome that encodes two or more ORFs and is translated to produce all viral proteins via diverse translation strategies. Positive sense (+) genomic RNA can be used as mRNA to translate the first ORF at the 5′ proximity. In some cases, the ribosome can avoid or skip the termination codon of the first ORF and continuously translate to produce a fusion protein via non-canonical translation strategies, such as readthrough and frameshifting [41,42]. Ribosomal frameshifting and readthrough are translation strategies commonly used by many viruses to produce an extended protein, which is usually viral RNA-dependent RNA polymerase [43–45]. Once ribosomes keep translation after frameshifting or readthrough, the ribosomal movement would remove the NMD components associated with the coding sequence, thus compromising NMD activation. For polycistronic genomic RNAs, the ORFs near the 3′ end are usually translated via subgenomic RNAs. RNA viruses employing subgenomic RNAs to produce viral proteins have the genomic RNA potentially targeted by NMD because the translation termination of the first 5′ ORF creates an extremely long untranslated sequence. Some viral genomes (e.g., *polioviruses*, HIV, *Dengue virus*, HCV, *potyviruses*) encode a large ORF that is translated to produce a polyprotein. The polyprotein is cleaved by cellular or viral protease enzymes to produce all functional viral proteins [46]. Such viruses use the polyprotein strategy to avoid the presence of any internal stop codon (iTC) in the large ORF, which would protect viral RNAs from NMD. A number of plant viruses or mycoviruses have segmented genomes, some of which are monocistronic. Thus, translation strategies of RNA viruses are closely associated with the activation or escaping of the host NMD defense.

2. NMD Defense against Viruses

The host NMD machinery can target genomic and sub-genomic RNAs of RNA viruses if these RNAs exhibit certain NMD-triggering features (Table 1). iTC is a primary feature in many (+) RNA viruses and is responsible for NMD activation [1]. Long 3′ UTRs, retained introns, and multiple ORFs are among the NMD-inducing features in mammalian retrovirus

RNAs. These features allow viral RNAs to interact with NMD machinery when translated in host cells (Figure 2) [47]. Some viral RNAs have several NMD-triggering features, leading to a strong NMD response. In addition, NMD-mediated degradation of (+) RNA viruses is independent of the EJC complex since viral RNAs do not undergo RNA splicing [48]. Either upstream ORF or long 3′ UTR plays a vital role in activating EJC-independent NMD [49–51]. Other distinctive features accountable for NMD activation might be present in viruses.

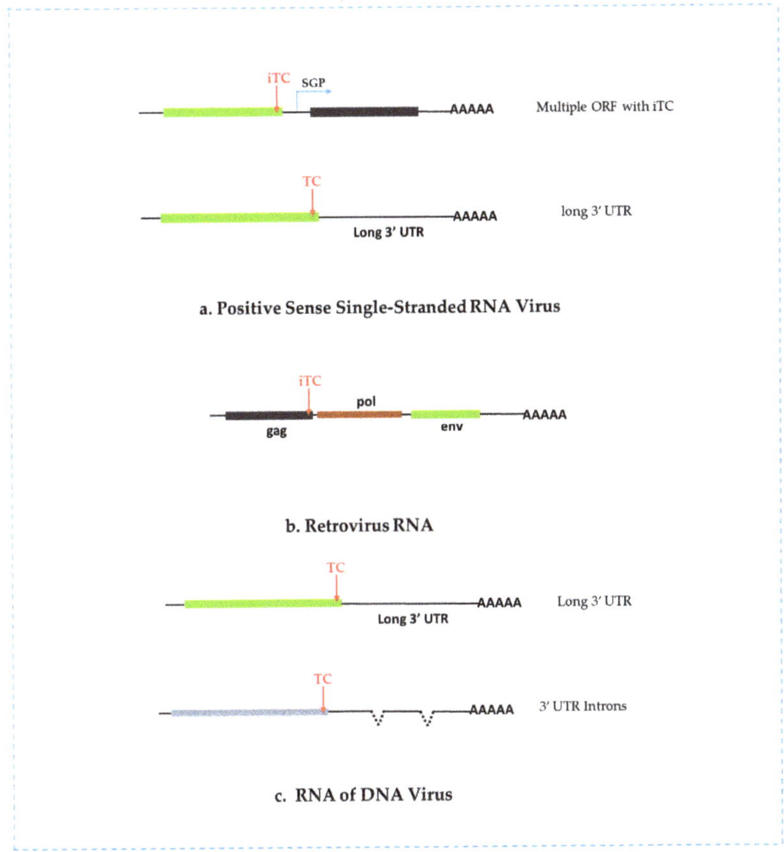

Figure 2. Schematic diagram of various viral RNAs targeted by host NMD. (a) Positive sense RNA virus possesses two or more separated open reading frames (e.g., *semliki forest virus* and *sindbis virus*) or a long 3′ UTR (e.g., *pea enation mosaic virus* 2); (**b**) Retroviral RNA (e.g., *HIV*, *Rous Sarcoma Virus*) harboring internal termination codons (iTC) in the *gag* gene; (**c**) Susceptible RNAs of DNA viruses (e.g., *Kaposi's sarcoma-associated herpesvirus*, *Epstein-Barr virus*) with an extended 3′ UTR and introns [SGP: Sub-Genomic Promoter].

2.1. NMD Targets Positive Sense RNA Viruses

(+) RNA viruses have their genome translated directly as mRNAs once infect host cells. A majority of RNA viruses have one or more distinctive features, such as polycistronic RNAs, iTC, long 3′ UTR, and uORF, which make (+) RNA viruses susceptible to NMD. For example, *Sindbis Virus* possesses a bicistronic gRNA with iTCs that confront host cellular NMD [52,53]. The unusually long 3′ UTR of *Semliki Forest Virus* (SFV) gRNA was supposed to make the viral gRNA susceptible to NMD, although its RNA has some other features that promote NMD without the long 3′ UTR [54]. *Turnip crinkle virus* (TCV) lacks the 5′ cap structure and a 3′ poly-A tail in its gRNA and uses a non-canonical mechanism to translate its genome. The iTC feature makes the gRNA of TCV sensitive to NMD [55]. iTC and long 3′ UTR in *Potato virus X* (PVX) gRNA and the extended 3′ UTR in sgRNA 1, 2, and 3 accounts for Arabidopsis's NMD targets [55]. In addition, PVX infection reduces endogenous NMD activity, indicating that PVX may produce anti-NMD protein or make the cellular state hostile to the cellular NMD response [55]. It is a common strategy for RNA viruses to use sub-genomic RNA for expressing viral proteins. Furthermore, compared with host mRNAs, higher GC content in viral 3′ UTR enforces *Tombusviridae* to NMD [56]. *Murine hepatitis virus* (MHV) is a (+) RNA virus belonging to one of the *coronaviruses* that possess NMD-inducing features, such as multiple ORFs with iTC that create a long 3′ UTR during translation [57]. Another case is SARS-CoV2 that is predicted to have a PTC at position 14 of ORF3b in 17.6% of isolates, suggesting that SARS-CoV-2 would be a potential target of NMD [58].

2.2. NMD Targets Double-Stranded RNA Viruses

Rotavirus is a double-stranded RNA virus displaying interplay with NMD. After entering the host cells, rotaviruses become active and produce 11 kinds of 5′ capped (+) ssRNAs [59]. The key NMD factor UPF1 seems to modulate the expression of viral proteins. The degradation kinetics of rotaviral RNAs is delayed in UPF1 and UPF2 depleted cells and UPF1 and UPF2 are associated with viral RNAs, indicating that viral RNA is sensitive to NMD, although the specific molecular mechanism of interaction between viral RNAs and UPF1 or UPF2 is not identified yet [59].

2.3. NMD Targets Retroviruses

Genome organization and lifecycle of Retroviruses differ from those of (+) RNA viruses. Retroviruses occupy multiple ORFs, long 3′ UTR, and retained intron, which can be predicted to trigger NMD (Figure 2b) [47,60]. The termination codon of the *gag* ORF and the iTC of the reverse transcriptase ORF would initiate NMD response for many retroviruses during their translation. *Rous Sarcoma Virus* (RSV) is a retrovirus, and its unspliced gRNA is targeted by NMD due to these NMD-triggering features [61–63]. Another retrovirus, *human T-cell leukemia virus type 1* (HTLV-1) RNA, exposes sensitivity to host NMD [64,65]. Still, the detailed interaction between a specific region of viral RNA and the NMD factor is not well understood. The interaction of the *human immunodeficiency virus* (HIV) with the RNA surveillance pathway is distinctive from other viruses [66]. Here, UPF1 plays an unexpected role in HIV infection [66]. In addition, a stop codon at the termination site of the *gag* gene in *Molony murine leukemia virus* (MoMLV) creates an extremely long 3′ UTR, which makes MoMLV RNA potential for NMD [67].

2.4. NMD Targets Viral RNAs of DNA Viruses

The *cauliflower mosaic virus* (CaMV) belongs to the family *Caulimoviridae* possessing a circular double-stranded DNA genome. The virus replicates via reverse transcription of 35S pre-genomic RNA under the control of the transactivator protein (TAV) [68]. 35S pre-genomic RNA holds six ORFs translated to produce six viral proteins (P1-P6) via the reinitiation mechanism. Alternative splicing is common for the 35S pre-genomic RNA [69], which produces spliced and alternatively spliced mRNAs potentially targeted by NMD [70]. Still, less is known about the mechanism by which NMD targets the CaMV RNA. It can be supposed that multiple uORFs in the 35S pre-genomic RNA may reduce the translation reinitiation efficacy of the original ORFs (P1–P6 coding ORF), which may be responsible for subsequent RNA decay by decay factors [70].

Recently, Zhao and colleagues [71] reported that NMD exerts an antiviral defense against the DNA virus *Kaposi's sarcoma-associated herpesvirus* (KSHV) by restricting viral lytic reactivation [71]. Using high-throughput transcriptome-wide analysis, they identified viral RNAs as an NMD substrate due to some NMD-eliciting features, such as long 3′ UTRs and introns [71]. These introns facilitate the deposition of EJC in the junction sites and further trigger the EJC-dependent NMD (Figure 2c). Another report suggested that NMD targets the *Rta* transactivator transcripts of KSHV and another herpesvirus, *Epstein-Barr virus* (EBV), for degradation through recognizing features in their $3'$ UTRs, which control the lytic reactivation of herpesviruses [72]. Thus, the host NMD system plays an important role in the regulation of the latent-lytic balance during herpesvirus infection.

3. Diversity of Viral Tactics Counteracts NMD-Mediated Host Defense

As summarized above, many viruses are targeted by host NMD for degradation. Some viruses have evolved into strategies to counteract the NMD-mediated antiviral defense (Table 1). The viral strategies can be distinct, which depends on viruses. It can be executed via *cis*-acting or *trans*-acting elements. *Cis*-acting elements are mainly viral RNA sequences or structures that impede host NMD. *Trans*-acting molecules are viral proteins via a molecular interaction, which restrain the NMD activating factors.

3.1. Viral Proteins: Targeting NMD Components and Interfering with their Function

3.1.1. Movement Proteins

Imamachi and colleagues reported that GC-rich motifs in UPF1 targets were indispensable for UPF1-mediated mRNA decay [56]. Thus, high GC content in the $3'$ UTR of viral RNAs could expose viral RNAs to NMD. *Umbravirus Pea enation mosaic virus* 2 (PEMV2) movement protein p26 is required for efficient virus accumulation irrespective of long-distance trafficking. Interestingly, it protects PEMV2 gRNA and NMD-sensitive host mRNAs bearing long, GC-rich 3'UTRs from NMD [73]. Additionally, p26 safeguards some host cellular mRNAs sensitive to NMD in *N. benthamiana* [73].

3.1.2. Tax and Rex Factors

The Tax factor of the *retrovirus* HTLV-1 attenuates UPF1 and translation commencement factor INT6 activity, both important for NMD [65]. Tax interacts with the central helicase core domain of UPF1 and might plug the RNA channel of UPF1, reducing its binding affinity of nucleic acids [64]. The authors also showed that in the single-molecule approach, Tax's sequential interaction with an RNA-bound UPF1 freezes and represses UPF1 translocation (Figure 3). In the presence of Tax, the affinity of UPF1 to nucleic acids is reduced by 10 times, and the mutation of UPF1 R843 residue decreases the Tax interaction with UPF1 [64]. The Tax-UPF1 complex accumulates in P-bodies, thus interfering with mRNA degradation [74]. The SMG5/SMG7 complex recycles UPF1 by dephosphorylating via PP2A in P bodies [75]. The Tax-UPF1 complex located in P-bodies indicates that Tax inhibits UPF1 recycling [74]. Tax interacts with other NMD factors, including UPF2 and INT6/eIF3E, preventing the degradation of some cellular and viral RNAs by NMD [65,76]. HTLV-1 Rex protein involved in the nuclear transport of spliced and unspliced viral protein also has an NMD inhibiting role and protects host and viral RNAs from being degraded [21,77]. Rex binds with multiple NMD proteins such as UPF1, SMG5, and SMG7 and selectively replaces UPF3B with UPF3A (a less active form of UPF3) to inhibit NMD [21].

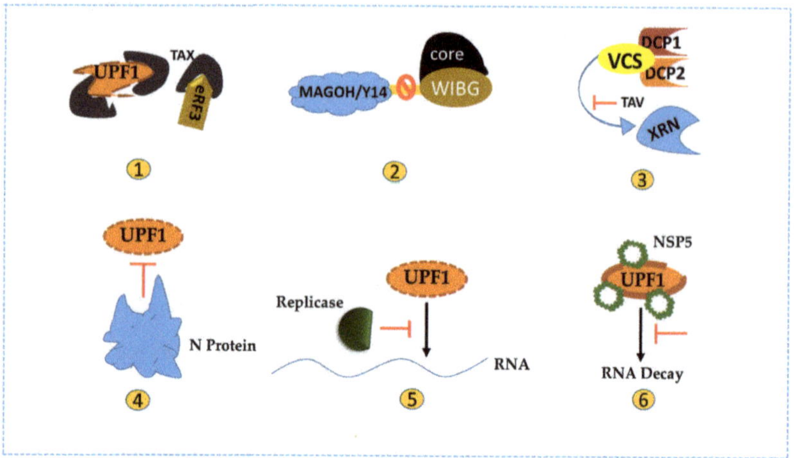

Figure 3. Viral protein-based strategies to modulate NMD activity: (**1**) Retroviral Tax factor antagonizes NMD by binding with eRF3 and UPF1 to prevent the SURF complex formation. By binding with UPF1, Tax disrupts its translocation and binding affinity to nucleic acids and other proteins; (**2**) HCV Core Protein disrupts NMD by binding with host protein WIBG to block the recycling interaction between WIBG and MAGOH/Y14; (**3**) Trans-acting activator (TAV) of CaMV interrupts decapping by interacting with VCS and subsequent decay of decapped transcripts by the XRN complex; (**4**) Coronavirus N-protein binds to UPF1 and diminishes its activity for NMD; (**5**) SFV replicase protein dislocates UPF1 from viral RNA to modulate host NMD response; (**6**) Non-structural protein (NSP5) promotes UPF1 degradation during *rotavirus* infection to inhibit the RNA decay pathway.

3.1.3. Capsid Protein (CP)

West Nile Virus (WNV) belongs to the genus *flavivirus*, closely related to *Dengue virus* and *Zika virus*. The WNV genome encodes some proteins that participate in the down-regulation of host NMD. RBM8A is an EJC protein directly associated with WNV RNA and induces its decay, while PYM1 is an EJC recycling factor that binds with RBM8A to sharpen the decay circumstances. The knockdown of PYM1 attenuates RBM8A binding with WNV RNA, indicating that the PYM1-RBM8A interaction plays a role in NMD against WNV [78]. Hence, the WNV CP protein interacts with PYM1 and restrains it from association with RBM8A, circumventing NMD [78]. RNAi screening-based study also demonstrates that the components (e.g., PYM1) associated with EJC-recycling and NMD pathway are functional in defense against WNV, *Dengue virus*, and *Zika virus*. *Zika virus* is a single-stranded positive-sense RNA virus responsible for severe neurological complications. It infects host cells by impairing endogenous NMD turnover. The CP protein of *Zika virus* depletes the level of UPF1, ultimately inactivating NMD [79]. By association with UPF1, *Zika virus* CP induces proteasome-mediated UPF1 degradation [80]. Recently, evidence has been found that SFV CP also suppresses NMD in translation-independent manner [81]. SFV CP binds with UPF1 and other viral proteins, but the detailed mechanism is not yet elucidated [81].

3.1.4. Core Protein

Core protein from *Hepatitis C Virus* (HCV) also down-regulates endogenous NMD turnover pathway. It disrupts the interaction between WIBG and the EJC component MAGOH/Y14, thus overwhelming NMD [82]. WIBG is a host protein participating in the cytoplasm's EJC disassembly and recycling process. Generally, WIBG binds MAGOH/Y14 to recycle EJC and accelerates the recycled EJC back to nuclei. By interacting with WIBG to interrupt EJC recycling, the HCV core protein perturbs EJC back to the nucleus from the cytoplasm (Figure 3) [82]. The WIBG–HCV core protein interaction possibly alters cellular transcriptome favoring HCV infection, which requires further study.

3.1.5. Trans-Activator Protein (TAV)

CaMV TAV acts as a suppressor of NMD by stimulating the reinitiation of viral RNA translation. As aforementioned (Section 3.2), failure to reinitiate translation of 35S pre-genomic RNA triggers RNA decay. TAV helps to accomplish translation reinitiation of 35S polycistronic mRNA of CaMV by recruiting the reinitiation supporting protein (RISP), eukaryotic translation initiation factor (eIF3), and other translation components at the translation site. To maintain the highest catalytic activity of RISP, eIF3, and other proteins involved in reinitiation, TAV activates rapamycin (TOR) targets [83–86], which represses the translation reinitiation barrier created by multiple uORFs in the 35S pre-genomic RNA [87]. In addition, TAV potentially interacts with the decapping complex scaffold protein VCS and protects NMD target transcripts containing uORFs within their leader regions (Figure 3) [70]. VCS contributes to more than 50% of short-live mRNA decay in *Arabidopsis* [88].

3.1.6. N-Protein

Evasion of NMD response is mandatory for Coronavirus (CoV) to establish successful infection. Several proteins of *Coronavirus* have been characterized to be involved in viral RNA package [89–92] and also help CoV to interact with host silencing factors for efficient replication. Coronavirus N protein plays multifunctional roles in the virus lifecycle. The MHV N protein represses the host NMD pathway and helps the virus to overcome the host defense [57]. In silico analysis suggests that the N protein of the chicken coronavirus *infectious bronchitis virus* (IBV) possibly binds to the NMD core factor UPF1 and compromises its activity, thus antagonizing NMD (Figure 3) [93]. Recently, protein interactome revealed that SARS-CoV-2 N protein interacts with mRNA decay factors UPF1 and MOV10, which might be one of the strategies for SARS-CoV-2 compromising NMD [58].

3.1.7. Replicase Protein

The genome of (+) RNA viruses is the template for translation to produce viral proteins, but it also is recruited into viral replicating compartments as the template for viral RNA replication. Thus, there must be a competition for viral gRNA between viral RNA translation and replication. NMD is dependent on RNA translation. Balistreri and colleagues found that the knockdown of NMD proteins (UPF1, Smg5, Smg7) markedly increased the concentration of SFV RNA and proteins, and the mutation or deletion of viral replicase enhanced NMD response [54]. Accordingly, a competition model was proposed for counteracting NMD-mediated antiviral defense, where SFV replicase protein competes with cellular UPF1 to bind (+) viral RNA by dislocating UPF1 from viral RNA inside the host cell (Figure 3) [54,80].

3.1.8. Non-Structural Protein

Recent evidence reported that non-structural protein from *rotavirus* circumvented the NMD pathway in a strain-independent manner [59]. Viral non-structural protein 5 (NSP5) binds with cellular UPF1 and activates proteasome-mediated degradation of UPF1 (Figure 3) [59].

3.2. Viral RNA Sequences or Structures

Besides *trans*-acting viral proteins, viruses have evolved to counteract the NMD-mediated antiviral defense via *cis*-acting RNA elements. These RNA elements include some RNA structures involved in viral non-canonical RNA translation, such as leaky scanning, readthrough, frameshifting and reinitiating, and some specialized RNA sequences engaged in RNA: protein interaction to restrict UPF1 binding [1,94]. However, some *cis*-acting RNA elements directly interact with core NMD components to overwhelm RNA decay [47,94,95].

3.2.1. *Rous Sarcoma Virus* RNA Stability Element (RSE)

Unspliced full-length RSV RNA is entirely resistant to NMD [63], but the insertion of PTC in the RSV molecular clone makes it an NMD substrate [61]. Several follow-up studies show that RSV evades NMD and expresses essential gag protein for virion synthesis [96]. A specific *cis*-acting element called RNA stability element (RSE) protects the RSV gRNA from host NMD. RSE restricts UPF1 to bind eRF3 by recruiting polypyrimidine tract binding protein 1 (PTBP1) [47,62]. Mutations or lack of RSE impose viral genome decay. This study also suggests that the 3′ UTRs of RSV *pol* and *src* may also function as stability elements. Increasing the stability of viral RNAs in cytoplasm also helps *vaccinia virus* and other poxviruses to overcome RNA decay [97]. Similarly, PTBP1 shields retroviral and cellular RNAs from NMD degradation [98]. PTBP1 restricts UPF1 binding and regulates gene expression by locating at the stop codon's vicinity.

3.2.2. Unstructured Element of TCV

TCV has a (+) single-stranded gRNA, which can be used as a template to express p33 and its readthrough product p94, which are viral replicases. TCV transcribes to create two sgRNAs that are translated to produce viral CP and MP proteins. Both gRNA or sgRNAs of TCV are resistant to NMD in *N. benthamiana* [1]. A 51 nt long unstructured region (USR) at the beginning of the TCV 3′ UTR was determined to be an RSE for sgRNAs and unrelated NMD-sensitive RNAs. Resistance was abolished while introducing two bases in the element to create a stable hairpin. Another RSE is the readthrough element located downstream of the TCV p28 termination codon, stabilizing an NMD-sensitive reporter transcript from decay [1]. This study also demonstrates that viral sgRNA1 lacking 5′cap is NMD resistant, while the addition of 5′cap makes the RNA sensitive to NMD.

3.2.3. Readthrough and Frameshifting Elements

Readthrough elements from *Moloney murine leukemia virus* (MoMLV) and *Colorado tick fever virus* (CTFV) protect sensitive transcripts from NMD [67,99]. The preliminary stage of NMD is initiated by eRF1 and eRF3, which then recruits UPF1. MoMLV reverse transcriptase (MoMLV-RT), interacts with the C terminal domain of eRF1 via its RNase H domain and blocks it from binding with eRF3. The RT-eRF1 interaction promotes mRNA's readthrough and escapes from NMD [67]. Inserting readthrough stimulatory element and MoMLV PseudoKnot (MLVPK) into an NMD reporter transcript protects it from NMD [5]. Further study reveals that MLVPK antagonizes UPF1 recruitment by eRF1 and subsequent NMD pathway. Another readthrough element is a hairpin structure from CTFV which allows ribosomes to pass the termination codon for continued scanning, thus preventing NMD [99,100].

Ribosomal frameshifting is the translation process where translating ribosomes move forward or step back by one or two nucleotides to escape in-frame stop codon, allowing ribosomes to translate continuously to produce a large fusion protein. Viruses commonly use programmed -1 ribosomal frameshifting to produce viral replicase. -1 frameshifting requires a slippery sequence just upstream of the in-frame stop codon and a structured RNA element several bases downstream of the slippery sequence. Ribosomal frameshifting has been demonstrated to be a means to increase the stability of PEMV2 RNA and its reporter transcript [1]. The frameshifting mechanism is also present in many important human viruses, such as SARS-CoV, HIV-1 and HIV-2, *Simian immunodeficiency virus*, and WNV [45], which are potential NMD targets. The underlying mechanism of ribosomal frameshifting is believed to remove viral RNA-bound UPF1 by the continuous movement of translating ribosomes.

3.3. Others

Alternative splicing is tightly controlled on HIV-1 genomic RNA when generating multiple mRNAs from single pre-mRNA species. Although HIV-1 viral RNA possesses NMD-inducing features, it can evade NMD by hijacking UPF1 [66,101]. UPF1 exerts multifaceted nature during HIV infection [102]. The viral protein pr55Gag is a structural protein that sharpens the nuclear transport of HIV-1 viral RNA. Interestingly, UPF1 co-localizes with the pr55Gag RNP complex, eventually intensifying viral replication [101]. Unexpectedly, UPF1 promotes HIV-1 viral RNA stability and translation [66], while UPF2 and SMG6 were found to inhibit HIV replication in CD4+ T cells [100] and primary monocyte-derived macrophages cells [102].

Table 1. Host NMD eliciting features and possible resistant factors of various viruses.

Virus Species	Types of Genome	Host	NMD-Eliciting Feature	NMD Resistant Factor	Reference
PVX	(+) ssRNA	*Arabidopsis thaliana*	iTC and long 3′ UTR	Unknown	[55]
TCV	(+) ssRNA	*Nicotiana benthamiana*	iTC and long 3′ UTR	Unstructured element	[55]
PEMV2	(+) ssRNA	*N. benthamiana*	Long 3′ UTR	p26 (MP), frameshifting element	[73]
Rotavirus	dsRNA	African green monkey kidney cell line	Unknown	non-structural protein 5 (NSP5)	[59]
CaMV	DNA	*N. benthamiana*	Polycistronic 35S pgRNA	Transactivator Protein (TAV)	[70]
HTLV-1	(+) ssRNA	Hela and HEK cell line	Unknown	TAX and REX factor	[21,64,65]
HCV	(+) ssRNA	hepatoma cells	Unknown	Core Protein	[82]
SFV	(+) ssRNA	Hela Cell line	Unknown	Replicase /nsp3 protein, capsid protein	[54,80,81,96]
MoMLV	(+) ssRNA	Mouse	Long 3′ UTR on *gag* gene	MoMLU-RT	[67]
CTFV	dsRNA	HEK293T cells	Unknown	Hairpin structure (CTFV-HP)	[99,100]
Zika Virus	(+) ssRNA	Human NPC cells	Unknown	Capsid Protein	[79]
RSV	(+) ssRNA	CEF cell line	uORF and long 3′ UTR	RNA Stability Element	[62]
HIV-1	(+) ssRNA	Hela cell line		REV	[101,103]
MHV	(+) ssRNA	17Cl-1 Cells, Mouse astrcytoma cell line	uORF and 3′ UTR	N Protein	[57]
KSHV	DNA	PEL cell line	Long 3′ UTR of *Orf50* mRNA	Unknown	[71,72]
EBV	DNA	HEK293T cells	Long 3′ UTR of BRLF1-encoding transcripts	Unknown	[71,72]

4. Concluding Remarks

The ample scenarios of the host NMD–virus interaction have been rooted out to a substantial measure. The diverse families of viruses evolve multifarious tactics to protect viral genomes from RNA decay and antiviral silencing during the infection cycle. Evolutionarily, to cope with host antiviral actions, (+) RNA viruses and retroviruses deploy *cis*- and *trans*-arm-based strategies to attenuate host antiviral immunity and facilitate viral replication [104]. Although the precise mechanism is not elucidated yet for most virus families, significant progress has been made in the molecular interaction of NMD and viruses in recent years.

Regarding DNA viruses that previously seemed unlikely to be associated with NMD, recent evidence shows that EBV and KSHV vigorously interact with host NMD components [71,72]. Long 3′ UTR and retained intron impose *Herpesvirus* RNAs to NMD. Whether DNA viruses encode anti-NMD factors or have any specific RNA sequence or structure that could resist NMD remains unsolved. On the other hand, questions can be raised that most (+) RNA viruses can potentially be an NMD target, but a significant number are still undiscovered. All (+) RNA viruses reported to be NMD targets have a monopartite viral genome that contains one or more iTC. Quite a portion of plant viruses is segmented in their genome, such as the family of *bromoviridea* that has a tripartite RNA genome. It would be interesting to investigate whether segmented viral genomes have the potential for NMD? Moreover, little is known about viruses whose genome is double-stranded RNA. Thus, numerous efforts will be required to shed light on taxonomically different viruses to create a whole picture of the NMD–virus molecular interaction.

NMD is a translation-dependent pathway. Alternative translation initiation factors such as VPg, 3′ cap-independent translation enhancer (CITE), and an internal ribosomal entry site (IRES) employed by viruses in various circumstances inside host cells may potentially impact the modulating or reshaping decay pathway, which have been overlooked till now. Viruses deploy virally induced cellular microenvironment to avoid the antiviral response [105]. What is the mechanism of action of NMD when a virus multiplies inside a virally encoded secured compartment in the cytoplasm or are there relations between NMD and viroplasms? Plenty of questions remain unanswered, which could expand our understanding of NMD and open new avenues for virus research in treating plant and animal diseases.

Author Contributions: Conceptualization, M.R.A. and Z.D.; writing—original; draft preparation, M.R.A.; writing—Z.D.; supervision, Z.D.; project administration, Z.D.; funding acquisition, Z.D. All authors have read and agreed to the published version of the manuscript.

Funding: This research was funded by the National Natural Science Foundation of China (31870144, 32070154 to Z.D.).

Institutional Review Board Statement: Not applicable.

Informed Consent Statement: Not applicable.

Data Availability Statement: Not applicable.

Conflicts of Interest: The authors declare no conflict of interest.

Abbreviations

QC: Quality Control; PTC: Premature Termination Codon; uORF: Upstream Open Reading Frame; vRNA: Viral RNA; UTR: Untranslated Region; EJC: Exon-Junction Complex; nt: Nucleotide; eRF: Eukaryotic Release Factor; iTC: Internal Termination Codon; sgRNA: Sub Genomic RNA; pgRNA: Pre Genomic RNA; PAPB: Poly Adenine Binding Protein; SMG: Suppressor with Morphogenetic effect on Genitalia; VPg: Viral Genome Linked Protein.

References

1. May, J.P.; Yuan, X.; Sawicki, E.; Simon, A.E. RNA Virus Evasion of Nonsense-Mediated Decay. *PLoS Pathog.* **2018**, *14*, 11. [CrossRef] [PubMed]
2. Kervestin, S.; Jacobson, A. NMD: A Multifaceted Response to Premature Translational Termination. *Nat. Rev. Mol. Cell Biol.* **2012**, *13*, 700–712. [CrossRef] [PubMed]
3. Schweingruber, C.; Rufener, S.C.; Zünd, D.; Yamashita, A.; Mühlemann, O. Nonsense-Mediated MRNA Decay—Mechanisms of Substrate MRNA Recognition and Degradation in Mammalian Cells. *Biochim. Biophys. Acta—Gene Regul. Mech.* **2013**, *1829*, 612–623. [CrossRef] [PubMed]
4. Singh, G.; Rebbapragada, I.; Lykke-Andersen, J. A Competition between Stimulators and Antagonists of Upf Complex Recruitment Governs Human Nonsense-Mediated MRNA Decay. *PLoS Biol.* **2008**, *6*, e111. [CrossRef] [PubMed]
5. Hogg, J.R.; Goff, S.P. Upf1 Senses 3′UTR Length to Potentiate MRNA Decay. *Cell* **2010**, *143*, 379–389. [CrossRef]
6. Hurt, J.A.; Robertson, A.D.; Burge, C.B. Global Analyses of UPF1 Binding and Function Reveal Expanded Scope of Nonsense-Mediated MRNA Decay. *Genome Res.* **2013**, *23*, 1636–1650. [CrossRef]
7. Colombo, M.; Karousis, E.D.; Bourquin, J.; Bruggmann, R.; Mühlemann, O. Transcriptome-Wide Identification of NMD-Targeted Human MRNAs Reveals Extensive Redundancy between SMG6- and SMG7-Mediated Degradation Pathways. *RNA* **2017**, *23*, 189–201. [CrossRef]
8. Wachter, A.; Hartmann, L. NMD: Nonsense-Mediated Defense. *Cell Host Microbe* **2014**, *16*, 273–275. [CrossRef]
9. He, F.; Jacobson, A. Nonsense-Mediated MRNA Decay: Degradation of Defective Transcripts Is Only Part of the Story. *Annu. Rev. Genet.* **2015**, *49*, 339–366. [CrossRef]
10. Dai, Y.; Li, W.; An, L. NMD Mechanism and the Functions of Upf Proteins in Plant. *Plant Cell Rep.* **2016**, *35*, 5–15. [CrossRef]
11. Karousis, E.D.; Mühlemann, O. Nonsense-Mediated MRNA Decay Begins Where Translation Ends. *Cold Spring Harb. Perspect. Biol.* **2019**, *11*, a032862. [CrossRef] [PubMed]
12. Peccarelli, M.; Kebaara, B.W. Regulation of Natural MRNAs by the Nonsense-Mediated MRNA Decay Pathway. *Eukaryot. Cell* **2014**, *13*, 1126–1135. [CrossRef] [PubMed]

13. Muhlrad, D.; Parker, R. Aberrant MRNAs with Extended 3′ UTRs Are Substrates for Rapid Degradation by MRNA Surveillance. *RNA* **1999**, *5*, 1299–1307. [CrossRef] [PubMed]
14. Vicente-Crespo, M.; Palacios, I.M. Nonsense-Mediated MRNA Decay and Development: Shoot the Messenger to Survive? *Biochem. Soc. Trans.* **2011**, *38*, 1500–1505. [CrossRef] [PubMed]
15. Jaffrey, S.R.; Wilkinson, M.F. Nonsense-Mediated RNA Decay in the Brain: Emerging Modulator of Neural Development and Disease. *Nat. Rev. Neurosci.* **2018**, *19*, 715–728. [CrossRef] [PubMed]
16. Metzstein, M.M.; Krasnow, M.A. Functions of the Nonsense-Mediated MRNA Decay Pathway in Drosophila Development. *PLoS Genet.* **2006**, *2*, e180. [CrossRef]
17. Nickless, A.; Bailis, J.M.; You, Z. Control of Gene Expression through the Nonsense-Mediated RNA Decay Pathway. *Cell Biosci.* **2017**, *7*, 26. [CrossRef]
18. Brogna, S.; Wen, J. Nonsense-Mediated MRNA Decay (NMD) Mechanisms. *Nat. Struct. Mol. Biol.* **2009**, *16*, 107–113. [CrossRef]
19. Maquat, L.E. Nonsense-Mediated MRNA Decay: Splicing, Translation and MRNP Dynamics. *Nat. Rev. Mol. Cell Biol.* **2004**, *5*, 89–99. [CrossRef]
20. Kashima, I.; Yamashita, A.; Izumi, N.; Kataoka, N.; Morishita, R.; Hoshino, S.; Ohno, M.; Dreyfuss, G.; Ohno, S. Binding of a Novel SMG-1-Upf1-ERF1-ERF3 Complex (SURF) to the Exon Junction Complex Triggers Upf1 Phosphorylation and Nonsense-Mediated MRNA Decay. *Genes Dev.* **2006**, *20*, 355–367. [CrossRef] [PubMed]
21. Nakano, K.; Karasawa, N.; Hashizume, M.; Tanaka, Y.; Ohsugi, T.; Uchimaru, K.; Watanabe, T. Elucidation of the Mechanism of Host NMD Suppression by HTLV-1 Rex: Dissection of Rex to Identify the NMD Inhibitory Domain. *Viruses* **2022**, *14*, 344. [CrossRef]
22. Lejeune, F.; Ishigaki, Y.; Li, X.; Maquat, L.E. The Exon Junction Complex Is Detected on CBP80-Bound but Not EIF4E-Bound MRNA in Mammalian Cells: Dynamics of MRNP Remodeling. *EMBO J.* **2002**, *21*, 3536–3545. [CrossRef]
23. Hwang, J.; Maquat, L.E. Nonsense-Mediated MRNA Decay (NMD) in Animal Embryogenesis: To Die or Not to Die, That Is the Question. *Curr. Opin. Genet. Dev.* **2011**, *21*, 422–430. [CrossRef]
24. Rigby, R.E.; Rehwinkel, J. RNA Degradation in Antiviral Immunity and Autoimmunity. *Trends Immunol.* **2015**, *36*, 179–188. [CrossRef] [PubMed]
25. Le Hir, H.; Gatfield, D.; Izaurralde, E.; Moore, M.J. The Exon-Exon Junction Complex Provides a Binding Platform for Factors Involved in MRNA Export and Nonsense-Mediated MRNA Decay. *EMBO J.* **2001**, *20*, 4987–4997. [CrossRef] [PubMed]
26. Yamashita, A.; Ohnishi, T.; Kashima, I.; Taya, Y.; Ohno, S. Human SMG-1, a Novel Phosphatidylinositol 3-Kinase-Related Protein Kinase, Associates with Components of the MRNA Surveillance Complex and Is Involved in the Regulation of Nonsense-Mediated MRNA Decay. *Genes Dev.* **2001**, *15*, 2215–2228. [CrossRef] [PubMed]
27. Okada-Katsuhata, Y.; Yamashita, A.; Kutsuzawa, K.; Izumi, N.; Hirahara, F.; Ohno, S. N-and C-Terminal Upf1 Phosphorylations Create Binding Platforms for SMG-6 and SMG-5:SMG-7 during NMD. *Nucleic Acids Res.* **2012**, *40*, 1251–1266. [CrossRef]
28. Chamieh, H.; Ballut, L.; Bonneau, F.; Le Hir, H. NMD Factors UPF2 and UPF3 Bridge UPF1 to the Exon Junction Complex and Stimulate Its RNA Helicase Activity. *Nat. Struct. Mol. Biol.* **2008**, *15*, 85–93. [CrossRef]
29. Yepiskoposyan, H.; Aeschimann, F.; Nilsson, D.; Okoniewski, M.; Mühlemann, O. Autoregulation of the Nonsense-Mediated MRNA Decay Pathway in Human Cells. *RNA* **2011**, *17*, 2108–2118. [CrossRef]
30. Gatfield, D.; Unterholzner, L.; Ciccarelli, F.D.; Bork, P.; Izaurralde, E. Nonsense-Mediated MRNA Decay in Drosophila: At the Intersection of the Yeast and Mammalian Pathways. *EMBO J.* **2003**, *22*, 3960–3970. [CrossRef]
31. Wen, J.; Brogna, S. Splicing-Dependent NMD Does Not Require the EJC in Schizosaccharomyces Pombe. *EMBO J.* **2010**, *29*, 1537–1551. [CrossRef]
32. Tian, M.; Yang, W.; Zhang, J.; Dang, H.; Lu, X.; Fu, C.; Miao, W. Nonsense-Mediated MRNA Decay in Tetrahymena Is EJC Independent and Requires a Protozoa-Specific Nuclease. *Nucleic Acids Res.* **2017**, *45*, 6848–6863. [CrossRef]
33. Metze, S.; Herzog, V.A.; Ruepp, M.D.; Mühlemann, O. Comparison of EJC-Enhanced and EJC-Independent NMD in Human Cells Reveals Two Partially Redundant Degradation Pathways. *RNA* **2013**, *19*, 1432–1448. [CrossRef] [PubMed]
34. Li, F.; Wang, A. RNA-Targeted Antiviral Immunity: More Than Just RNA Silencing. *Trends Microbiol.* **2019**, *27*, 792–805. [CrossRef] [PubMed]
35. Loh, B.; Jonas, S.; Izaurralde, E. The SMG5-SMG7 Heterodimer Directly Recruits the CCR4-NOT Deadenylase Complex to MRNAs Containing Nonsense Codons via Interaction with POP2. *Genes Dev.* **2013**, *27*, 2125–2138. [CrossRef] [PubMed]
36. Eberle, A.B.; Lykke-Andersen, S.; Mühlemann, O.; Jensen, T.H. SMG6 Promotes Endonucleolytic Cleavage of Nonsense MRNA in Human Cells. *Nat. Struct. Mol. Biol.* **2009**, *16*, 49–55. [CrossRef] [PubMed]
37. Nagarajan, V.K.; Jones, C.I.; Newbury, S.F.; Green, P.J. XRN 5′→3′ Exoribonucleases: Structure, Mechanisms and Functions. *Biochim. Biophys. Acta-Gene Regul. Mech.* **2013**, *1829*, 590–603. [CrossRef] [PubMed]
38. Kerényi, Z.; Mérai, Z.; Hiripi, L.; Benkovics, A.; Gyula, P.; Lacomme, C.; Barta, E.; Nagy, F.; Silhavy, D. Inter-Kingdom Conservation of Mechanism of Nonsense-Mediated MRNA Decay. *EMBO J.* **2008**, *27*, 1585–1595. [CrossRef]
39. Causier, B.; Li, Z.; De Smet, R.; Lloyd, J.P.B.; Van De Peer, Y.; Davies, B. Conservation of Nonsense-Mediated MRNA Decay Complex Components Throughout Eukaryotic Evolution. *Sci. Rep.* **2017**, *7*, 16692. [CrossRef]
40. Lloyd, J.P.B. The Evolution and Diversity of the Nonsense-Mediated MRNA Decay Pathway [Version 1; Peer Review: 4 Approved]. *F1000Research* **2018**, *7*, 1299. [CrossRef]

41. Miras, M.; Allen Miller, W.; Truniger, V.; Aranda, M.A. Non-Canonical Translation in Plant RNA Viruses. *Front. Plant Sci.* **2017**, *8*, 1385–1409. [CrossRef]

42. Ho, J.S.Y.; Zhu, Z.; Marazzi, I. Unconventional Viral Gene Expression Mechanisms as Therapeutic Targets. *Nature* **2021**, *593*, 362–371. [CrossRef] [PubMed]

43. Skuzeski, J.M.; Nichols, L.M.; Gesteland, R.F.; Atkins, J.F. The Signal for a Leaky UAG Stop Codon in Several Plant Viruses Includes the Two Downstream Codons. *J. Mol. Biol.* **1991**, *218*, 365–373. [CrossRef] [PubMed]

44. Dinman, J.D. Mechanisms and Implications of Programmed Translational Frameshifting. *Wiley Interdiscip. Rev. RNA* **2012**, *3*, 661–673. [CrossRef]

45. Jaafar, Z.A.; Kieft, J.S. Viral RNA Structure-Based Strategies to Manipulate Translation. *Nat. Rev. Microbiol.* **2019**, *17*, 110–123. [CrossRef] [PubMed]

46. Yost, S.A.; Marcotrigiano, J. Viral Precursor Polyproteins: Keys of Regulation from Replication to Maturation. *Curr. Opin. Virol.* **2013**, *3*, 137–142. [CrossRef]

47. Withers, J.B.; Beemon, K.L. Structural Features in the Rous Sarcoma Virus RNA Stability Element Are Necessary for Sensing the Correct Termination Codon. *Retrovirology* **2010**, *7*, 65. [CrossRef]

48. Kim, D.; Lee, J.Y.; Yang, J.S.; Kim, J.W.; Kim, V.N.; Chang, H. The Architecture of SARS-CoV-2 Transcriptome. *Cell* **2020**, *181*, 914–921. [CrossRef]

49. Bühler, M.; Steiner, S.; Mohn, F.; Paillusson, A.; Mühlemann, O. EJC-Independent Degradation of Nonsense Immunoglobulin-μ MRNA Depends on 3′ UTR Length. *Nat. Struct. Mol. Biol.* **2006**, *13*, 462–464. [CrossRef]

50. Kertész, S.; Kerényi, Z.; Mérai, Z.; Bartos, I.; Pálfy, T.; Barta, E.; Silhavy, D. Both Introns and Long 3′-UTRs Operate as Cis-Acting Elements to Trigger Nonsense-Mediated Decay in Plants. *Nucleic Acids Res.* **2006**, *34*, 6147–6157. [CrossRef]

51. Karousis, E.D.; Nasif, S.; Mühlemann, O. Nonsense-Mediated MRNA Decay: Novel Mechanistic Insights and Biological Impact. *Wiley Interdiscip. Rev. RNA* **2016**, *7*, 661–682. [CrossRef]

52. Strauss, J.H.; Strauss, E.G. The Alphaviruses: Gene Expression, Replication, and Evolution. *Microbiol. Rev.* **1994**, *58*, 491–562. [CrossRef]

53. Wernet, M.F.; Klovstad, M.; Clandinin, T.R. Generation of Infectious Virus Particles from Inducible Transgenic Genomes. *Curr. Biol.* **2014**, *24*, 107–108. [CrossRef]

54. Balistreri, G.; Horvath, P.; Schweingruber, C.; Zünd, D.; McInerney, G.; Merits, A.; Mühlemann, O.; Azzalin, C.; Helenius, A. The Host Nonsense-Mediated MRNA Decay Pathway Restricts Mammalian RNA Virus Replication. *Cell Host Microbe* **2014**, *16*, 403–411. [CrossRef]

55. Garcia, D.; Garcia, S.; Voinnet, O. Nonsense-Mediated Decay Serves as a General Viral Restriction Mechanism in Plants. *Cell Host Microbe* **2014**, *16*, 391–402. [CrossRef]

56. Imamachi, N.; Salam, K.A.; Suzuki, Y.; Akimitsu, N. A GC-Rich Sequence Feature in the 3′ UTR Directs UPF1-Dependent MRNA Decay in Mammalian Cells. *Genome Res.* **2017**, *27*, 407–418. [CrossRef]

57. Wada, M.; Lokugamage, K.G.; Nakagawa, K.; Narayanan, K.; Makino, S. Interplay between Coronavirus, a Cytoplasmic RNA Virus, and Nonsense-Mediated MRNA Decay Pathway. *Proc. Natl. Acad. Sci. USA* **2018**, *115*, E10157–E10166. [CrossRef] [PubMed]

58. Gordon, D.E.; Jang, G.M.; Bouhaddou, M.; Xu, J.; Obernier, K.; White, K.M.; O'Meara, M.J.; Rezelj, V.V.; Guo, J.Z.; Swaney, D.L.; et al. A SARS-CoV-2 Protein Interaction Map Reveals Targets for Drug Repurposing. *Nature* **2020**, *583*, 459–468. [CrossRef]

59. Sarkar, R.; Banerjee, S.; Mukherjee, A.; Chawla-Sarkar, M. Rotaviral Nonstructural Protein 5 (NSP5) Promotes Proteasomal Degradation of up-Frameshift Protein 1 (UPF1), a Principal Mediator of Nonsense-Mediated MRNA Decay (NMD) Pathway, to Facilitate Infection. *Cell Signal.* **2022**, *89*, 110180. [CrossRef] [PubMed]

60. Bolinger, C.; Boris-Lawrie, K. Mechanisms Employed by Retroviruses to Exploit Host Factors for Translational Control of a Complicated Proteome. *Retrovirology* **2009**, *6*, 8. [CrossRef] [PubMed]

61. LeBlanc, J.J.; Beemon, K.L. Unspliced Rous Sarcoma Virus Genomic RNAs Are Translated and Subjected to Nonsense-Mediated MRNA Decay before Packaging. *J. Virol.* **2004**, *78*, 5139–5146. [CrossRef]

62. Quek, B.L.; Beemon, K. Retroviral Strategy to Stabilize Viral RNA. *Curr. Opin. Microbiol.* **2014**, *18*, 78–82. [CrossRef]

63. Barker, G.F.; Beemon, K. Rous Sarcoma Virus RNA Stability Requires an Open Reading Frame in the Gag Gene and Sequences Downstream of the Gag-Pol Junction. *Mol. Cell. Biol.* **1994**, *14*, 1986–1996. [CrossRef]

64. Fiorini, F.; Robin, J.P.; Kanaan, J.; Borowiak, M.; Croquette, V.; Le Hir, H.; Jalinot, P.; Mocquet, V. HTLV-1 Tax Plugs and Freezes UPF1 Helicase Leading to Nonsense-Mediated MRNA Decay Inhibition. *Nat. Commun.* **2018**, *9*, 431. [CrossRef]

65. Mocquet, V.; Neusiedler, J.; Rende, F.; Cluet, D.; Robin, J.-P.; Terme, J.-M.; Duc Dodon, M.; Wittmann, J.; Morris, C.; Le Hir, H.; et al. The Human T-Lymphotropic Virus Type 1 Tax Protein Inhibits Nonsense-Mediated MRNA Decay by Interacting with INT6/EIF3E and UPF1. *J. Virol.* **2012**, *86*, 7530–7543. [CrossRef] [PubMed]

66. Ajamian, L.; Abrahamyan, L.; Milev, M.; Ivanov, P.V.; Kulozik, A.E.; Gehring, N.H.; Mouland, A.J. Unexpected Roles for UPF1 in HIV-1 RNA Metabolism and Translation. *RNA* **2008**, *14*, 914–927. [CrossRef] [PubMed]

67. Tang, X.; Zhu, Y.; Baker, S.L.; Bowler, M.W.; Chen, B.J.; Chen, C.; Hogg, J.R.; Goff, S.P.; Song, H. Structural Basis of Suppression of Host Translation Termination by Moloney Murine Leukemia Virus. *Nat. Commun.* **2016**, *7*, 12070. [CrossRef]

68. Hohn, T.; Rothnie, H. Plant Pararetroviruses: Replication and Expression. *Curr. Opin. Virol.* **2013**, *3*, 621–628. [CrossRef]

69. Bouton, C.; Geldreich, A.; Ramel, L.; Ryabova, L.A.; Dimitrova, M.; Keller, M. Cauliflower Mosaic Virus Transcriptome Reveals a Complex Alternative Splicing Pattern. *PLoS ONE* **2015**, *10*, e0132665. [CrossRef]

70. Lukhovitskaya, N.; Ryabova, L.A. Cauliflower Mosaic Virus Transactivator Protein (TAV) Can Suppress Nonsense-Mediated Decay by Targeting VARICOSE, a Scaffold Protein of the Decapping Complex. *Sci. Rep.* **2019**, *9*, 7042. [CrossRef] [PubMed]

71. Zhao, Y.; Ye, X.; Shehata, M.; Dunker, W.; Xie, Z.; Karijolich, J. The RNA Quality Control Pathway Nonsense-Mediated MRNA Decay Targets Cellular and Viral RNAs to Restrict KSHV. *Nat. Commun.* **2020**, *11*, 3345. [CrossRef]

72. Van Gent, M.; Reich, A.; Velu, S.E.; Gack, M.U. Nonsense-Mediated Decay Controls the Reactivation of the Oncogenic Herpesviruses Ebv and Kshv. *PLoS Biol.* **2021**, *19*, e3001097. [CrossRef] [PubMed]

73. May, J.P.; Johnson, P.Z.; Ilyas, M.; Gao, F.; Simon, A.E. The Multifunctional Long-Distance Movement Protein of Pea Enation Mosaic Virus 2 Protects Viral and Host Transcripts from Nonsense-Mediated Decay. *MBio* **2020**, *11*, e00204–e00220. [CrossRef]

74. Franks, T.M.; Singh, G.; Lykke-Andersen, J. Upf1 ATPase-Dependent MRNP Disassembly Is Required for Completion of Nonsense-Mediated MRNA Decay. *Cell* **2010**, *143*, 938–950. [CrossRef] [PubMed]

75. Lemasson, I.; Lewis, M.R.; Polakowski, N.; Hivin, P.; Cavanagh, M.-H.; Thébault, S.; Barbeau, B.; Nyborg, J.K.; Mesnard, J.-M. Human T-Cell Leukemia Virus Type 1 (HTLV-1) BZIP Protein Interacts with the Cellular Transcription Factor CREB To Inhibit HTLV-1 Transcription. *J. Virol.* **2007**, *81*, 1543–1553. [CrossRef] [PubMed]

76. Morris, C.; Wittmann, J.; Jäck, H.M.; Jalinot, P. Human INT6/EIF3e Is Required for Nonsense-Mediated MRNA Decay. *EMBO Rep.* **2007**, *8*, 596–602. [CrossRef] [PubMed]

77. Nakano, K.; Ando, T.; Yamagishi, M.; Yokoyama, K.; Ishida, T.; Ohsugi, T.; Tanaka, Y.; Brighty, D.W.; Watanabe, T. Viral Interference with Host MRNA Surveillance, the Nonsense-Mediated MRNA Decay (NMD) Pathway, through a New Function of HTLV-1 Rex: Implications for Retroviral Replication. *Microbes Infect.* **2013**, *15*, 491–505. [CrossRef]

78. Li, M.; Johnson, J.R.; Truong, B.; Kim, G.; Weinbren, N.; Dittmar, M.; Shah, P.S.; Von Dollen, J.; Newton, B.W.; Jang, G.M.; et al. Identification of Antiviral Roles for the Exon–Junction Complex and Nonsense-Mediated Decay in Flaviviral Infection. *Nat. Microbiol.* **2019**, *4*, 985–995. [CrossRef]

79. Fontaine, K.A.; Leon, K.E.; Khalid, M.M.; Tomar, S.; Jimenez-Morales, D.; Dunlap, M.; Kaye, J.A.; Shah, P.S.; Finkbeiner, S.; Krogan, N.J.; et al. The Cellular NMD Pathway Restricts Zika Virus Infection and Is Targeted by the Viral Capsid Protein. *MBio* **2018**, *9*, e02126-18. [CrossRef]

80. Wei-lin Popp, M.; Cho, H.; Maquat, L.E. Viral Subversion of Nonsense-Mediated MRNA Decay nonsense-mediated mrna decay in plants and animals: An overview. *RNA* **2020**, *26*, 1509–1518. [CrossRef]

81. Contu, L.; Balistreri, G.; Domanski, M.; Uldry, A.C.; Mühlemann, O. Characterisation of the Semliki Forest Virus-Host Cell Interactome Reveals the Viral Capsid Protein as an Inhibitor of Nonsense-Mediated MRNA Decay. *PLoS Pathog.* **2021**, *17*, 1–29. [CrossRef]

82. Ramage, H.R.; Kumar, G.R.; Verschueren, E.; Johnson, J.R.; VonDollen, J.; Johnson, T.; Newton, B.; Shah, P.; Horner, J.; Krogan, N.J.; et al. A Combined Proteomics/Genomics Approach Links Hepatitis C Virus Infection with Nonsense-Mediated MRNA Decay. *Mol. Cell* **2015**, *57*, 329–340. [CrossRef] [PubMed]

83. Thiébeauld, O.; Schepetilnikov, M.; Park, H.S.; Geldreich, A.; Kobayashi, K.; Keller, M.; Hohn, T.; Ryabova, L.A. A New Plant Protein Interacts with EIF3 and 60S to Enhance Virus-Activated Translation Re-Initiation. *EMBO J.* **2009**, *28*, 3171–3184. [CrossRef] [PubMed]

84. Park, H.-S.; Himmelbach, A.; Browning, K.S.; Hohn, T.; Ryabova, L.A. A Plant Viral "Reinitiation" Factor Interacts with the Host Translational Machinery. *Cell* **2001**, *106*, 723–733. [CrossRef]

85. Schepetilnikov, M.; Kobayashi, K.; Geldreich, A.; Caranta, C.; Robaglia, C.; Keller, M.; Ryabova, L.A. Viral Factor TAV Recruits TOR/S6K1 Signalling to Activate Reinitiation after Long ORF Translation. *EMBO J.* **2011**, *30*, 1343–1356. [CrossRef]

86. Ryabova, L.A.; Pooggin, M.M.; Hohn, T. Translation Reinitiation and Leaky Scanning in Plant Viruses. *Virus Res.* **2006**, *119*, 52–62. [CrossRef]

87. Schepetilnikov, M.; Dimitrova, M.; Mancera-Martínez, E.; Geldreich, A.; Keller, M.; Ryabova, L.A. TOR and S6K1 Promote Translation Reinitiation of UORF-Containing MRNAs via Phosphorylation of EIF3h. *EMBO J.* **2013**, *32*, 1087–1102. [CrossRef] [PubMed]

88. Sorenson, R.S.; Deshotel, M.J.; Johnson, K.; Adler, F.R.; Sieburth, L.E. Arabidopsis MRNA Decay Landscape Arises from Specialized RNA Decay Substrates, Decapping-Mediated Feedback, and Redundancy. *Proc. Natl. Acad. Sci. USA* **2018**, *115*, E1485–E1494. [CrossRef]

89. Masters, P.S. The Molecular Biology of Coronaviruses. *Adv. Virus Res.* **2006**, *65*, 193–292. [CrossRef]

90. Hsin, W.C.; Chang, C.H.; Chang, C.Y.; Peng, W.H.; Chien, C.L.; Chang, M.F.; Chang, S.C. Nucleocapsid Protein-Dependent Assembly of the RNA Packaging Signal of Middle East Respiratory Syndrome Coronavirus. *J. Biomed. Sci.* **2018**, *25*, 47. [CrossRef]

91. Chang, C.K.; Lo, S.C.; Wang, Y.S.; Hou, M.H. Recent Insights into the Development of Therapeutics against Coronavirus Diseases by Targeting N Protein. *Drug Discov. Today* **2016**, *21*, 562–572. [CrossRef] [PubMed]

92. Kuo, L.; Koetzner, C.A.; Hurst, K.R.; Masters, P.S. Recognition of the Murine Coronavirus Genomic RNA Packaging Signal Depends on the Second RNA Binding Domain of the Nucleocapsid Protein. *J. Virol.* **2014**, *88*, 4451–4465. [CrossRef] [PubMed]

93. Emmott, E.; Munday, D.; Bickerton, E.; Britton, P.; Rodgers, M.A.; Whitehouse, A.; Zhou, E.-M.; Hiscox, J.A. The Cellular Interactome of the Coronavirus Infectious Bronchitis Virus Nucleocapsid Protein and Functional Implications for Virus Biology. *J. Virol.* **2013**, *87*, 9486–9500. [CrossRef]

94. Narayanan, K.; Makino, S. Interplay between Viruses and Host MRNA Degradation. *Biochim. Biophys. Acta-Gene Regul. Mech.* **2013**, *1829*, 732–741. [CrossRef] [PubMed]

95. Weil, J.E.; Hadjithomas, M.; Beemon, K.L. Structural Characterization of the Rous Sarcoma Virus RNA Stability Element. *J. Virol.* **2009**, *83*, 2119–2129. [CrossRef]

96. Weil, J.E.; Beemon, K.L. A 3′ UTR Sequence Stabilizes Termination Codons in the Unspliced RNA of Rous Sarcoma Virus. *RNA* **2006**, *12*, 102–110. [CrossRef]

97. Liu, S.-W.; Wyatt, L.S.; Orandle, M.S.; Minai, M.; Moss, B. The D10 Decapping Enzyme of Vaccinia Virus Contributes to Decay of Cellular and Viral MRNAs and to Virulence in Mice. *J. Virol.* **2014**, *88*, 202–211. [CrossRef]

98. Ge, Z.; Quek, B.L.; Beemon, K.L.; Hogg, J.R. Polypyrimidine Tract Binding Protein 1 Protects MRNAs from Recognition by the Nonsense-Mediated MRNA Decay Pathway. *Elife* **2016**, *5*, e11155. [CrossRef]

99. Baker, S.L.; Hogg, J.R. A System for Coordinated Analysis of Translational Readthrough and Nonsensemediated MRNA Decay. *PLoS ONE* **2017**, *12*, e0173980. [CrossRef]

100. Napthine, S.; Yek, C.; Powell, M.L.; Brown, D.T.K.; Brierley, I. Characterization of the Stop Codon Readthrough Signal of Colorado Tick Fever Virus Segment 9 RNA. *RNA* **2012**, *18*, 241–252. [CrossRef]

101. Ajamian, L.; Abel, K.; Rao, S.; Vyboh, K.; García-de-Gracia, F.; Soto-Rifo, R.; Kulozik, A.E.; Gehring, N.H.; Mouland, A.J. HIV-1 Recruits UPF1 but Excludes UPF2 to Promote Nucleocytoplasmic Export of the Genomic RNA. *Biomolecules* **2015**, *5*, 2808–2839. [CrossRef] [PubMed]

102. Rao, S.; Amorim, R.; Niu, M.; Breton, Y.; Tremblay, M.J.; Mouland, A.J. Host MRNA Decay Proteins Influence HIV-1 Replication and Viral Gene Expression in Primary Monocyte-Derived Macrophages. *Retrovirology* **2019**, *16*, 3. [CrossRef] [PubMed]

103. Toro-Ascuy, D.; Rojas-Araya, B.; Valiente-Echeverría, F.; Soto-Rifo, R. Interactions between the HIV-1 Unspliced MRNA and Host MRNA Decay Machineries. *Viruses* **2016**, *8*, 320. [CrossRef]

104. Leon, K.; Ott, M. An 'Arms Race' between the Nonsense-Mediated MRNA Decay Pathway and Viral Infections. *Semin. Cell Dev. Biol.* **2021**, *111*, 101–107. [CrossRef] [PubMed]

105. Schmid, M.; Speiseder, T.; Dobner, T.; Gonzalez, R.A. DNA Virus Replication Compartments. *J. Virol.* **2014**, *88*, 1404–1420. [CrossRef]

 *viruses*

Review

Host Cell Targets for Unconventional Antivirals against RNA Viruses

Vicky C. Roa-Linares [1,*], Manuela Escudero-Flórez [1], Miguel Vicente-Manzanares [2] and Juan C. Gallego-Gómez [1,*]

[1] Molecular and Translation Medicine Group, University of Antioquia, Medellin 050010, Colombia
[2] Molecular Mechanisms Program, Centro de Investigación del Cáncer, Instituto de Biología Molecular y Celular del Cáncer, Consejo Superior de Investigaciones Científicas (CSIC), University of Salamanca, 37007 Salamanca, Spain; miguel.vicente@csic.es
* Correspondence: vicky.roa@udea.edu.co (V.C.R.-L.); carlos.gallego@udea.edu.co (J.C.G.-G.)

Abstract: The recent COVID-19 crisis has highlighted the importance of RNA-based viruses. The most prominent members of this group are SARS-CoV-2 (coronavirus), HIV (human immunodeficiency virus), EBOV (Ebola virus), DENV (dengue virus), HCV (hepatitis C virus), ZIKV (Zika virus), CHIKV (chikungunya virus), and influenza A virus. With the exception of retroviruses which produce reverse transcriptase, the majority of RNA viruses encode RNA-dependent RNA polymerases which do not include molecular proofreading tools, underlying the high mutation capacity of these viruses as they multiply in the host cells. Together with their ability to manipulate the immune system of the host in different ways, their high mutation frequency poses a challenge to develop effective and durable vaccination and/or treatments. Consequently, the use of antiviral targeting agents, while an important part of the therapeutic strategy against infection, may lead to the selection of drug-resistant variants. The crucial role of the host cell replicative and processing machinery is essential for the replicative cycle of the viruses and has driven attention to the potential use of drugs directed to the host machinery as therapeutic alternatives to treat viral infections. In this review, we discuss small molecules with antiviral effects that target cellular factors in different steps of the infectious cycle of many RNA viruses. We emphasize the repurposing of FDA-approved drugs with broad-spectrum antiviral activity. Finally, we postulate that the ferruginol analog (18-(phthalimide-2-yl) ferruginol) is a potential host-targeted antiviral.

Keywords: host-targeted antivirals; RNA viruses; drug repositioning

Citation: Roa-Linares, V.C.; Escudero-Flórez, M.; Vicente-Manzanares, M.; Gallego-Gómez, J.C. Host Cell Targets for Unconventional Antivirals against RNA Viruses. *Viruses* **2023**, *15*, 776. https://doi.org/10.3390/v15030776

Academic Editors: Yiping Li and Yuliang Liu

Received: 23 December 2022
Revised: 12 February 2023
Accepted: 28 February 2023
Published: 17 March 2023

1. Introduction

Viral diseases are an important focus of study in biomedical sciences due to their impact on human health. Many RNA and DNA viruses infect vertebrates, including humans. RNA viruses have a notable impact as they cause severe diseases with an important socio-economic burden, e.g., acquired human immunodeficiency (HIV), hemorrhagic diseases (Ebola, Zika, DENV), hepatitis C (HCV), or influenza [1]. Human intervention in natural ecosystems and the continuous growth of the human population have led to increased human contact with zoonotic virus reservoirs and an increased frequency of interspecies infection, as illustrated by the COVID-19 pandemic, caused by coronavirus SARS-CoV-2 [2].

Although antiviral drugs and vaccines for human RNA viruses do exist, a major challenge is the emergence of drug-resistant variants [3]. This phenomenon is due to several factors, particularly that RNA produced by RNA virus-encoded RNA polymerases have a very high mutation rate (one per each 100,000 nucleotides) [4]. In addition, although reverse transcriptase has a lower probability to induce mutations (five in the full genome) [5], together both phenomena produce a very high rate of spontaneous mutation compared to the rate of mutation in, for example, eukaryotic cells [6]. In most cases, these mutations put the mutated virus at an evolutionary disadvantage by, for example, decreasing its rate of replication or impairing the proper assembly of the capsid. However, random mutation sometimes leads to the selection of antiviral-resistant variants (quasispecies) in which viral replication and/or infective capability remain unaffected or even improve [7]. Coronaviruses (CoVs) display a lower rate of random mutation, as they express nonstructural protein (nsp) 14. nsp14 is a $3'$-$5'$ exoribonuclease with proofreading activity [8]. The rapid emergence of SARS-CoV-2 variants (e.g., the omicron variants and its subvariants) insensitive to vaccines suggests that this mechanism is not as efficient as originally proposed. Another compensating factor is the infectivity and replicative speed of SARS-CoV-2. By jumping quickly from cell to cell and propagating to multiple individuals in a population, the rate of mutation is elevated despite the existence of proofreading mechanisms [9]. Dengue, Zika, and Chikungunya have no currently approved treatments or effective prevention alternatives.

At present, antiviral drug development has focused on the identification of viral proteins and/or structures as potential targets for different compounds and small molecules, termed direct-acting antivirals (DAAs) [10]. DAAs are useful against DNA viruses due to the comparatively low rate of spontaneous mutations [11]. However, their efficacy against RNA viruses is less consistent [12], and each DAA needs to be evaluated independently. Nevertheless, there are DAAs with documented efficacy against RNA viruses, e.g., transcriptase reverse inhibitors against HIV (reviewed in [13]) or remdesivir, which targets the RNA polymerase of SARS-CoV-2 [14].

An alternative approach is to identify antivirals that do not select drug-resistant strains and are directed against the host factors required for productive viral infection. Host-targeted antivirals (HTAs) could be more effective against RNA viruses because host factors are genetically more stable than viral factors, potentially overcoming viral heterogeneity and the emergence of drug-resistant mutants [15]. Additionally, exploring FDA-approved drugs with HTA properties (drug repositioning) reduces time and optimizes economic resources [16]. For example, antitumor small molecules are an attractive option due to the convergence of some signaling pathways involved in cancer and infection caused by RNA viruses [17].

Here, we review recent evidence positioning several HTAs for the treatment of human diseases caused by RNA viruses. As a proof of principle, we focus on the case of an abietane ferruginol analog endowed with a broad antiviral spectrum against dengue, Zika, and herpesvirus [18,19]. These approaches suggest that repurposing current HTAs is a viable strategy that may be useful for the treatment of RNA viral infections [20]. We illustrate this point by describing the cases of several antitumor drugs that have displayed promising results against SARS-CoV-2 infection [20–22] as well as other RNA viruses [23,24] by targeting host cell mechanisms.

2. Cellular Factors Used by RNA Viruses in Their Replicative Cycle

Viruses are obligate intracellular parasites that require infection of living organisms (cells) to replicate [25]. In the course of their replicative cycle, viruses can modulate a wide variety of cellular processes that encompass the remodeling of the endomembrane system [26], cytoskeletal polymerization and organization [27] dynamics [19], modulation of gene and host protein expression [28], apoptosis and autophagy [29,30], cell division [31], evasion of the immune response [32], induction of epithelial to mesenchymal transition (EMT) [17], and regulation of lipid metabolism [33].

Viruses use diverse cellular receptors to attach and enter the host cell. Heparan sulfate receptors and other sulphated glycans are widely used by several families of viruses, including DENV [34] and some alphaviruses [35]. DENV, HCV, HIV, and Ebola viruses trigger clathrin-dependent receptor endocytosis [36–39]. Influenza and Ebola viruses also enter cells through pinocytosis [39,40], and SARS-CoV-2 infects cells via pH- and receptor-dependent/lysosomal entry [41]. After virion internalization and the release of the viral genome into the cytoplasm, positive-strand RNA viruses modify the cellular endomembrane system [42–44], forming structures known as "viral factories" [45].

The flavivirus genus produces at least two types of intracellular membrane structures: vesicle packets (VPs) and convoluted membranes (CMs). Replication of the viral genome happens at VPs, whereas RNA translation/polyprotein processing occurs at CMs [42,43]. CMs are formed through the membrane remodeling of the endoplasmic reticulum, whereas VPs seem to be derived from the Golgi apparatus [46]. The development of these structures is induced by changes in the lipid composition, the influence of integral membrane proteins, the activity of diverse cytoskeletal proteins and microtubule motors, and scaffolding by peripheral and integral membrane proteins [47]. Additionally, the endoplasmic reticulum and Golgi apparatus provide viruses with different sets of host proteins required for the processing, folding, and function of viral glycoproteins. Examples include ER α-glucosidases [48] and other cellular proteins that cleave viral proteins into their mature/active forms, such as furins [49].

The cytoskeleton is also heavily involved in the development of the viral cycle [50,51]. For example, Ebola virus hijacks microfilaments to transport viral nucleocapsids from viral replication centers to membrane budding sites [52], similar to vaccinia virus [53]. Actin filaments are also crucial during DENV internalization and the release of new virions [54]. Many viruses, e.g., influenza, HIV, and Ebola, co-opt microtubule-associated molecular motors for intracellular motility [55]. Influenza virus uses microtubules to move ribonucleoproteins from the nucleus to plasma membranes [56]. HIV uses microtubules to facilitate group-specific antigen (gag) trafficking and virus particle production [57,58], while Ebola virions hijack microtubules within membranous compartments to travel to the acidified vesicular compartment, where viral and cellular membrane fusion occurs [59].

RNA viruses not only hijack pre-existing cellular machinery to propagate infection but also trigger cellular modifications that favor viral dissemination. One crucial example is the virus-induced epithelial-to-mesenchymal transition (EMT).

The EMT is a trans-differentiation process that has been classically associated to development and the metastasis cascade [60]. It is an epigenetic program characterized by a progressive loss of epithelial polarity, the acquisition of individual cellular motility, and an invasive capacity [61]. During this process, epithelial cells adopt a fibroblastic, mesenchymal-like morphology. Viral triggers of the EMT include the AKT-mediated phosphorylation of β-catenin (HCV), which triggers its translocation to the nucleus and the induction of mesenchymal genes as well as the repression of E-cadherin transcription [62]. Recent data demonstrated that DENV infection activates the PI3K/Akt/Rho GTPase pathway, resulting in the activation of Rac1 and Cdc42 Rho GTPases, which in turn induce a lamellipodia and filopodia extension [63]. These cytoskeletal rearrangements highlight the convergence of virus-induced actin remodeling with an EMT induction.

Figure 1 summarizes the key role of the host mechanisms for viral infection and propagation, as discussed above and elsewhere [54], also postulating the role of a collection of host-targeting drugs as antivirals.

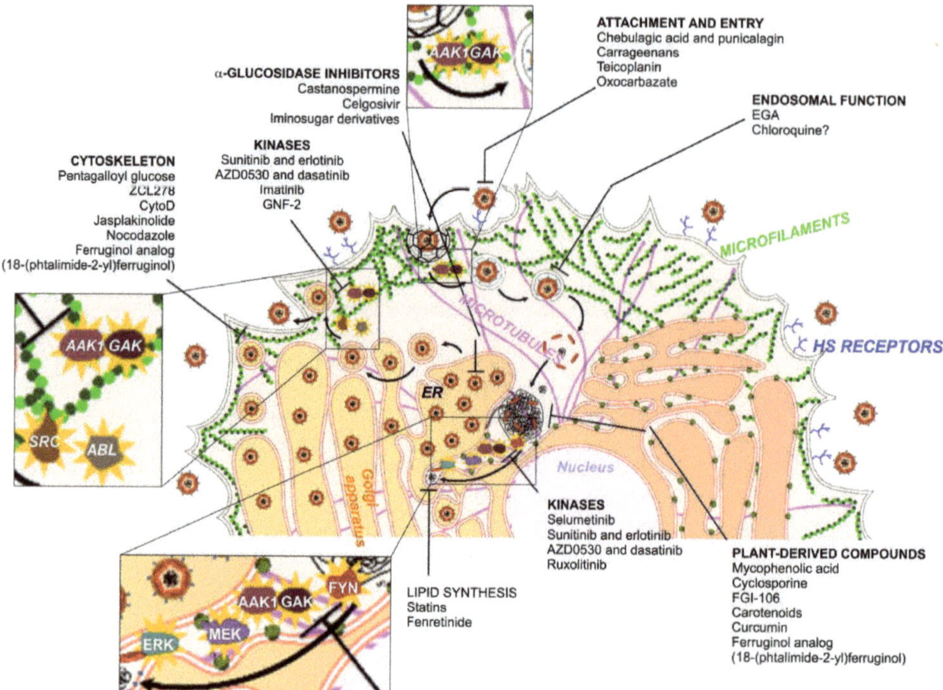

Figure 1. Host factors required for RNA virus entry and propagation. Host-targeted antivirals are indicated in the figure according to their presumed point of interference with the viral infectious cycle. ER: endoplasmic reticulum; EGA: bromobenzaldehyde N-(2,6-dimethylphenyl) semicarbazone.

3. Host-Targeted Antivirals against RNA Viruses

From a phylogenetic perspective, RNA viruses evolve at a much faster rate than their hosts (Figure 2). As discussed above, this is due to the lack of proofreading of the RNA-dependent RNA polymerases (RdRp) and reverse transcriptases (RT) [64], resulting in an error frequency approximately three orders of magnitude higher than that of DNA-dependent DNA polymerases [65].

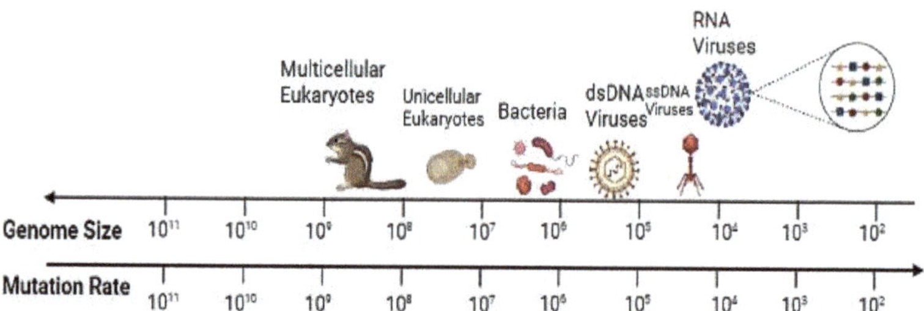

Figure 2. Biological mutation rate compared with genome size. RNA viruses have higher mutation rates. Adaptive mutations are highlighted by colored symbols, illustrating the genetic variation present in the RNA virus quasispecies. Data are adapted from [66].

RNA viruses exist in nature as quasispecies, which can be defined as complex distributions of genomes that exhibit genetic variations that endow the mutation carriers with advantages to survive and thrive over strains that do not carry the mutation. Genetic variations emerge during non-proofed replication into host organisms, enhancing the ability of specific quasispecies to proliferate, infect more easily, evade immunity, and become resistant to treatment and/or vaccination [7]. Quasispecies that do not enjoy such an advantage are eventually eliminated. While replication fidelity is a crucial element in the survival of one quasispecies over another, additional factors, e.g., the genome architecture and speed of genome replication, may influence these evolutionary adaptations [67].

Direct-acting antivirals (DAAs) are employed to treat RNA virus infections. However, quasispecies emergence triggers the rapid appearance of resistant populations. When the RdRp inhibitor balapiravir was used to treat DENV infection, no differences were found in the plasma viral load, cytokine concentrations, and fever clearance time between the treated group and the control, indicating the resistance of the DENV quasispecies to its action [68]. On the other hand, host-targeted antivirals (HTAs) constitute a novel strategy that may overcome this issue. The main reason is that host factors are genetically more stable than viral factors, and the mutation rate of cellular genes is very low in comparison with the genes of RNA viruses [69]. Moreover, HTAs have pan-genotype and serotype antiviral activity and complementary mechanisms of action to DAAs in clinical development; thus, they could be used jointly [70]. Additionally, many HTAs are already FDA-approved drugs for the treatment of other pathologies, e.g., cardiovascular diseases, inflammatory diseases, and cancer. This means that toxicity and other adverse effects were assessed and reported beforehand [71]. However, this does not rule out the emergence of quasispecies that bypass the cellular effector targeted. This is a distinct possibility that needs to be addressed on a case-by-case basis.. Figure 3 displays a graphic representation of a scale that compares the advantages and disadvantages of HTAs with those of DAAs. In our view, the net balance of HTAs outweighs that of DAAs, hence tipping the scale toward the former.

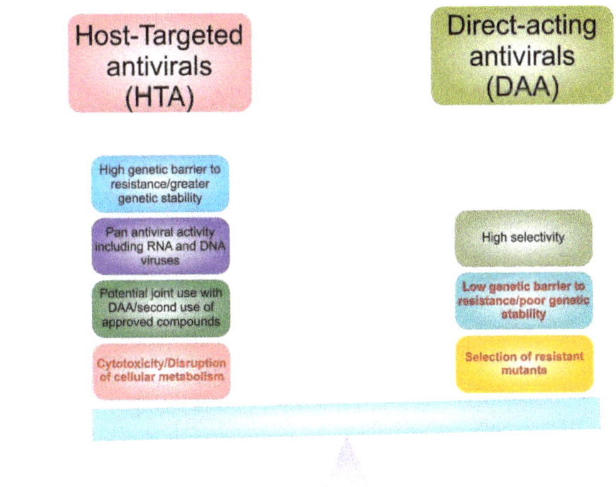

Antiviral features	Host-Targeted Antivirals (HTA)	Direct-Acting Antivirals (DAA)
Genetic barrier to resistance	High	Low
Genetic stability	High	Low
Spectrum of antiviral activity	High (Pan-activity)	Low (Virus-specific)
Synergism with other antivirals	Reported use with DAA	Reported use with HTA
Second use of approved compounds (repurposing)	Common	Restricted to antiviral treatments
Cytotoxicity/disruption of cellular metabolism	Potentially high, depends on doses and treatment plan	Low
Selection of resistant mutants	Scarce	Frequent for chronic diseases produced by RNA viruses and less frequent for acute infections

Figure 3. Graphic summary and summary table of the most salient features of HTAs and DAAs. Red text indicates the disadvantages of the indicated approach.

3.1. Attachment and Entry Inhibitors

The endpoint of the virion entry for most RNA viruses is the release of the viral genome into the host cell cytoplasm; hence, the disruption of the initial attachment and entry of viruses can prevent the next steps of the viral replication cycle. There are a few genetic examples of this mechanism. For example, naturally occurring mutations of the chemokine receptor and HIV-1 entry co-factor CCR5 prevent infection [72–74]. Thereby, targeting mechanisms of entry into the host cell could be a viable antiviral strategy. A key example is HIV-1, which enters lymphocytes, macrophages, and neurons via CD4 (main receptor) and CCR5/CXCR4 (co-receptors) [75]. Maraviroc is a CCR5 antagonist that can block HIV infection [76]. However, HIV-1 can also use CXCR4 as a co-receptor. In fact, plerixafor (AMD3100), which is currently used for stem cell mobilization in MM patients undergoing autologous treatment [77,78], was originally conceived as a form of anti-HIV treatment [79]. Some examples are already available in the literature. Chebulagic acid and punicalagin compounds inhibit viral glycoprotein interactions with cell surface glycosaminoglycans in the 10–100 µM range [80], which are employed by many viruses as primary entry factors, including DENV and HCV. Similarly, λ- and ι-carrageenans act as heparan sulfate (HS)-mimetic compounds and are potent (IC_{50} = 0.14 to 4.1 µg/mL) inhibitors of DENV-2 and DENV-3 in monkey and human cells [81]. Likewise, other sulfated polysaccharides display antiviral activity against HIV and HCV [82]. Bromobenzaldehyde N-(2,6-dimethylphenyl) semicarbazone (EGA) inhibits influenza A entry into host cells by blocking the acidification of endosomes [83].

Importantly, SARS-CoV-2 infection of the host cell is enabled by the interaction of the Receptor-Binding Domain (RBD) of the SARS-CoV-2 spike (S) protein with the cellular ACE2 receptor. This makes ACE2 a potential antiviral target. Chloroquine (CQ), a 4-aminoquinoline base that causes an increase in the lysosomal pH, was suggested as a potential antiviral agent for SARS-CoV at a relatively low dose ($IC_{50} \approx 1$ µM) [84], although its real-life efficacy is questionable at best [85]. Its potential mechanism of action is based on the fact that CQ inhibits the acidification and maturation of the endosome, thereby blocking the pathway in the intermediate stages of endocytosis and preventing the further transport of virus particles to the final release site [86]. CQ could also impair the terminal glycosylation of the ACE2 receptor used by SARS-CoV-2 [87], reducing the affinity of the S protein for ACE2. Additional HTAs targeting this mechanism were screened, including NAAE (N-(2-amino-ethyl)-1 aziridine-ethanamine). NAEE displayed a dose-dependent inhibition of the ACE2 catalytic activity, blocking cell fusion with the S protein. MLN-4760 is a potent (IC_{50} = 0.44 nM) inhibitor for ACE2 that forms a stable complex with ACE2 [88]. However, there is no well-established consensus on the feasibility of ACE2 receptor inhibition due to its important physiological role, including its protective effect on lung injury in acute respiratory diseases [89].

Cathepsin L and TMPRSS are suitable targets to block the entry of some viruses, e.g., SARS-CoV-2 [90,91], into cells. Teicoplanin, a glycopeptide antibiotic, inhibits cathepsin L, blocking the entry of SARS-CoV pseudotyped viruses and SARS-CoV-2 in vitro with a very low IC_{50}, approximately 300 nM [92]. Likewise, oxocarbazate suppresses cathepsin L and inhibits the entry of SARS-CoV and Ebola virus with an IC_{50} also in the 300 nM range [93].

3.2. ER α-Glucosidase Inhibitors

Endoplasmic reticulum α-glucosidase inhibitors efficiently disrupt the morphogenesis and assembly of multiple enveloped viruses. The main reason is that these enzymes are essential for the processing, proper folding, and function of many viral glycoproteins that are part of the viral capsid.

The naturally occurring iminosugar castanospermine inhibits all serotypes of DENV in vitro with various IC_{50} values [94], also decreasing mortality in an in vivo model [95]; this compound also inhibits cell adhesion and cell-to-cell spread, glycoprotein processing, and the replication of HIV [96]. Celgosivir, a pro-drug stemming from castanospermine, conferred full protection to mice infected with DENV (IC_{50} = 5 μM) [94]. A pre-clinical trial in mice showed that celgosivir enhanced survival, reduced viremia, and promoted a vigorous immune response [97]. Celgosivir also displayed antiviral properties against HCV [98].

More recently, some iminosugar derivatives were shown to display activity against DENV and influenza virus. UV-4B protected mice from a DENV-2 lethal challenge, with IC_{50} = 17 μM [99]. It also promoted survival in a lethal influenza virus mouse model [100].

To address the advantages or disadvantages of HTA therapy, Plummer and coworkers conducted the first evolutionary study to investigate the evolution of DENV-2 under selective pressure by UV-4B [15]. This study revealed that DENV does not acquire mutations that increase fitness during in vivo replication in the presence of the compound, indicating that host factors display a high genetic barrier during the treatment of DENV infections.

3.3. Lipid Synthesis Inhibitors

During infection of enveloped viruses, the host's lipid synthesis pathways are co-opted in several steps of the virus replicative cycle. Statins inhibit the mevalonate pathway, blocking the enzyme HMG-CoA reductase, which plays a central role in the production of cholesterol and is employed regularly in patients with hypercholesterolemia [101]. Previous studies performed in our laboratory showed that lovastatin inhibited DENV particle assembly (IC_{50} = 10–50 μM) in a cell culture via the inhibition of the prenylation of Rho and Rab GTPases, which are important in DENV morphogenesis [102]. A delay in infection and an increase in survival rates was observed in AG129 mice infected with DENV-2; these results were the basis of a randomized, double-blind, placebo-controlled trial in adults with dengue that demonstrated the potential use of lovastatin in anti-DENV therapy [103]. Likewise, lovastatin was effective in clinical trials to control HIV-1 replication in chronically infected individuals who were not receiving antiretroviral medication [104].

The antiviral activity of statins on the HCV subgenomic replicon showed that mevastatin and simvastatin exhibited the strongest anti-HCV activity, whereas fluvastatin and lovastatin had moderate inhibitory effects. Moreover, the combination of statins with several selective HCV inhibitors resulted in a pronounced antiviral effect in a cell culture that prevented or delayed the emergence of drug-resistant variants [105]. Simvastatin exhibited antiviral effects against influenza A virus by modulating common cellular pathways, especially during endocytosis and lysosomal activity, affecting the autophagosome function [106]. Fenretinide (4-HPR) is a synthetic retinoid that alters ceramide homeostasis. 4-HPR displayed in vitro antiviral activity against some flaviviruses, including DENV, West Nile, Modoc, and HCV viruses, and in vivo activity against DENV, with the IC_{50} in the 1–50 μM range [107].

3.4. Antagonists of Cytoskeletal Polymerization

Several studies have shown that actin filaments and microtubules interact with viral particles at several steps of the replicative cycle. Pentagalloyl glucose decreased the intracellular levels of cofilin [108], blocking the reorganization of the actin cytoskeleton and influenza virus assembly and budding with an IC_{50} around 10 μM [109]. Importantly, protruding actin-based structures such as Cdc42-dependent filopodia [110] seem particularly important in viral dissemination. In this regard, ZCL278 had potent antiviral effects against junin virus (JUNV), vesicular stomatitis virus (VSV), lymphocytic choriomeningitis (LCM) virus, and dengue virus [111], by inhibiting the Cdc42 function in host cells ($IC_{50} \approx$ 50 μM) [112]. During DENV-2 infection, nocodazole and cytochalasin D significantly inhibited the production of infectious particles both before and after infection [54]. Similar results were reported by Wang et al. [113], who showed that cytochalasin D and jasplakinolide reduced virion ingress during treatment in early events of infection and reported an accumulation of the E viral protein later in infection. Other researchers have emphasized the importance of the interactions between the cytoskeleton and flavivirus replication cycles, suggesting the cytoskeleton as a viable target for the design of new antiviral therapies [50].

3.5. Approved Small Antitumor Molecules with Antiviral Effects

Mutations to multiple protein kinases underlie the initiation or development of various forms of cancer [114]. Interestingly, some of these kinases are also important mediators of viral infection; therefore, kinase antagonists (e.g., small organic molecules) developed to treat cancer could be repurposed as a therapeutic alternative to treat infections caused by RNA viruses. Some protein kinase inhibitors have antiviral activity against one or more DENV serotypes. For example, selumetinib (AZD6244) is a MEK/ERK inhibitor that displays antiviral activity against DENV-2, DENV-3, and St. Louis encephalitis virus (SLEV) by altering the virion morphogenesis in live mice (dose 100 mg/kg/day against 10^5 CFU) [115]. Sunitinib and erlotinib interfere with the intracellular trafficking of HCV-infective particles by inhibiting AP2-associated protein kinase 1 (AAK1) and cyclin G-associated kinase (GAK). Furthermore, they produce a decrease in DENV and EBOV infection, demonstrating that these molecules could be useful as broad-spectrum antivirals [116].

Small molecules such as AZD0530 and dasatinib target the Tyr kinase Fyn with very high affinity ($IC_{50} \approx 1$ nM). This kinase is involved in DENV-2 replication, thereby possessing potential anti-DENV activity [117]. Dasatinib also inhibits the activity of other Src-like kinases, exhibiting a potent inhibitory effect on dengue virus (serotypes 1–4) by preventing the assembly of dengue virions within the virus-induced membranous replication complex. Indeed, dasatinib efficiently ($IC_{50} \approx 1$–5 µM) blocked DENV infection in vitro [118]. Furthermore, dasatinib prevents permeability alterations during DENV infection [119], likely by targeting the RhoA/ROCK axis [120]. ABL kinases regulate several cellular pathways, including cell migration, adhesion, and actin reorganization [121]. A prepandemic study identified imatinib as an antagonist ($IC_{50} \approx 10$–50 µM) of both SARS-CoV and MERS-CoV replication in vitro [122]. Imatinib inhibited the early stages of the coronavirus life cycle by blocking viral fusion with the endosomal membrane, thus inhibiting subsequent viral genome replication. In Ebola infection, imatinib prevents the phosphorylation of the viral VP40 protein, which is necessary for virion egress from the host cell [123].

More recently, GNF-2 displayed a potent and specific effect (IC_{50} similar to that of imatinib) against DENV by inhibiting the kinase activity of Abl by allosterically binding to the myristoyl-binding pocket of Abl. However, it also displayed an Abl-independent mechanism, targeting the interaction with the viral E protein in the prefusion state [124]. Finally, ruxolitinib, an inhibitor of Janus Kinases 1/2 and an FDA-approved drug for the treatment of proliferative neoplastic myelofibrosis, reduces HIV-1 replication in human macrophages at a very low IC_{50} (0.1–0.4 µM) [125]. Together, these data provide evidence of the broad-spectrum activity of FDA-approved antitumor drugs.

3.6. Plant-Derived Natural Compounds

Plant-derived natural products are an important form of antiviral treatment, particularly in developing countries due to the ancient social roots of traditional medicine [126]. Many potential antivirals derived from natural products were described, of which we discuss only the best characterized. Importantly, some of these can act as direct antivirals. For example, curcumin (see below) can inhibit the DENV NS2B/NS3 protease system [127]. However, they can also target host-dependent mechanisms. This section focuses on the latter.

Immunosuppressive drugs such as mycophenolic acid (IMP dehydrogenase inhibitor) exhibit anti-DENV activity, preventing the synthesis and accumulation of viral RNA [128]. Cyclosporine, a cyclophilin protein inhibitor and immunosuppressant agent used conventionally to prevent graft rejection, displayed antiviral activity against DENV, HCV, and HIV at the step of RNA synthesis [129,130]. FGI-106, a low-molecular-weight inhibitor, has shown broad-spectrum antiviral activity against Ebola virus both in vitro and in vivo.

Although its mechanism of action has yet to be elucidated, this compound was shown to also be effective against DENV, HIV, and HCV infections [131].

Carotenoid pigments have antitumor and antiviral effects due to the activation of caspases that inhibit metalloproteinases. Treating HCV- and HBV-infected mononuclear cells with these pigments decreased the activity of viral polymerases, thereby inhibiting viral replication.

Curcumin has effects on several cellular processes and/or structures, e.g., the cytoskeleton and the ubiquitin–proteasome system [132]. It also induces apoptosis [133]. All these effects interfere with dengue virus infection [134]. On the other hand, ferruginol derivatives are active against DENV-2 in post-infection stages, resulting in a dramatic reduction in the viral plaque size [18].

A summary of host-targeted antiviral with a non-comprehensive list of compounds appears in Table 1.

Table 1. Host-targeted antivirals against some RNA viruses.

Category	Compound	Cellular Target	Virus	References
Entry and attachment inhibitors	Chebulagic acid and punicalagin	Cellular surface glycosaminoglycans	DENV, HCV	[80]
	Carrageenans	HS-imitative compounds	DENV, HIV, HCV	[81]
	Teicoplanin	Cathepsin L	SARS-CoV-2	[92]
	Oxocarbazate		SARS-CoV, EBOV	[93]
Endosomal function inhibitors	EGA	Endosome acidification	Influenza A	[83]
	Chloroquine		SARS-CoV-2, SARS-CoV	[84,86,87]
α-glucosidase inhibitors	Castanospermine	ER glucosidases/ disruption of glycoprotein processing	DENV, HIV	[94–96]
	Celgosivir		DENV, HCV	[99,100]
	Iminosugar derivatives		DENV, Influenza	[15,100]
Lipid synthesis inhibitors	Lovastatin	Prenylation of Rho and Rab GTPases	DENV, HCV, HIV	[102–104]
	Mevastatin Simvastatin	HMG-CoA reductase	HCV, Influenza A	[71,106]
	Fenretinide	Ceramide homeostasis	DENV, West Nile, Modoc, HCV	[107]
Agents acting on cytoskeleton	Pentagalloyl glucose	Cofilin	Influenza	[108,109]
	ZCL278	Cdc42	DENV, JUNV, VSV, LCM	[111]
	Cytochalasin D Jasplakinolide	Actin filaments	DENV	[54,113]
	Nocodazol	Microtubules		
Antitumor molecules	Selumetinib (AZD6244)	MEK/ERK	DENV, SLEV	[115]
	Sunitinib and erlotinib	AAK1/GAK	HCV, DENV, EBOV	[116]
	AZD0530 and dasatinib	Fyn/Src kinases	DENV, Modoc	[117,118]
	Imatinib	c-ABL	DENV, SARS-CoV, MERS-CoV, EBOV	[122,123]
	GNF-2		DENV	[124]
	Ruxolitinib	JAK 1/2	HIV	[125]
Plant-derived natural compounds	Mycophenolic acid	IMP dehydrogenase	DENV	[128]
	Cyclosporine	Cyclophilin protein	DENV, HCV, HIV	[129,130]
	FGI-106	Unknown	DENV, EBOV, HIV, HCV	[131]
	Carotenoid pigments	Metalloproteinases	HCV	[135]
	Curcumin	Cytoskeleton, ubiquitin–proteasome-proteasome system, apoptosis	DENV	[134]
	18-(phthalimide-2-yl) ferruginol	Actin remodeling, polyprotein translation, replicative complexes	DENV, ZIKV, CHIKV	[18]

HS: heparan sulfate; ER: endoplasmic reticulum; HMG CoA: 3 hydroxy 3 methylglutaryl coenzyme A reductase; IMP dehydrogenase: Inosine-5′-monophosphate dehydrogenase; EGA: bromobenzaldehyde N-(2,6-dimethylphenyl) semicarbazone; MEK/ERK: mitogen-activated protein kinase kinase/extracellular signal-regulated kinase; AAK1/GAK: AP2-associated protein kinase 1/cyclin G-associated kinase; Fyn/Src: proto-oncogene tyrosine-protein kinase; JAK: Janus kinase.

4. Ferruginol Analogs as Potential Host-Targeted Antiviral

Ferruginol is a diterpenoid phenol isolated from plants of the *Podocarpaceae, Cupressaceae, Lamiaceae,* and *Verbenaceae* families. Ferruginol has a wide spectrum of biological activities, such as antibacterial, antifungal, antimicrobial, acaricide, cardioactive, antioxidant, anti-Leishmania, antiplasmodium, nematicide, anti-ulcerous, and cytotoxic activities on tumor cells. Previously, we reported that two ferruginol analogs, 18-(phthalimide-2-yl) ferruginol (compound 8) and 18-oxoferruginol (compound 9, shown in Figure 4), reduced the in vitro infection of human herpesvirus type 1 and 2 and dengue serotype 2 when added in post-infection stages. In addition, compound 8 significantly reduced the size of the viral plaques when DENV-2-infected cells were treated [18].

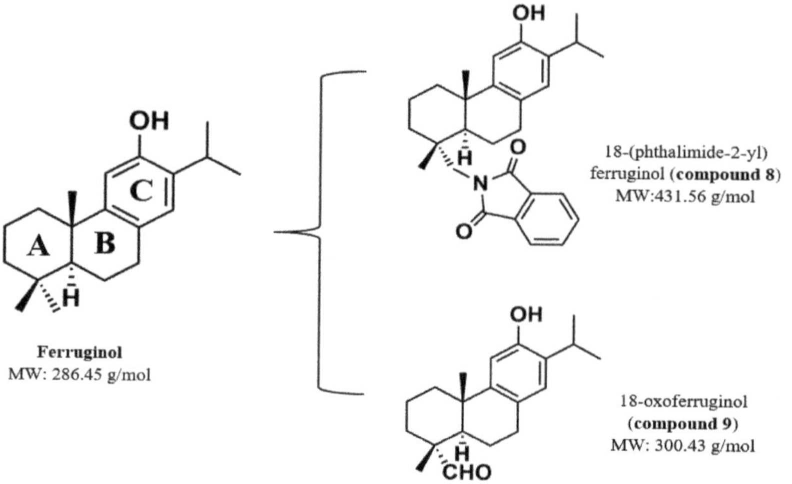

Figure 4. Chemical structure of ferruginol and two of its analogs, 18-(phthalimide-2-yl) ferruginol (compound 8) and 18-oxoferruginol (compound 9). MW: molecular weight.

18-(phthalimide-2-yl) ferruginol displays a phthalimide group in ring A of the ferruginol backbone, specifically in carbon 18, as indicated in Figure 4. Phthalimide is an imide derived from phthalic acid with two carbonyl groups joined to a secondary amine. These compounds are hydrophobic, endowing the molecule with the ability to traverse biological membranes in vivo [136]. In addition, these molecules and some of their derivatives are endowed with antibacterial, antifungal, analgesic, antitumor, anxiolytic, hypolipemic, and analgesic activities and have low in vivo toxicity [137].

Our recent data suggest that compound 8 is endowed with potential broad-spectrum antiviral activity. It does inhibit in vitro infection by Zika virus in Vero [19], PC3, and HeLa cells at concentrations below 10 μM. Likewise, this molecule has antiviral activity in Vero cells infected with CHIKV (Alphavirus genus) [138]. Further unpublished studies related to the antiviral mechanism of action of this molecule strongly suggest that 18-(phthalimide-2-yl)-ferruginol has an HTA-related mechanism of action by disrupting the DENV-2 polyprotein translation via the alteration of actin remodeling and other related cellular and viral processes involved in the replicative complex formation. Additionally, in vitro and in silico evidence indicates that this compound has few cytotoxic and potentially reversible effects on host cells.

5. Integration of Bioinformatics with the Search for Host-Targeted Antivirals

As discussed above, new therapeutic strategies for RNA viruses could be based on targeting host proteins and processes that disturb the virus–cell interactomes that emerge during virus colonization of the host cell. This view integrates the fact that viruses can manipulate and modify cellular protein networks essential for the viral cycle [139,140]. Databases such as VirusMINT (https://bio.tools/virusmint accessed on 25 November 2022) and VirHostNet (https://virhostnet.prabi.fr/ accessed on 25 November 2022) and drug-target databases such as DrugBank (https://go.drugbank.com/ accessed on 25 November 2022) are powerful tools to obtain preliminary information regarding the interactions between virus and host proteins.

Experimentally, the high-throughput screening of virus-host protein–protein inter-action methods, such as yeast two-hybrid assays, co-affinity purification/MS techniques, protein arrays, and protein complementation protocols such as mass spectrometry, offers great benefits for expanding virus–cell interactome knowledge and provides the oppor-tunity to discover new targets in humans [141]. Another promising technology for novel drug discovery is image-based profiling, which includes computational equipment, such as deep learning and single-cell methods; in short, these tools collect relevant biological infor-mation present in an image, reducing it to a multidimensional profile (see Figure 5B) [142]. Altogether, the focus of this outlook must be the host cell targets improving the repurposing of drugs or finding new ones, because actual treatments remain limited, inefficient, and incapable of challenging drug resistance. The workflow proposed for such a search strategy is depicted in Figure 5A.

(A)

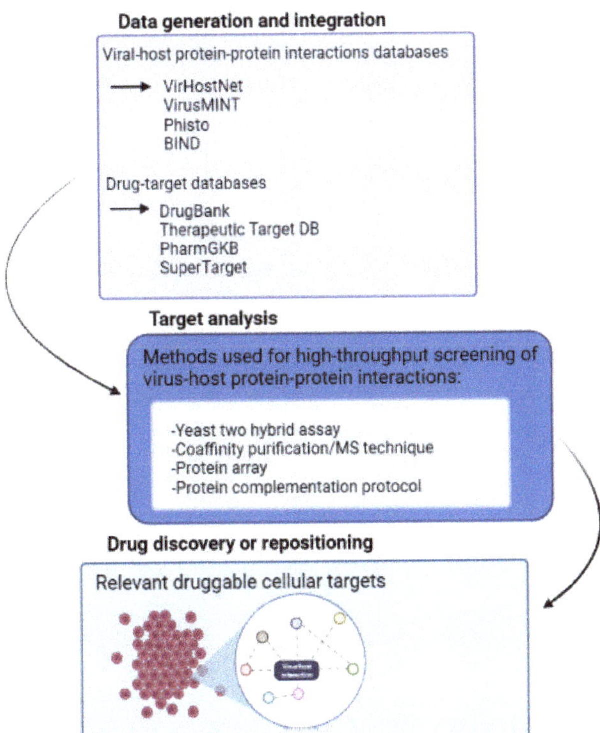

Figure 5. *Cont.*

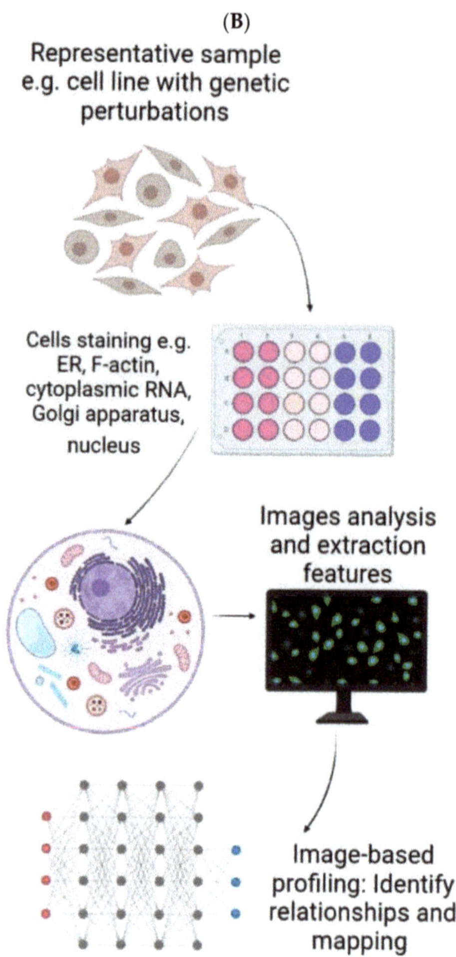

Figure 5. Potential pipeline for the discovery of host-targeted antivirals. (**A**) Workflow to search for cellular targets involved in virus–cell interactions. Databases provide initial identification of possible targets in the interactions between viruses and cells that need to be confirmed by experimental techniques. The next step includes repositioning existing drugs or discovering new inhibitors that may act as host-targeted antivirals. (**B**) Image-based profiling overview from biological samples.

6. Conclusions and Perspectives

The goal of an antiviral drug is to avoid or resolve the infection to prevent the development of severe disease or the death of the host. This kind of therapy must be effective, ensuring complete resolution of the infection. Likewise, understanding the mechanisms by which molecules exert their antiviral action not only on the virus but also on the host cell allows an estimation of the risk–benefit balance.

Moreover, there is overwhelming proof of the impact of viral escape mutants on global human health. The COVID-19 pandemic is a relevant example of the appearance of viral strains resistant to vaccination that can become massively contagious.

All these lessons show that HTA therapy is a valid alternative for the new millennium in research, not only against viruses but also in general for microorganisms that evolve faster than their hosts. It is true that HTA therapy has many potential adverse effects, and many clinical trials are ongoing to ascertain this aspect among others. Doubts that remain in this research field are more numerous than the data provided thus far but generate new hope in the field of pharmacology to design more rational drugs.

Author Contributions: V.C.R.-L., M.E.-F. and J.C.G.-G., conceptualization, manuscript writing, figure preparation, and bibliographic compilation. M.V.-M., additional conceptualization, and rebuttal. All authors have read and agreed to the published version of the manuscript.

Funding: This research was funded by Ministerio de Ciencia, Tecnología e Innovación (Minciencias), grant number 11584466951 and projects Sustainability-CODI-UdeA-2018-9.

Acknowledgments: The authors thank to Camilo Hernández-Cuellar for additional graphic design. This research is funded by the University of Antioquia, grant CODI-2020-34137, Sustainability Program 2018-9 of CODI-UdeA, and Exclusivity Program-UdeA (2022-2023), to J.C.G.-G. Moreover, V.R.-L thanks the financial support from CODI (Comité para el Desarrollo de la Investigación-Universidad de Antioquia)/Grant 2014-1041.

Conflicts of Interest: The authors declare no conflict of interest.

References

1. Carrasco-Hernandez, R.; Jácome, R.; López Vidal, Y.; Ponce de León, S. Are RNA Viruses Candidate Agents for the Next Global Pandemic? A Review. *ILAR J.* **2017**, *58*, 343–358. [CrossRef] [PubMed]
2. Dhama, K.; Patel, S.K.; Sharun, K.; Pathak, M.; Tiwari, R.; Yatoo, M.I.; Malik, Y.S.; Sah, R.; Rabaan, A.A.; Panwar, P.K.; et al. SARS-CoV-2 jumping the species barrier: Zoonotic lessons from SARS, MERS and recent advances to combat this pandemic virus. *Travel Med. Infect. Dis.* **2020**, *37*, 101830. [CrossRef] [PubMed]
3. Irwin, K.K.; Renzette, N.; Kowalik, T.F.; Jensen, J.D. Antiviral drug resistance as an adaptive process. *Virus Evol.* **2016**, *2*, vew014. [CrossRef] [PubMed]
4. Knippa, K.; Peterson, D.O. Fidelity of RNA polymerase II transcription: Role of Rpb9 [corrected] in error detection and proofreading. *Biochemistry* **2013**, *52*, 7807–7817. [CrossRef] [PubMed]
5. Preston, B.D.; Poiesz, B.J.; Loeb, L.A. Fidelity of HIV-1 reverse transcriptase. *Science* **1988**, *242*, 1168–1171. [CrossRef]
6. Cuevas, J.M.; Geller, R.; Garijo, R.; Lopez-Aldeguer, J.; Sanjuan, R. Extremely High Mutation Rate of HIV-1 In Vivo. *PLoS Biol.* **2015**, *13*, e1002251. [CrossRef]
7. Novella, I.S.; Domingo, E.; Holland, J.J. Rapid viral quasispecies evolution: Implications for vaccine and drug strategies. *Mol. Med. Today* **1995**, *1*, 248–253. [CrossRef]
8. Minskaia, E.; Hertzig, T.; Gorbalenya, A.E.; Campanacci, V.; Cambillau, C.; Canard, B.; Ziebuhr, J. Discovery of an RNA virus $3' \rightarrow 5'$ exoribonuclease that is critically involved in coronavirus RNA synthesis. *Proc. Natl. Acad. Sci. USA* **2006**, *103*, 5108–5113. [CrossRef]
9. Wang, R.; Chen, J.; Wei, G.W. Mechanisms of SARS-CoV-2 Evolution Revealing Vaccine-Resistant Mutations in Europe and America. *J. Phys. Chem. Lett.* **2021**, *12*, 11850–11857. [CrossRef]
10. Paintsil, E.; Cheng, Y.C. Antiviral Agents. *Encycl. Microbiol.* **2019**, 176–225. [CrossRef]
11. Vere Hodge, A.; Field, H.J. General Mechanisms of Antiviral Resistance. In *Genetics and Evolution of Infectious Disease*; Elsevier: Amsterdam, The Netherlands, 2011; pp. 339–362. [CrossRef]
12. Gonzalez-Hernandez, M.J.; Pal, A.; Gyan, K.E.; Charbonneau, M.E.; Showalter, H.D.; Donato, N.J.; O'Riordan, M.; Wobus, C.E. Chemical derivatives of a small molecule deubiquitinase inhibitor have antiviral activity against several RNA viruses. *PLoS ONE* **2014**, *9*, e94491. [CrossRef] [PubMed]
13. Adamson, C.S.; Freed, E.O. Anti-HIV-1 therapeutics: From FDA-approved drugs to hypothetical future targets. *Mol. Interv.* **2009**, *9*, 70–74. [CrossRef] [PubMed]
14. Takashita, E.; Kinoshita, N.; Yamayoshi, S.; Sakai-Tagawa, Y.; Fujisaki, S.; Ito, M.; Iwatsuki-Horimoto, K.; Halfmann, P.; Watanabe, S.; Maeda, K.; et al. Efficacy of Antiviral Agents against the SARS-CoV-2 Omicron Subvariant BA.2. *N. Engl. J. Med.* **2022**, *386*, 1475–1477. [CrossRef] [PubMed]
15. Plummer, E.; Buck, M.D.; Sanchez, M.; Greenbaum, J.A.; Turner, J.; Grewal, R.; Klose, B.; Sampath, A.; Warfield, K.L.; Peters, B.; et al. Dengue Virus Evolution under a Host-Targeted Antiviral. *J. Virol.* **2015**, *89*, 5592–5601. [CrossRef] [PubMed]
16. Ashburn, T.T.; Thor, K.B. Drug repositioning: Identifying and developing new uses for existing drugs. *Nat. Rev. Drug Discov.* **2004**, *3*, 673–683. [CrossRef] [PubMed]
17. Hofman, P.; Vouret-Craviari, V. Microbes-induced EMT at the crossroad of inflammation and cancer. *Gut Microbes* **2012**, *3*, 176–185. [CrossRef]
18. Roa Linares, V.C.; Brand, Y.M.; Agudelo-Gomez, L.S.; Tangarife-Castano, V.; Betancur-Galvis, L.A.; Gallego-Gomez, J.C.; Gonzalez, M.A. Anti-herpetic and anti-dengue activity of abietane ferruginol analogues synthesized from (+) dehydroabietylamine. *Eur. J. Med. Chem.* **2016**, *108*, 79–88. [CrossRef]
19. Sousa, F.T.G.; Nunes, C.; Romano, C.M.; Sabino, E.C.; Gonzalez-Cardenete, M.A. Anti-Zika virus activity of several abietane-type ferruginol analogues. *Rev. Inst. Med. Trop. Sao Paulo* **2020**, *62*, e97. [CrossRef]

20. Sachse, M.; Tenorio, R.; Fernandez de Castro, I.; Munoz-Basagoiti, J.; Perez-Zsolt, D.; Raich-Regue, D.; Rodon, J.; Losada, A.; Aviles, P.; Cuevas, C.; et al. Unraveling the antiviral activity of plitidepsin against SARS-CoV-2 by subcellular and morphological analysis. *Antivir. Res.* **2022**, *200*, 105270. [CrossRef]
21. Strobelt, R.; Adler, J.; Paran, N.; Yahalom-Ronen, Y.; Melamed, S.; Politi, B.; Shulman, Z.; Schmiedel, D.; Shaul, Y. Imatinib inhibits SARS-CoV-2 infection by an off-target-mechanism. *Sci. Rep.* **2022**, *12*, 5758. [CrossRef]
22. Boytz, R.; Slabicki, M.; Ramaswamy, S.; Patten, J.J.; Zou, C.; Meng, C.; Hurst, B.L.; Wang, J.; Nowak, R.P.; Yang, P.L.; et al. Anti-SARS-CoV-2 activity of targeted kinase inhibitors: Repurposing clinically available drugs for COVID-19 therapy. *J. Med. Virol.* **2023**, *95*, e28157. [CrossRef]
23. Garcia-Serradilla, M.; Risco, C.; Pacheco, B. Drug repurposing for new, efficient, broad spectrum antivirals. *Virus Res.* **2019**, *264*, 22–31. [CrossRef] [PubMed]
24. Mercorelli, B.; Palu, G.; Loregian, A. Drug Repurposing for Viral Infectious Diseases: How Far Are We? *Trends Microbiol.* **2018**, *26*, 865–876. [CrossRef]
25. Flint, J.; Racaniello, V.R.; Rall, G.F.; Hatziioannou, T.; Skalka, A.M. *Principles of Virology, Multi-Volume*, 5th ed.; American Society for Microbiology: Washington, DC, USA, 2020.
26. Arakawa, M.; Morita, E. Flavivirus Replication Organelle Biogenesis in the Endoplasmic Reticulum: Comparison with Other Single-Stranded Positive-Sense RNA Viruses. *Int. J. Mol. Sci.* **2019**, *20*, 2336. [CrossRef] [PubMed]
27. Taylor, M.P.; Koyuncu, O.O.; Enquist, L.W. Subversion of the actin cytoskeleton during viral infection. *Nat. Rev. Microbiol.* **2011**, *9*, 427–439. [CrossRef] [PubMed]
28. Walsh, D.; Mohr, I. Viral subversion of the host protein synthesis machinery. *Nat. Rev. Microbiol.* **2011**, *9*, 860–875. [CrossRef]
29. Jackson, W.T. Viruses and the autophagy pathway. *Virology* **2015**, *479-480*, 450–456. [CrossRef]
30. Orozco-García, E.; Gallego-Gómez, J.C. Autophagy and Lipid Metabolism—A Cellular Platform where Molecular and Metabolic Pathways Converge to Explain Dengue Viral Infection. In *Cell Biology-New Insights*; Najman, S., Ed.; IntechOpen: Rijeka, Croatia, 2016; Volume 1.
31. Fan, Y.; Sanyal, S.; Bruzzone, R. Breaking Bad: How Viruses Subvert the Cell Cycle. *Front. Cell. Infect. Microbiol.* **2018**, *8*, 396. [CrossRef]
32. Nelemans, T.; Kikkert, M. Viral Innate Immune Evasion and the Pathogenesis of Emerging RNA Virus Infections. *Viruses* **2019**, *11*, 961. [CrossRef]
33. Martin-Acebes, M.A.; Jimenez de Oya, N.; Saiz, J.C. Lipid Metabolism as a Source of Druggable Targets for Antiviral Discovery against Zika and Other Flaviviruses. *Pharmaceuticals* **2019**, *12*, 97. [CrossRef]
34. Artpradit, C.; Robinson, L.N.; Gavrilov, B.K.; Rurak, T.T.; Ruchirawat, M.; Sasisekharan, R. Recognition of heparan sulfate by clinical strains of dengue virus serotype 1 using recombinant subviral particles. *Virus Res.* **2013**, *176*, 69–77. [CrossRef] [PubMed]
35. Sahoo, B.; Chowdary, T.K. Conformational changes in Chikungunya virus E2 protein upon heparan sulfate receptor binding explain mechanism of E2-E1 dissociation during viral entry. *Biosci. Rep.* **2019**, *39*, 1–14. [CrossRef] [PubMed]
36. Blanchard, E.; Belouzard, S.; Goueslain, L.; Wakita, T.; Dubuisson, J.; Wychowski, C.; Rouille, Y. Hepatitis C virus entry depends on clathrin-mediated endocytosis. *J. Virol.* **2006**, *80*, 6964–6972. [CrossRef] [PubMed]
37. Piccini, L.E.; Castilla, V.; Damonte, E.B. Dengue-3 Virus Entry into Vero Cells: Role of Clathrin-Mediated Endocytosis in the Outcome of Infection. *PLoS ONE* **2015**, *10*, e0140824. [CrossRef] [PubMed]
38. Daecke, J.; Fackler, O.T.; Dittmar, M.T.; Kräusslich, H.G. Involvement of clathrin-mediated endocytosis in human immunodeficiency virus type 1 entry. *J. Virol.* **2005**, *79*, 1581–1594. [CrossRef] [PubMed]
39. Aleksandrowicz, P.; Marzi, A.; Biedenkopf, N.; Beimforde, N.; Becker, S.; Hoenen, T.; Feldmann, H.; Schnittler, H.J. Ebola virus enters host cells by macropinocytosis and clathrin-mediated endocytosis. *J. Infect. Dis.* **2011**, *204* (Suppl. S3), S957–S967. [CrossRef]
40. Rossman, J.S.; Leser, G.P.; Lamb, R.A. Filamentous influenza virus enters cells via macropinocytosis. *J. Virol.* **2012**, *86*, 10950–10960. [CrossRef]
41. Mahmoud, I.S.; Jarrar, Y.B.; Alshaer, W.; Ismail, S. SARS-CoV-2 entry in host cells-multiple targets for treatment and prevention. *Biochimie* **2020**, *175*, 93–98. [CrossRef]
42. den Boon, J.A.; Ahlquist, P. Organelle-like membrane compartmentalization of positive-strand RNA virus replication factories. *Annu. Rev. Microbiol.* **2010**, *64*, 241–256. [CrossRef]
43. Miller, S.; Krijnse-Locker, J. Modification of intracellular membrane structures for virus replication. *Nat. Rev. Microbiol.* **2008**, *6*, 363–374. [CrossRef]
44. Spuul, P.; Balistreri, G.; Hellstrom, K.; Golubtsov, A.V.; Jokitalo, E.; Ahola, T. Assembly of alphavirus replication complexes from RNA and protein components in a novel trans-replication system in mammalian cells. *J. Virol.* **2011**, *85*, 4739–4751. [CrossRef] [PubMed]
45. Fernández de Castro, I.; Tenorio, R.; Risco, C. Virus Factories. *Encycl. Virol.* **2021**, 495–500. [CrossRef]
46. Blanchard, E.; Roingeard, P. Virus-induced double-membrane vesicles. *Cell. Microbiol.* **2015**, *17*, 45–50. [CrossRef] [PubMed]
47. McMahon, H.T.; Gallop, J.L. Membrane curvature and mechanisms of dynamic cell membrane remodelling. *Nature* **2005**, *438*, 590–596. [CrossRef]
48. Chang, J.; Block, T.M.; Guo, J.T. Antiviral therapies targeting host ER alpha-glucosidases: Current status and future directions. *Antivir. Res.* **2013**, *99*, 251–260. [CrossRef]

49. Becker, G.L.; Lu, Y.; Hardes, K.; Strehlow, B.; Levesque, C.; Lindberg, I.; Sandvig, K.; Bakowsky, U.; Day, R.; Garten, W.; et al. Highly potent inhibitors of proprotein convertase furin as potential drugs for treatment of infectious diseases. *J. Biol. Chem.* **2012**, *287*, 21992–22003. [CrossRef]
50. Foo, K.Y.; Chee, H.Y. Interaction between Flavivirus and Cytoskeleton during Virus Replication. *Biomed. Res. Int.* **2015**, *2015*, 427814. [CrossRef]
51. Zhang, Y.; Gao, W.; Li, J.; Wu, W.; Jiu, Y. The Role of Host Cytoskeleton in Flavivirus Infection. *Virol. Sin.* **2019**, *34*, 30–41. [CrossRef]
52. Schudt, G.; Kolesnikova, L.; Dolnik, O.; Sodeik, B.; Becker, S. Live-cell imaging of Marburg virus-infected cells uncovers actin-dependent transport of nucleocapsids over long distances. *Proc. Natl. Acad. Sci. USA* **2013**, *110*, 14402–14407. [CrossRef]
53. Cudmore, S.; Cossart, P.; Griffiths, G.; Way, M. Actin-based motility of vaccinia virus. *Nature* **1995**, *378*, 636–638. [CrossRef]
54. Orozco-García, E.; Trujillo-Correa, A.; Gallego-Gómez, J.C. Cell Biology of Virus Infection. The Role of Cytoskeletal Dynamics Integrity in the Effectiveness of Dengue Virus Infection. In *Cell Biology-New Insights*; Najman, S., Ed.; IntechOpen: Rijeka, Croatia, 2016; Volume 1.
55. Greber, U.F.; Way, M. A superhighway to virus infection. *Cell* **2006**, *124*, 741–754. [CrossRef] [PubMed]
56. Amorim, M.J.; Bruce, E.A.; Read, E.K.; Foeglein, A.; Mahen, R.; Stuart, A.D.; Digard, P. A Rab11- and microtubule-dependent mechanism for cytoplasmic transport of influenza A virus viral RNA. *J. Virol.* **2011**, *85*, 4143–4156. [CrossRef] [PubMed]
57. McDonald, D.; Vodicka, M.A.; Lucero, G.; Svitkina, T.M.; Borisy, G.G.; Emerman, M.; Hope, T.J. Visualization of the intracellular behavior of HIV in living cells. *J. Cell Biol.* **2002**, *159*, 441–452. [CrossRef]
58. Nishi, M.; Ryo, A.; Tsurutani, N.; Ohba, K.; Sawasaki, T.; Morishita, R.; Perrem, K.; Aoki, I.; Morikawa, Y.; Yamamoto, N. Requirement for microtubule integrity in the SOCS1-mediated intracellular dynamics of HIV-1 Gag. *FEBS Lett.* **2009**, *583*, 1243–1250. [CrossRef]
59. Yonezawa, A.; Cavrois, M.; Greene, W.C. Studies of ebola virus glycoprotein-mediated entry and fusion by using pseudotyped human immunodeficiency virus type 1 virions: Involvement of cytoskeletal proteins and enhancement by tumor necrosis factor alpha. *J. Virol.* **2005**, *79*, 918–926. [CrossRef] [PubMed]
60. Lambert, A.W.; Weinberg, R.A. Linking EMT programmes to normal and neoplastic epithelial stem cells. *Nat. Rev. Cancer* **2021**, *21*, 325–338. [CrossRef]
61. Yang, J.; Antin, P.; Berx, G.; Blanpain, C.; Brabletz, T.; Bronner, M.; Campbell, K.; Cano, A.; Casanova, J.; Christofori, G.; et al. Guidelines and definitions for research on epithelial-mesenchymal transition. *Nat. Rev. Mol. Cell Biol.* **2020**, *21*, 341–352. [CrossRef]
62. Bose, S.K.; Meyer, K.; Di Bisceglie, A.M.; Ray, R.B.; Ray, R. Hepatitis C virus induces epithelial-mesenchymal transition in primary human hepatocytes. *J. Virol.* **2012**, *86*, 13621–13628. [CrossRef]
63. Cuartas-Lopez, A.M.; Hernandez-Cuellar, C.E.; Gallego-Gomez, J.C. Disentangling the role of PI3K/Akt, Rho GTPase and the actin cytoskeleton on dengue virus infection. *Virus Res.* **2018**, *256*, 153–165. [CrossRef]
64. Domingo, E.; Garcia-Crespo, C.; Lobo-Vega, R.; Perales, C. Mutation Rates, Mutation Frequencies, and Proofreading-Repair Activities in RNA Virus Genetics. *Viruses* **2021**, *13*, 1882. [CrossRef]
65. Moya, A.; Elena, S.F.; Bracho, A.; Miralles, R.; Barrio, E. The evolution of RNA viruses: A population genetics view. *Proc. Natl. Acad. Sci. USA* **2000**, *97*, 6967–6973. [CrossRef] [PubMed]
66. Duffy, S. Why are RNA virus mutation rates so damn high? *PLoS Biol.* **2018**, *16*, e3000003. [CrossRef]
67. Domingo, E.; Garcia-Crespo, C.; Perales, C. Historical Perspective on the Discovery of the Quasispecies Concept. *Annu. Rev. Virol.* **2021**, *8*, 51–72. [CrossRef] [PubMed]
68. Botta, L.; Rivara, M.; Zuliani, V.; Radi, M. Drug repurposing approaches to fight Dengue virus infection and related diseases. *Front. Biosci. (Landmark Ed.)* **2018**, *23*, 997–1019. [CrossRef] [PubMed]
69. Troost, B.; Smit, J.M. Recent advances in antiviral drug development towards dengue virus. *Curr. Opin. Virol.* **2020**, *43*, 9–21. [CrossRef]
70. Koonin, E.V.; Wolf, Y.I. Evolution of microbes and viruses: A paradigm shift in evolutionary biology? *Front. Cell Infect. Microbiol.* **2012**, *2*, 119. [CrossRef]
71. Delang, L.; Vliegen, I.; Froeyen, M.; Neyts, J. Comparative study of the genetic barriers and pathways towards resistance of selective inhibitors of hepatitis C virus replication. *Antimicrob. Agents Chemother.* **2011**, *55*, 4103–4113. [CrossRef]
72. Liu, R.; Paxton, W.A.; Choe, S.; Ceradini, D.; Martin, S.R.; Horuk, R.; MacDonald, M.E.; Stuhlmann, H.; Koup, R.A.; Landau, N.R. Homozygous defect in HIV-1 coreceptor accounts for resistance of some multiply-exposed individuals to HIV-1 infection. *Cell* **1996**, *86*, 367–377. [CrossRef]
73. Samson, M.; Libert, F.; Doranz, B.J.; Rucker, J.; Liesnard, C.; Farber, C.M.; Saragosti, S.; Lapoumeroulie, C.; Cognaux, J.; Forceille, C.; et al. Resistance to HIV-1 infection in caucasian individuals bearing mutant alleles of the CCR-5 chemokine receptor gene. *Nature* **1996**, *382*, 722–725. [CrossRef]

74. Dean, M.; Carrington, M.; Winkler, C.; Huttley, G.A.; Smith, M.W.; Allikmets, R.; Goedert, J.J.; Buchbinder, S.P.; Vittinghoff, E.; Gomperts, E.; et al. Genetic restriction of HIV-1 infection and progression to AIDS by a deletion allele of the CKR5 structural gene. Hemophilia Growth and Development Study, Multicenter AIDS Cohort Study, Multicenter Hemophilia Cohort Study, San Francisco City Cohort, ALIVE Study. *Science* **1996**, *273*, 1856–1862. [CrossRef]

75. Chen, B. Molecular Mechanism of HIV-1 Entry. *Trends Microbiol.* **2019**, *27*, 878–891. [CrossRef] [PubMed]

76. Woollard, S.M.; Kanmogne, G.D. Maraviroc: A review of its use in HIV infection and beyond. *Drug Des. Devel. Ther.* **2015**, *9*, 5447–5468. [CrossRef] [PubMed]

77. Keating, G.M. Plerixafor: A review of its use in stem-cell mobilization in patients with lymphoma or multiple myeloma. *Drugs* **2011**, *71*, 1623–1647. [CrossRef] [PubMed]

78. Bilgin, Y.M. Use of Plerixafor for Stem Cell Mobilization in the Setting of Autologous and Allogeneic Stem Cell Transplantations: An Update. *J. Blood Med.* **2021**, *12*, 403–412. [CrossRef] [PubMed]

79. De Clercq, E.; Yamamoto, N.; Pauwels, R.; Balzarini, J.; Witvrouw, M.; De Vreese, K.; Debyser, Z.; Rosenwirth, B.; Peichl, P.; Datema, R.; et al. Highly potent and selective inhibition of human immunodeficiency virus by the bicyclam derivative JM3100. *Antimicrob. Agents Chemother.* **1994**, *38*, 668–674. [CrossRef] [PubMed]

80. Lin, L.T.; Chen, T.Y.; Lin, S.C.; Chung, C.Y.; Lin, T.C.; Wang, G.H.; Anderson, R.; Lin, C.C.; Richardson, C.D. Broad-spectrum antiviral activity of chebulagic acid and punicalagin against viruses that use glycosaminoglycans for entry. *BMC Microbiol.* **2013**, *13*, 187. [CrossRef] [PubMed]

81. Talarico, L.B.; Damonte, E.B. Interference in dengue virus adsorption and uncoating by carrageenans. *Virology* **2007**, *363*, 473–485. [CrossRef] [PubMed]

82. Bouhlal, R.; Haslin, C.; Chermann, J.C.; Colliec-Jouault, S.; Sinquin, C.; Simon, G.; Cerantola, S.; Riadi, H.; Bourgougnon, N. Antiviral activities of sulfated polysaccharides isolated from Sphaerococcus coronopifolius (Rhodophyta, Gigartinales) and Boergeseniella thuyoides (Rhodophyta, Ceramiales). *Mar. Drugs* **2011**, *9*, 1187–1209. [CrossRef]

83. Gillespie, E.J.; Ho, C.L.; Balaji, K.; Clemens, D.L.; Deng, G.; Wang, Y.E.; Elsaesser, H.J.; Tamilselvam, B.; Gargi, A.; Dixon, S.D.; et al. Selective inhibitor of endosomal trafficking pathways exploited by multiple toxins and viruses. *Proc. Natl. Acad. Sci. USA* **2013**, *110*, E4904–E4912. [CrossRef]

84. Vincent, M.J.; Bergeron, E.; Benjannet, S.; Erickson, B.R.; Rollin, P.E.; Ksiazek, T.G.; Seidah, N.G.; Nichol, S.T. Chloroquine is a potent inhibitor of SARS coronavirus infection and spread. *Virol. J.* **2005**, *2*, 69. [CrossRef]

85. Axfors, C.; Schmitt, A.M.; Janiaud, P.; Van't Hooft, J.; Abd-Elsalam, S.; Abdo, E.F.; Abella, B.S.; Akram, J.; Amaravadi, R.K.; Angus, D.C.; et al. Mortality outcomes with hydroxychloroquine and chloroquine in COVID-19 from an international collaborative meta-analysis of randomized trials. *Nat. Commun.* **2021**, *12*, 2349. [CrossRef] [PubMed]

86. Al-Bari, M.A.A. Targeting endosomal acidification by chloroquine analogs as a promising strategy for the treatment of emerging viral diseases. *Pharm. Res. Perspect.* **2017**, *5*, e00293. [CrossRef] [PubMed]

87. Savarino, A.; Di Trani, L.; Donatelli, I.; Cauda, R.; Cassone, A. New insights into the antiviral effects of chloroquine. *Lancet Infect. Dis.* **2006**, *6*, 67–69. [CrossRef] [PubMed]

88. Ahmad, I.; Pawara, R.; Surana, S.; Patel, H. The Repurposed ACE2 Inhibitors: SARS-CoV-2 Entry Blockers of Covid-19. *Top. Curr. Chem. (Cham.)* **2021**, *379*, 40. [CrossRef]

89. Cheng, H.; Wang, Y.; Wang, G.Q. Organ-protective effect of angiotensin-converting enzyme 2 and its effect on the prognosis of COVID-19. *J. Med. Virol.* **2020**, *92*, 726–730. [CrossRef]

90. Zhao, M.M.; Yang, W.L.; Yang, F.Y.; Zhang, L.; Huang, W.J.; Hou, W.; Fan, C.F.; Jin, R.H.; Feng, Y.M.; Wang, Y.C.; et al. Cathepsin L plays a key role in SARS-CoV-2 infection in humans and humanized mice and is a promising target for new drug development. *Signal Transduct. Target. Ther.* **2021**, *6*, 134. [CrossRef]

91. Hoffmann, M.; Kleine-Weber, H.; Schroeder, S.; Kruger, N.; Herrler, T.; Erichsen, S.; Schiergens, T.S.; Herrler, G.; Wu, N.H.; Nitsche, A.; et al. SARS-CoV-2 Cell Entry Depends on ACE2 and TMPRSS2 and Is Blocked by a Clinically Proven Protease Inhibitor. *Cell* **2020**, *181*, 271–280 e278. [CrossRef]

92. Zhou, N.; Pan, T.; Zhang, J.; Li, Q.; Zhang, X.; Bai, C.; Huang, F.; Peng, T.; Zhang, J.; Liu, C.; et al. Glycopeptide Antibiotics Potently Inhibit Cathepsin L in the Late Endosome/Lysosome and Block the Entry of Ebola Virus, Middle East Respiratory Syndrome Coronavirus (MERS-CoV), and Severe Acute Respiratory Syndrome Coronavirus (SARS-CoV). *J. Biol. Chem.* **2016**, *291*, 9218–9232. [CrossRef]

93. Shah, P.P.; Wang, T.; Kaletsky, R.L.; Myers, M.C.; Purvis, J.E.; Jing, H.; Huryn, D.M.; Greenbaum, D.C.; Smith, A.B., 3rd; Bates, P.; et al. A small-molecule oxocarbazate inhibitor of human cathepsin L blocks severe acute respiratory syndrome and ebola pseudotype virus infection into human embryonic kidney 293T cells. *Mol. Pharmacol.* **2010**, *78*, 319–324. [CrossRef]

94. Sayce, A.C.; Alonzi, D.S.; Killingbeck, S.S.; Tyrrell, B.E.; Hill, M.L.; Caputo, A.T.; Iwaki, R.; Kinami, K.; Ide, D.; Kiappes, J.L.; et al. Iminosugars Inhibit Dengue Virus Production via Inhibition of ER Alpha-Glucosidases–Not Glycolipid Processing Enzymes. *PLoS Negl. Trop. Dis.* **2016**, *10*, e0004524. [CrossRef]

95. Whitby, K.; Pierson, T.C.; Geiss, B.; Lane, K.; Engle, M.; Zhou, Y.; Doms, R.W.; Diamond, M.S. Castanospermine, a potent inhibitor of dengue virus infection in vitro and in vivo. *J. Virol.* **2005**, *79*, 8698–8706. [CrossRef] [PubMed]

96. Sunkara, P.S.; Kang, M.S.; Bowlin, T.L.; Liu, P.S.; Tyms, A.S.; Sjoerdsma, A. Inhibition of glycoprotein processing and HIV replication by castanospermine analogues. *Ann. N. Y. Acad. Sci.* **1990**, *616*, 90–96. [CrossRef] [PubMed]

97. Rathore, A.P.; Paradkar, P.N.; Watanabe, S.; Tan, K.H.; Sung, C.; Connolly, J.E.; Low, J.; Ooi, E.E.; Vasudevan, S.G. Celgosivir treatment misfolds dengue virus NS1 protein, induces cellular pro-survival genes and protects against lethal challenge mouse model. *Antivir. Res.* **2011**, *92*, 453–460. [CrossRef] [PubMed]

98. Qu, X.; Pan, X.; Weidner, J.; Yu, W.; Alonzi, D.; Xu, X.; Butters, T.; Block, T.; Guo, J.T.; Chang, J. Inhibitors of endoplasmic reticulum alpha-glucosidases potently suppress hepatitis C virus virion assembly and release. *Antimicrob. Agents Chemother.* **2011**, *55*, 1036–1044. [CrossRef] [PubMed]

99. Perry, S.T.; Buck, M.D.; Plummer, E.M.; Penmasta, R.A.; Batra, H.; Stavale, E.J.; Warfield, K.L.; Dwek, R.A.; Butters, T.D.; Alonzi, D.S.; et al. An iminosugar with potent inhibition of dengue virus infection in vivo. *Antivir. Res.* **2013**, *98*, 35–43. [CrossRef] [PubMed]

100. Warfield, K.L.; Barnard, D.L.; Enterlein, S.G.; Smee, D.F.; Khaliq, M.; Sampath, A.; Callahan, M.V.; Ramstedt, U.; Day, C.W. The Iminosugar UV-4 is a Broad Inhibitor of Influenza A and B Viruses ex Vivo and in Mice. *Viruses* **2016**, *8*, 71. [CrossRef] [PubMed]

101. Davidson, M.H. Safety profiles for the HMG-CoA reductase inhibitors: Treatment and trust. *Drugs* **2001**, *61*, 197–206. [CrossRef] [PubMed]

102. Martinez-Gutierrez, M.; Castellanos, J.E.; Gallego-Gomez, J.C. Statins reduce dengue virus production via decreased virion assembly. *Intervirology* **2011**, *54*, 202–216. [CrossRef]

103. Martinez-Gutierrez, M.; Correa-Londono, L.A.; Castellanos, J.E.; Gallego-Gomez, J.C.; Osorio, J.E. Lovastatin delays infection and increases survival rates in AG129 mice infected with dengue virus serotype 2. *PLoS ONE* **2014**, *9*, e87412. [CrossRef]

104. Montoya, C.J.; Jaimes, F.; Higuita, E.A.; Convers-Paez, S.; Estrada, S.; Gutierrez, F.; Amariles, P.; Giraldo, N.; Penaloza, C.; Rugeles, M.T. Antiretroviral effect of lovastatin on HIV-1-infected individuals without highly active antiretroviral therapy (The LIVE study): A phase-II randomized clinical trial. *Trials* **2009**, *10*, 41. [CrossRef]

105. Delang, L.; Paeshuyse, J.; Vliegen, I.; Leyssen, P.; Obeid, S.; Durantel, D.; Zoulim, F.; Op de Beeck, A.; Neyts, J. Statins potentiate the in vitro anti-hepatitis C virus activity of selective hepatitis C virus inhibitors and delay or prevent resistance development. *Hepatology* **2009**, *50*, 6–16. [CrossRef] [PubMed]

106. Mehrbod, P.; Omar, A.R.; Hair-Bejo, M.; Haghani, A.; Ideris, A. Mechanisms of action and efficacy of statins against influenza. *BioMed. Res. Int.* **2014**, *2014*, 872370. [CrossRef] [PubMed]

107. Carocci, M.; Hinshaw, S.M.; Rodgers, M.A.; Villareal, V.A.; Burri, D.J.; Pilankatta, R.; Maharaj, N.P.; Gack, M.U.; Stavale, E.J.; Warfield, K.L.; et al. The bioactive lipid 4-hydroxyphenyl retinamide inhibits flavivirus replication. *Antimicrob. Agents Chemother.* **2015**, *59*, 85–95. [CrossRef] [PubMed]

108. Liu, G.; Xiang, Y.; Guo, C.; Pei, Y.; Wang, Y.; Kitazato, K. Cofilin-1 is involved in regulation of actin reorganization during influenza A virus assembly and budding. *Biochem. Biophys. Res. Commun.* **2014**, *453*, 821–825. [CrossRef]

109. Liu, G.; Xiong, S.; Xiang, Y.F.; Guo, C.W.; Ge, F.; Yang, C.R.; Zhang, Y.J.; Wang, Y.F.; Kitazato, K. Antiviral activity and possible mechanisms of action of pentagalloylglucose (PGG) against influenza A virus. *Arch. Virol.* **2011**, *156*, 1359–1369. [CrossRef]

110. Van den Broeke, C.; Jacob, T.; Favoreel, H.W. Rho'ing in and out of cells: Viral interactions with Rho GTPase signaling. *Small GTPases* **2014**, *5*, e28318. [CrossRef]

111. Chou, Y.Y.; Cuevas, C.; Carocci, M.; Stubbs, S.H.; Ma, M.; Cureton, D.K.; Chao, L.; Evesson, F.; He, K.; Yang, P.L.; et al. Identification and Characterization of a Novel Broad-Spectrum Virus Entry Inhibitor. *J. Virol.* **2016**, *90*, 4494–4510. [CrossRef]

112. Friesland, A.; Zhao, Y.; Chen, Y.H.; Wang, L.; Zhou, H.; Lu, Q. Small molecule targeting Cdc42-intersectin interaction disrupts Golgi organization and suppresses cell motility. *Proc. Natl. Acad. Sci. USA* **2013**, *110*, 1261–1266. [CrossRef]

113. Wang, J.L.; Zhang, J.L.; Chen, W.; Xu, X.F.; Gao, N.; Fan, D.Y.; An, J. Roles of small GTPase Rac1 in the regulation of actin cytoskeleton during dengue virus infection. *PLoS Negl. Trop. Dis.* **2010**, *4*. [CrossRef]

114. Bhullar, K.S.; Lagaron, N.O.; McGowan, E.M.; Parmar, I.; Jha, A.; Hubbard, B.P.; Rupasinghe, H.P.V. Kinase-targeted cancer therapies: Progress, challenges and future directions. *Mol. Cancer.* **2018**, *17*, 48. [CrossRef]

115. de Oliveira, L.C.; Ribeiro, A.M.; Albarnaz, J.D.; Torres, A.A.; Guimaraes, L.F.Z.; Pinto, A.K.; Parker, S.; Doronin, K.; Brien, J.D.; Buller, M.R.; et al. The small molecule AZD6244 inhibits dengue virus replication in vitro and protects against lethal challenge in a mouse model. *Arch. Virol.* **2020**, *165*, 671–681. [CrossRef]

116. Bekerman, E.; Neveu, G.; Shulla, A.; Brannan, J.; Pu, S.Y.; Wang, S.; Xiao, F.; Barouch-Bentov, R.; Bakken, R.R.; Mateo, R.; et al. Anticancer kinase inhibitors impair intracellular viral trafficking and exert broad-spectrum antiviral effects. *J. Clin. Investig.* **2017**, *127*, 1338–1352. [CrossRef] [PubMed]

117. de Wispelaere, M.; LaCroix, A.J.; Yang, P.L. The small molecules AZD0530 and dasatinib inhibit dengue virus RNA replication via Fyn kinase. *J. Virol.* **2013**, *87*, 7367–7381. [CrossRef] [PubMed]

118. Chu, J.J.; Yang, P.L. c-Src protein kinase inhibitors block assembly and maturation of dengue virus. *Proc. Natl. Acad. Sci. USA* **2007**, *104*, 3520–3525. [CrossRef]

119. Singh, S.; Anupriya, M.G.; Modak, A.; Sreekumar, E. Dengue virus or NS1 protein induces trans-endothelial cell permeability associated with VE-Cadherin and RhoA phosphorylation in HMEC-1 cells preventable by Angiopoietin-1. *J. Gen. Virol.* **2018**, *99*, 1658–1670. [CrossRef] [PubMed]

120. Kroutzman, A.; Colom-Fernandez, B.; Jimenez, A.M.; Ilander, M.; Cuesta-Mateos, C.; Perez-Garcia, Y.; Arevalo, C.D.; Bruck, O.; Hakanen, H.; Saarela, J.; et al. Dasatinib Reversibly Disrupts Endothelial Vascular Integrity by Increasing Non-Muscle Myosin II Contractility in a ROCK-Dependent Manner. *Clin. Cancer Res. Off. J. Am. Assoc. Cancer Res.* **2017**, *23*, 6697–6707. [CrossRef] [PubMed]

121. Bradley, W.D.; Koleske, A.J. Regulation of cell migration and morphogenesis by Abl-family kinases: Emerging mechanisms and physiological contexts. *J. Cell. Sci.* **2009**, *122*, 3441–3454. [CrossRef] [PubMed]

122. Coleman, C.M.; Sisk, J.M.; Mingo, R.M.; Nelson, E.A.; White, J.M.; Frieman, M.B. Abelson Kinase Inhibitors Are Potent Inhibitors of Severe Acute Respiratory Syndrome Coronavirus and Middle East Respiratory Syndrome Coronavirus Fusion. *J. Virol.* **2016**, *90*, 8924–8933. [CrossRef]

123. Garcia, M.; Cooper, A.; Shi, W.; Bornmann, W.; Carrion, R.; Kalman, D.; Nabel, G.J. Productive replication of Ebola virus is regulated by the c-Abl1 tyrosine kinase. *Sci. Transl. Med.* **2012**, *4*, 123ra24. [CrossRef]

124. Clark, M.J.; Miduturu, C.; Schmidt, A.G.; Zhu, X.; Pitts, J.D.; Wang, J.; Potisopon, S.; Zhang, J.; Wojciechowski, A.; Hann Chu, J.J.; et al. GNF-2 Inhibits Dengue Virus by Targeting Abl Kinases and the Viral E Protein. *Cell Chem. Biol.* **2016**, *23*, 443–452. [CrossRef]

125. Haile, W.B.; Gavegnano, C.; Tao, S.; Jiang, Y.; Schinazi, R.F.; Tyor, W.R. The Janus kinase inhibitor ruxolitinib reduces HIV replication in human macrophages and ameliorates HIV encephalitis in a murine model. *Neurobiol. Dis.* **2016**, *92*, 137–143. [CrossRef] [PubMed]

126. Ernst, E. The efficacy of herbal medicine—An overview. *Fundam. Clin. Pharm.* **2005**, *19*, 405–409. [CrossRef] [PubMed]

127. Balasubramanian, A.; Pilankatta, R.; Teramoto, T.; Sajith, A.M.; Nwulia, E.; Kulkarni, A.; Padmanabhan, R. Inhibition of dengue virus by curcuminoids. *Antivir. Res.* **2019**, *162*, 71–78. [CrossRef] [PubMed]

128. Diamond, M.S.; Zachariah, M.; Harris, E. Mycophenolic acid inhibits dengue virus infection by preventing replication of viral RNA. *Virology* **2002**, *304*, 211–221. [CrossRef]

129. Ciesek, S.; Steinmann, E.; Wedemeyer, H.; Manns, M.P.; Neyts, J.; Tautz, N.; Madan, V.; Bartenschlager, R.; von Hahn, T.; Pietschmann, T. Cyclosporine A inhibits hepatitis C virus nonstructural protein 2 through cyclophilin A. *Hepatology* **2009**, *50*, 1638–1645. [CrossRef]

130. Qing, M.; Yang, F.; Zhang, B.; Zou, G.; Robida, J.M.; Yuan, Z.; Tang, H.; Shi, P.Y. Cyclosporine inhibits flavivirus replication through blocking the interaction between host cyclophilins and viral NS5 protein. *Antimicrob. Agents Chemother.* **2009**, *53*, 3226–3235. [CrossRef]

131. Aman, M.J.; Kinch, M.S.; Warfield, K.; Warren, T.; Yunus, A.; Enterlein, S.; Stavale, E.; Wang, P.; Chang, S.; Tang, Q.; et al. Development of a broad-spectrum antiviral with activity against Ebola virus. *Antivir. Res.* **2009**, *83*, 245–251. [CrossRef]

132. Holy, J. Curcumin inhibits cell motility and alters microfilament organization and function in prostate cancer cells. *Cell Motil. Cytoskelet.* **2004**, *58*, 253–268. [CrossRef]

133. Leu, T.H.; Maa, M.C. The molecular mechanisms for the antitumorigenic effect of curcumin. *Curr. Med. Chem. Anticancer Agents* **2002**, *2*, 357–370. [CrossRef]

134. Padilla, S.L.; Rodriguez, A.; Gonzales, M.M.; Gallego, G.J.; Castano, O.J. Inhibitory effects of curcumin on dengue virus type 2-infected cells in vitro. *Arch. Virol.* **2014**, *159*, 573–579. [CrossRef]

135. Hegazy, G.E.; Abu-Serie, M.M.; Abo-Elela, G.M.; Ghozlan, H.; Sabry, S.A.; Soliman, N.A.; Abdel-Fattah, Y.R. In vitro dual (anticancer and antiviral) activity of the carotenoids produced by haloalkaliphilic archaeon Natrialba sp. M6. *Sci. Rep.* **2020**, *10*, 5986. [CrossRef] [PubMed]

136. Sharma, U.; Kumar, P.; Kumar, N.; Singh, B. Recent advances in the chemistry of phthalimide analogues and their therapeutic potential. *Mini Rev. Med. Chem.* **2010**, *10*, 678–704. [CrossRef] [PubMed]

137. Guedes da Silva Jr, J.; Nogueira Holanda, V.; Rodrigues Gambôa, D.S.; Siqueira do Monte, T.V.; Andrade de Araújo, H.D.; Alves do Nascimento Jr, J.A.; da Silva Araújo, V.F.; Macedo Callôu, M.A.; Pôrto de Oliveira Assis, S.; Menezes Lima, V.L. Therapeutic Potential of Phthalimide Derivatives: A Review. *Am. J. Biomed. Sci. Res.* **2019**, *3*, 378–384. [CrossRef]

138. Gonzalez-Cardenete, M.A.; Hamulic, D.; Miquel-Leal, F.J.; Gonzalez-Zapata, N.; Jimenez-Jarava, O.J.; Brand, Y.M.; Restrepo-Mendez, L.C.; Martinez-Gutierrez, M.; Betancur-Galvis, L.A.; Marin, M.L. Antiviral Profiling of C-18- or C-19-Functionalized Semisynthetic Abietane Diterpenoids. *J. Nat. Prod.* **2022**, *85*, 2044–2051. [CrossRef]

139. de Chassey, B.; Meyniel-Schicklin, L.; Aublin-Gex, A.; Andre, P.; Lotteau, V. New horizons for antiviral drug discovery from virus-host protein interaction networks. *Curr. Opin. Virol.* **2012**, *2*, 606–613. [CrossRef]

140. Dyer, M.D.; Murali, T.M.; Sobral, B.W. The landscape of human proteins interacting with viruses and other pathogens. *PLoS Pathog.* **2008**, *4*, e32. [CrossRef]

141. de Chassey, B.; Meyniel-Schicklin, L.; Vonderscher, J.; Andre, P.; Lotteau, V. Virus-host interactomics: New insights and opportunities for antiviral drug discovery. *Genome Med.* **2014**, *6*, 115. [CrossRef]

142. Chandrasekaran, S.N.; Ceulemans, H.; Boyd, J.D.; Carpenter, A.E. Image-based profiling for drug discovery: Due for a machine-learning upgrade? *Nat. Rev. Drug Discov.* **2021**, *20*, 145–159. [CrossRef]

MDPI

St. Alban-Anlage 66

4052 Basel

Switzerland

www.mdpi.com

Viruses Editorial Office

E-mail: viruses@mdpi.com

www.mdpi.com/journal/viruses